Communications
in Computer and Information Science 2376

Series Editors

Gang Li , *School of Information Technology, Deakin University, Burwood, VIC, Australia*
Joaquim Filipe , *Polytechnic Institute of Setúbal, Setúbal, Portugal*
Zhiwei Xu, *Chinese Academy of Sciences, Beijing, China*

Rationale

The CCIS series is devoted to the publication of proceedings of computer science conferences. Its aim is to efficiently disseminate original research results in informatics in printed and electronic form. While the focus is on publication of peer-reviewed full papers presenting mature work, inclusion of reviewed short papers reporting on work in progress is welcome, too. Besides globally relevant meetings with internationally representative program committees guaranteeing a strict peer-reviewing and paper selection process, conferences run by societies or of high regional or national relevance are also considered for publication.

Topics

The topical scope of CCIS spans the entire spectrum of informatics ranging from foundational topics in the theory of computing to information and communications science and technology and a broad variety of interdisciplinary application fields.

Information for Volume Editors and Authors

Publication in CCIS is free of charge. No royalties are paid, however, we offer registered conference participants temporary free access to the online version of the conference proceedings on SpringerLink (http://link.springer.com) by means of an http referrer from the conference website and/or a number of complimentary printed copies, as specified in the official acceptance email of the event.

CCIS proceedings can be published in time for distribution at conferences or as post-proceedings, and delivered in the form of printed books and/or electronically as USBs and/or e-content licenses for accessing proceedings at SpringerLink. Furthermore, CCIS proceedings are included in the CCIS electronic book series hosted in the SpringerLink digital library at http://link.springer.com/bookseries/7899. Conferences publishing in CCIS are allowed to use Online Conference Service (OCS) for managing the whole proceedings lifecycle (from submission and reviewing to preparing for publication) free of charge.

Publication process

The language of publication is exclusively English. Authors publishing in CCIS have to sign the Springer CCIS copyright transfer form, however, they are free to use their material published in CCIS for substantially changed, more elaborate subsequent publications elsewhere. For the preparation of the camera-ready papers/files, authors have to strictly adhere to the Springer CCIS Authors' Instructions and are strongly encouraged to use the CCIS LaTeX style files or templates.

Abstracting/Indexing

CCIS is abstracted/indexed in DBLP, Google Scholar, EI-Compendex, Mathematical Reviews, SCImago, Scopus. CCIS volumes are also submitted for the inclusion in ISI Proceedings.

How to start

To start the evaluation of your proposal for inclusion in the CCIS series, please send an e-mail to ccis@springer.com.

Mukta Majumder · J. K. M. Sadique Uz Zaman ·
Mili Ghosh · Samarjit Chakraborty

Editors

Computational Technologies and Electronics

First International Conference, ICCTE 2023
Siliguri, India, November 23–25, 2023
Proceedings, Part I

 Springer

Editors
Mukta Majumder (iD)
University of North Bengal
Darjeeling, West Bengal, India

J. K. M. Sadique Uz Zaman (iD)
University of North Bengal
Darjeeling, West Bengal, India

Mili Ghosh (iD)
University of North Bengal
Darjeeling, West Bengal, India

Samarjit Chakraborty (iD)
University of North Carrolina
Chapel Hill, NC, USA

ISSN 1865-0929   ISSN 1865-0937 (electronic)
Communications in Computer and Information Science
ISBN 978-3-031-81934-6   ISBN 978-3-031-81935-3 (eBook)
https://doi.org/10.1007/978-3-031-81935-3

This Springer imprint is published by the registered company Springer Nature Switzerland AG
The registered company address is: Gewerbestrasse 11, 6330 Cham, Switzerland

If disposing of this product, please recycle the paper.

Preface

We are thrilled to share the proceedings of the International Conference on Computational Technologies and Electronics (ICCTE 2023), hosted by the Department of Computer Science and Technology at the University of North Bengal, Siliguri, India, held from November 23–25, 2023. ICCTE aims to bring together intellectuals, leading scientists, research scholars, and students from around the world to exchange and share their research ideas on device intelligence, computing technologies, next-generation communication, and networking. The motivation behind this conference stems from the pressing need to accommodate the ever-increasing number of intelligent devices while maintaining highly efficient communication systems.

ICCTE 2023 featured distinguished keynote speakers, renowned experts in their respective fields. We were honoured to welcome Samarjit Chakraborty from the University of North Carolina, Chapel Hill, USA, Subarna Shakya from Tribhuvan University, Nepal, Sudan Jha from Kathmandu University, Nepal, Sansanee Auephanwiriyakul, Chiang Mai University, Thailand, and Rajkumar Upadhayay, Enterprise Architect Fellow at Navistar International, Lisle, Illinois, USA.

In technical collaboration with IEEE CIS Kolkata Chapter, IEEE EDS Kolkata Chapter, and CSI Siliguri Chapter, ICCTE 2023 attracted the international research community to a large extent. Our prestigious publication partner, Springer's CCIS Series, made sure that the conference proceedings are of the highest calibre and widely distributed.

With 157 initial submissions, we are happy to provide the ICCTE 2023 submission statistics, which demonstrate the conference's increasing importance. Each article was thoroughly reviewed by at least three esteemed reviewers in a double-blind format to ensure impartial evaluation. Out of these 46 papers were accepted and registered, indicating a remarkable 29% acceptance rate.

The conference proceedings are meticulously organized into distinct tracks. These proceedings, readers will find 28 insightful papers in the "Pattern Recognition & AI" track, 11 thought-provoking papers in the "Data Communication & Security" track, and 7 informative articles in the "Applied Electronics" track. These contributions represent the forefront of research in computational technologies and electronics, covering a diverse array of application areas.

As a debut international conference, we are delighted by its resounding success and impact. ICCTE 2023 successfully drew top-notch research from around the world. We would like to extend our sincere gratitude to the organizing committee members, evaluators, and distinguished keynote speakers whose efforts helped to make ICCTE an exceptional forum for information sharing.

We would like to thank all of the authors who contributed to the proceedings, the esteemed reviewers who went over the submissions, and the attendees who added to the conversations throughout the conference. The scholarly conversation surrounding computational technologies and applied electronics has been enhanced by your combined efforts.

November 2023

Mukta Majumder
J. K. M. Sadique Uz Zaman
Mili Ghosh
Samarjit Chakraborty

Organization

Chief Patron

C. M. Ravindran   University of North Bengal, India

Patron

Subhas Chandra Roy   University of North Bengal, India

Advisory Committee

Alvaro Rocha   University of Lisbon, Portugal
Subarna Shakya   Tribhuvan University, Nepal
Samarjit Chakraborty   University of North Carolina, Chapel Hill, USA
Xiao-Zhi Gao   University of Eastern Finland, Finland
Hadj Bourdoucen   Sultan Qaboos University, Sultanate of Oman
Esteban Alfaro-Cortés   University of Castilla - La Mancha, Spain
Sudan Jha   Kathmandu University, Nepal
Tharinda Nishantha Vidanagama   Wayamba University of Sri Lanka, Sri Lanka
Sansanee Auephanwiriyakul   Chiang Mai University, Thailand
Rajkumar Upadhayay   Navistar International, USA
Koushik Roy   West Bengal State University, India
Rajat Kumar Pal   University of Calcutta, India
Gautam Kumar Das   IIT Guwahati, India
Utpal Biswas   Kalyani University, India
Atal Chowdhury   Jadavpur University, India
Sunil Karforma   University of Burdwan, India
Tandra Pal   NIT Durgapur, India
Jaya Shil   IIEST Shibpur, India
Samarjit Bohra   Sikkim Manipal Institute of Technology, India
Utpal Garain   ISI Kolkata, India
Avishek Adhikari   Presidency University, India
Sudipta Roy   Assam University, India
Shikhar Kumar Sarma   Gauhati University, India
Angshuman Sarkar   Kalyani Government Engineering College, India
Manash Chanda   Meghnad Saha Institute of Technology, India

Arpan Deyasi   RCC Institute of Information Technology, India
Diptendu Sinha Roy   NIT Meghalaya, India
Swaraj Kumar Biswas   NIT Silchar, India
Santanu Phadikar   MAKAUT, India
Biswapati Jana   Vidyasagar University, India
Shashanka Roy   ISI-Kolkata, India
Sk Md Obaidullah   Aliah University, India
Siddhartha Bhattacharya   Siuri Government College, India
Abhijit Mustafi   Birla Institute of Technology Mesra, India
Sujan Kumar Saha   NIT Durgapur, India

General Chairs

Rakesh Kumar Mandal   University of North Bengal, India
Sharad Sinha   University of North Bengal, India

Organizing Chairs

J. K. M. Sadique Uz Zaman   University of North Bengal, India
Mukta Majumder   University of North Bengal, India

Convenor

Rakesh Kumar Mandal   University of North Bengal, India

Programme Committee Chairs

Mukta Majumder   University of North Bengal, India
J. K. M. Sadique Uz Zaman   University of North Bengal, India
Mili Ghosh   University of North Bengal, India
Samarjit Chakraborty   University of North Carolina, Chapel Hill, USA

Publication Committee

Mili Ghosh   University of North Bengal, India
Mukta Majumder   University of North Bengal, India

Publicity Committee

Anil Tudu	University of North Bengal, India
Md Ajij	University of North Bengal, India
Ardhendu Mandal	University of North Bengal, India

ICT Committee

Swarup Das	University of North Bengal, India
Md. Ajij	University of North Bengal, India
Anil Tudu	University of North Bengal, India
Anirban Biswas	University of North Bengal, India
Arghya Das	University of North Bengal, India
Sudeep Basu	University of North Bengal, India
Amitava Deb Barman	University of North Bengal, India
Utpal Mandi	University of North Bengal, India

Technical Programme Committee

Ardhendu Mandal	University of North Bengal, India
Rakesh Kumar Mandal	University of North Bengal, India
Swarup Das	University of North Bengal, India
Mukta Majumder	University of North Bengal, India
J. K. M. Sadique Uz Zaman	University of North Bengal, India
Mili Ghosh	University of North Bengal, India
Marcin Paprzycki	Polish Academy of Sciences, Poland
Ke-Lin Du	Concordia University, Canada
Xiao-Zhi Gao	University of Eastern Finland, Finland
Hedayat Omidvar	National Iranian Gas Company, Iran
Abdel-Badeeh M.Salem	Ain Shams University, Egypt
Pradeep Kumar	University of KwaZulu-Natal, South Africa
Sara Paiva	University of Vigo, Spain
Varun Kumar Ojha	University of Reading, UK
Hamid Abdi	Deakin University, Australia
Subhas Mukhopadhyay	Macquarie University, Australia
Sangeet Saha	University of Essex, UK
Rourab Paul	University of Pisa, Italy
Nimisha Ghosh	Siksha 'O' Anusandhan University, India
Debaditya Barman	Visva Bharati University, India
Uttam Mandal	Vidyasagar University, India

K. L. Hassan	Aliah University, India
Sanjoy Pratihar	IIIT Kalyani, India
Jadav Das	MAKAUT, India
Bhagyashri R. Hanji	Global Academy of Technology, India
Dibya Ranjan Dasadhikari	Siksha O Anusandhan University, India
A. F. Mollah	Aliah University, India
Abu Sufian	University of Gour Banga, India
Kakali Datta	Visva Bharati University, India
Asit Barman	Siliguri Institute of Technology, India
Arnab Sadhu	Vidyasagar University, India
Dipanjan Moitra	University of North Bengal, India
Sk. Arif Ahmed	XIM University, India
Bidyut Das	Haldia Institute of Technology, India
Swalpa Kumar Ray	Jalpaiguri Government Engineering College, India
Subhas Barman	Jalpaiguri Government Engineering College, India
Bishnu Prasad De	Kalinga Institute of Industrial Technology, India
Jamimamul Bakas	Birla Institute of Technology Mesra, India
Mahabub Hasan Mahalat	Siksha 'O' Anusandhan University, India
Vineeta Shukla	MathWorks, India
Pabitra Pal	MAKAUT, India
Wriddhi Bhowmik	Kalinga Institute of Industrial Technology, India
Abhijit Mustafi	Birla Institute of Technology Mesra, India
Debjani Mustafi	Birla Institute of Technology Mesra, India
Apash Roy	LPU, India
Dilip Kumar Choubey	IIIT Bhagalpur, India
Shivlal Mawada	Holkar Science College, India
Indrajit Ghosh	A.C. College, India
Sandip Dey	Sukanta Mahavidyalaya, India
Jayati Lahiri	Raiganj University, India
Bipul Shyam Purkayastha	Assam University, India
Anirban Chakraborty	Barrackpore Rastraguru Surendranath College, India
Sudakshina Dasgupta	Govt. College of Engg. & Textile Technology, India
Subhasis Banerjee	Visva Bharati University, India
Suvamoy Changder	NIT Durgapur, India
Debayani Ghosh	Thapar University, India
Amar Singh	LPU, India
Manmohan Sharma	LPU, India
Sofhia Seik	LPU, India
Kartick Chandra Mondal	Jadavpur University, India
Indrajit Bhattacharya	Kalyani Government Engineering College, India

Parikshit N. Mahalle   Smt. Kashibai Navale College of Engineering,
                                 India
Uddalak Mitra   Siliguri Institute of Technology, India
Jyoti Prakash Singh   NIT Patna, India
Arindam Sarkar   Ramkrishna Mission Vidyamandir, India
Debarka Mukhopadhyay   Christ University, India
Saurabh Ghosh   SRM-IST, India
Saptarsi Goswami   Calcutta University, India
Papri Ghosh   Narula Institute of Technology, India
Ritam Dutta   Poornima University, India
Sandip Karar   DeepSell (Augmented SCM Pvt. Ltd.), India
Soumen Nandi   Netaji Subhas Open University, India
Manoranjan Singha   University of North Bengal, India
Siddhartha Biswas   University of North Bengal, India
Subhrangsu Mandal   IIT Indore, India
Kousik Dasgupta   Kalyani Government Engineering College, India
Debjyoti Misra   Siliguri Institute of Technology, India

Additional Reviewers

Amit Ball   Pennsylvania State University, USA
Abhijit Dasgupta   St Jude Children's Hospital, USA
Somshuvra Bhattacharya   Jackson Laboratory, USA
Prakash Singh   IBM India Pvt Ltd., India
Sourav Kaity   DRDO, India
Koushik Mondal   ISM Dhanbad, India
Manas Parai   Siliguri Institute of Technology, India
Tamal Sarkar   University of North Bengal, India
Anagha Bhattacharya   NIT Mizoram, India
Chinmoy Ghorai   Jadavpur University, India
Swanirbhar Majumder   Tripura University, India
Dibyendu Mukherjee   NSHM Knowledge Campus, India
Dattatray Takale   Vishwakarma Institute of Information Tech., India
Amrita Sarkar   Birla Institute of Technology Mesra, India
Sweta Srivastava   Amity University, India
Suchandra Banerjee   NSHM Knowledge Campus, India
Sujay Biswas   University of North Bengal, India
Ipsita Saha   Guru Nanak Institute of Technology, India
Suchismita Maiti   Narula Institute of Technology, India
Utpal Nandi   Vidyasagar University, India
Dhrubasish Sarkar   Supreme Institute of Management & Tech., India

Rajesh Gundlapalle	New Horizon College of Engineering, India
Dhawaleswar Rao	Centurion University of Tech. & Management, India
Saikat Maity	Sister Nivedita University, India
Swarup Kr Ghosh	Sister Nivedita University, India
Rakesh Patra	Graphic Era Hill University, India
Krishna Dhal	Midnapore College, India
Bijoy Mandal	NSHM Knowledge Campus, India
Indrajit Chowdhury	University of North Bengal, India
Avijit Datta	Cooch Behar Panchanan Barma University, India
Barnali Choudhury	Netaji Subhas Open University, India
Rupnar Dutta	University of North Bengal, India
Sidhartha Laha	University of North Bengal, India
Subhas Pal	University of North Bengal, India
Md Nasir	Gitam University, India
Kasturi Mukherjee	Adamas University, India
Anirban Bhowmick	VIT, Bhopal Campus, India
Mir Wajahat Hussain	Alliance University, India
Rajat Dey	University of North Bengal, India
Rashidul Islam	Mathabhanga College, India
Abhilash Pati	Siksha O Anusandhan University, India
R. K. Samanta	University of North Bengal, India
Partha Chowdhuri	Vidyasagar University, India
Prajnamita Dasgupta	University of North Bengal, India
Ravi Raushan	National Institute of Technology Surathkal, India
Prashant G. K.	VIT, Bhopal Campus, India
Rajiv Kumar	Birla Institute of Technology Mesra, India
Dharmveer Yadav	Katihar Engineering College, India
Samit Biswas	IIEST, Shibpur, India
Ajay Kumar	Galgotias University, India
Dhananjoy Bhakta	IIIT Ranchi, India
Sandip Mal	VIT, Bhopal Campus, India
Chiranjit Changdar	Belda College, India
Debasis Dhal	University of Calcutta
Oishila Bandyopadhyay	IIIT Kalyani, India
Suvra Choudhury	ISI Kolkata, India
Preetam Mukherjee	Digital University, India
Sudakshina Mandal	Techno India University, India
Ankita Dhar	Sister Nivedita University, India
Arindam Roy	Prabhat Kumar College, India
Chandrashekhar Azad	NIT Jamshedpur, India

Amit Kumar   Jaypee University of Information Technology,
                             India
Prabhash Singh   Vidyasagar University, India
Chiranjib Sarkar   University of North Bengal, India
Rizwan Alam   Aligarh Muslim University, India

Keynote Lectures

Lecture 1: Issues and Challenges of adopting Cloud Computing in Government

Prof. Subarna Shakya

Tribhuvan University, Nepal

Cloud Computing is an internet based computing where all the shared resources, software and information are provided to the computers and devices on demand. Users can access the information from anywhere and anytime. E-Governance plays a vital role in any organization and clouds with different layers are helpful to the e-Governance services.

Cloud services are helpful to reduce the cost of infrastructure and software cost. This keynote concentrates on the adopt of cloud computing in governments to reduce infrastructure, and platform cost, to increase network security, to increase scalability and quick implementation. The results reveal that cloud computing adoption could help solve problems such as infrastructure issues, cost issues, and improve service delivery and transparency. Propose that cloud computing has the potential to offer benefits to both implementation of efficient e-government services and to users of those service. Explore the issues and challenges affecting the adoption of cloud computing in government services. At the end of this key note some recommendation of adoption of cloud computing in government services.

Lecture 2: Fostering Synergy: Bridging the Gap between the Internet of Computing and the Internet of Things for a Smarter World

Prof. Sudan Jha

Kathmandu University, Nepal

The keynote address explores the evolving landscape of the Internet of Things (IoT) and its integration with the Internet of Computing (IoC) to create a more interconnected and intelligent world. It delves into the evolution from IoT to IoC, emphasizing the benefits and challenges of this transition. The presentation covers the framework for Internet of Computing, the handshaking model between IoT and IoC, and highlights real-world applications and future prospects. By examining the synergy between IoT and IoC, the speech aims to shed light on the significant potential for innovation, efficiency, and sustainability in various sectors, ultimately contributing to the development of a smarter and more cohesive digital environment.

Lecture 3: Efficient and Certifiable Cyber-Physical Systems

Prof. Samarjit Chakraborty

The University of North Carolina, USA

The keynote address explores the evolving landscape of the Internet of Things (IoT) and its integration with the Internet of Computing (IoC) to create a more interconnected and intelligent world. It delves into the evolution from IoT to IoC, emphasizing the benefits and challenges of this transition. The presentation covers the framework for Internet of Computing, the handshaking model between IoT and IoC, and highlights real-world applications and future prospects. By examining the synergy between IoT and IoC, the speech aims to shed light on the significant potential for innovation, efficiency, and sustainability in various sectors, ultimately contributing to the development of a smarter and more cohesive digital environment.

Lecture 4: Computational Intelligence in Biomedical Engineering

Dr. Sansanee Auephanwiriyakul

Chiang Mai University, Thailand

Computational Intelligence (CI) relies on and combines several algorithms in fuzzy systems, neural networks, evolutionary computation, swarm intelligence, fractals, chaos theory, artificial immune systems, wavelets, etc., to produce an algorithm that is intelligent somehow. CI has been utilized in many applications for several years. One of the areas that CI has an impact on is the area of biomedical engineering, e.g., medical image processing, medical signal processing and biometrics. One of the CI tools used in those mentioned application is classification or sometimes called decision making. The major area in the classification is to develop a classifier, including, feature generation and selection. In this talk, feature generation methods and classifier methods based on the CI will be presented. We also show those methods on real application, including, medical image diagnosis, medical signal diagnosis, and data analysis in health care system.

Lecture 5: Bridging the Gap: Emerging Technologies and Industry Expertise in IT World (Automobile Industry)

Mr. Rajkumar Upadhyay

Enterprise Architect Fellow Navistar International, Lisle, USA

Computational Intelligence (CI) relies on and combines several algorithms in fuzzy systems, neural networks, evolutionary computation, swarm intelligence, fractals, chaos theory, artificial immune systems, wavelets, etc., to produce an algorithm that is intelligent somehow. CI has been utilized in many applications for several years. One of the areas that CI has an impact on is the area of biomedical engineering, e.g., medical image processing, medical signal processing and biometrics. One of the CI tools used in those mentioned application is classification or sometimes called decision making. The major area in the classification is to develop a classifier, including, feature generation and selection. In this talk, feature generation methods and classifier methods based on the CI will be presented. We also show those methods on real application, including, medical image diagnosis, medical signal diagnosis, and data analysis in health care system.

Contents – Part I

Pattern Recognition and AI

Resource Description Framework Statement Generation Using Soft
Attention Based Hybrid Resnet-Bidirectional Long Short Term Memory
Model 3
   Rubaya Khatun and Arup Sarkar

Evaluating the Tri-Script Writer Verification System Using a Handcrafted
Features and Vision Transformer Learning Approach 18
   Jaya Paul, Kalpita Dutta, Anasua Sarkar, Kaushik Roy, and Nibaran Das

Wildlife Detection Using ANN and Other Modern Technology: A Survey
of Literatures 31
   Priyodarshini Dhar and Rakesh Kumar Mandal

Enhancing Learning Outcomes Through the Use of Conducive Learning
Spaces 45
   Shabina Modi, Deepali Sale, Vishal Borate, and Yogesh Kisan Mali

Smart Diagnosis Using Symptoms for Seeking a Specialist Doctor 54
   Bidyut Das and Rishu Kumar

A Comprehensive Review and Future Prospects of Lie Detection Using
Machine Learning 64
   Debanil Chanda and R. K. Mandal

Searching Optimizers for Deep Learning Based Hyperspectral Image
Classification 78
   Anish Sarkar, Utpal Nandi, Bachchu Paul, Sudipta Kr. Ghosal,
   Moirangthem Marjit Singh, Jyotsna Kumar Mandal,
   and Nuruzzaman Faruqui

Efficient Crop Recommendation System: A Machine Learning Based
Approach 93
   Swagatika Tripathy, Dibya Ranjan Das Adhikary,
   and Premansu Sekhara Rath

Human Fall Detection Using Transfer Learning-Based 3D CNN 102
   Ekram Alam, Abu Sufian, Paramartha Dutta, and Marco Leo

Anomaly Detection in Respiratory Events Using Machine Learning 114
   Arundhati Roy and Sriparna Saha

Federated Learning to Speed Up Pre-processing of Large Data Sets 127
   Lhamu Sherpa and Nandan Banerji

Flood Susceptibility Zonation Using Geospatial Frequency Ratio
and Artificial Neural Network Techniques within Himalayan Terai Region:
A Comparative Exploration 136
   Deepanjan Sen, Swarup Das, Sumon Dey, and Arindam Sarkar

A Fertilizer Recommendation System Using an Assembly of Regressors
Coupled with Nature-Inspired Optimization Algorithms 149
   Uditendu Sarkar, Gouravmoy Banerjee, and Indrajit Ghosh

State-of-the-Art in Feature Selection: Applications of the Slime Mould
Algorithm 160
   Taniya Chatterjee, Puja Bhakta, Mili Ghosh, and Debaditya Barman

MOODBYTBLB: Impact of Covid-19 Among Indians: A Sentiment
Analysis Using Textblob 173
   Sanchita Neogi and Rahul Karmakar

MythBuster: A Comparative Analysis of Few Machine Learning and Deep
Learning Models for Fake News Detection 184
   Barsha Pattanaik, Pratyush Mukherjee, Sourav Mandal, Rohini Basak,
   and Rudra M. Tripathy

Unsupervised Approach for Word Sense Disambiguation in Bengali 196
   Ratul Das, Alok Ranjan Pal, and Diganta Saha

A Hybrid Method for Bengali Word Segmentation from Handwritten
Copies of School Students 207
   Moumita Moitra, Souvik Ganguly, and Sujan Kumar Saha

Securing Social Spaces: Harnessing Deep Learning to Eradicate
Cyberbullying 219
   Rohan Biswas, Kasturi Ganguly, Arijit Das, and Diganta Saha

Exploring Intonation Patterns in Nepali Speech: A Phonetic and Linguistic
Analysis for Text-to-Speech System 229
   Pratika Rai, Suraj Gurung, Sharad Sinha, Ranjit Subba,
   and Roshan Kumar Prasad

Sentiment Analysis on Airline Customer Review Using Language Model
and Capsule Network 240
Nilanjana Das, Rakesh Dutta, Uttam Kumar Mondal,
and Jyotsna Kumar Mandal

Implementation of Digital Healthcare for Improving Maternal Care:
A Systematic Review 251
Sarika Kumari Shaw and Jayati Lahiri Dey

Video Content Analysis and Classification Based on Human Activity
Recognition 263
K. C. Hari, Manish Pokharel, and Sushil Shrestha

Machine Learning Based Expert System for Breast Cancer Prediction
(MLESBCP) 275
Akhil Kumar Das, Saroj Kr. Biswas, Ardhendu Mandal,
Arijit Bhattacharya, and Debasmita Saha

ARU NET: Follicle Segmentation from Ultrasound Images of Ovaries
Using Attention Residual U-NET Model 287
Debasmita Saha, Ardhendu Mandal, Saroj Kr. Biswas,
Arijit Bhattacharya, and Akhil Kumar Das

d-RIMNet: RIMNet with Depthwise Separable Convolutional Layer
for Retinal OCTA Image Segmentation 299
Farhana Sultana, Abu Sufian, and Paramartha Dutta

Multi-modal Biometric Authentication: Harnessing Human Gait
and Keystroke Dynamics for Enhanced Security 311
Sandip Dutta, Utpal Roy, and Soumen Roy

Author Index 323

Contents – Part II

Data Communication and Security

Performance Evaluation of PAM-Based Optical Communication Link
Using Eye Diagram Analysis 3
    Manjit Singh, Himali Sarangal, Butta Singh, Harmandar Kaur,
    Satveer Kour, and Vikramjeet Singh

Simulative Analysis of Free Space Optical Link by Incorporating
Modernization Technique 14
    Himali Sarangal, Manjit Singh, Satveer Kour, Harmandar Kaur,
    and Butta Singh

Effect of Increasing Multiple-Interface Enabled Nodes in an Opportunistic
Network: A Case Study 26
    Ritwik Mondal, Subhasis Dutta, Sushmita Panda,
    Indrajit Bhattacharya, and Priya Ranjan Sinha Mahapatra

Design of an Application Authentication Framework to Secure Controller
Access via Northbound Interface in SDN 38
    Tinku Adhikari, Ajoy Kumar Khan, and Malay Kule

A Dual-Image Based Robust Reversible Data Hiding Scheme Based
on Weighted Matrix and LZW Algorithm 48
    Kankana Datta, Biswapati Jana, and Mamata Dalui Chakraborty

Looking for Variant of S-box Used in AES Algorithm 61
    Md. Hasanujjaman and J. K. M. Sadique Uz Zaman

Evaluating SMT Solvers on Schedulability Checking Instances 71
    Ravindra Metta, Anand Yeolekar, and Samarjit Chakraborty

An SMT Toolbox for Adversarial Robustness Evaluation for Spiking
Neural Networks 82
    Soham Banerjee, Sumana Ghosh, Ansuman Banerjee,
    and Swarup K. Mohalik

Designing a Context-Aware Virtual Machine (CWVM) to Enhance Data
Integration in Pervasive Wireless Sensor Networks (PWSN) 94
    Sushovan Das and Uttam Kr. Mondal

Applied Electronics

Energy Efficient Lossless Audio Encoder for IoT Enable Devices   109
  Sushovan Das, Uttam Kr. Mondal, and Bibek Bikash Roy

Position Finding of Airborne Object with Dynamic Electro Optics Using
Triangulation   121
  J. S. Bharti, Jana Biswapati, Kaity Sourav, and Vardhan Aditya

Smart Electronic Voting: Emergence, Challenges, and Strategies in India   130
  Arnab Sadhu, Arindam Ghosh, and Avishek Nandi

Security Challenges in Autonomous Systems Design   142
  Mohammad Hamad and Sebastian Steinhorst

Indrajit: A Collection-Task Scheduling Algorithm to Mitigate Sensor
Occlusions for Small Satellite Constellations   155
  Himadri Sekhar Paul and Swagata Biswas

Denoising Audio Signal for Restoring Audible Quality in Wireless
Acoustic Sensor Network (WASN)   171
  Utpal Ghosh and Uttam Kr. Mondal

Discrete-Time Realization of Fractional-Order Proportional Integral
Derivative Controller Using Modified Ziegler-Nichols Method   185
  Bhanita Adhikary and Jaydeep Swarnakar

Multiple Faults Detection Technique by an Efficient Traversal Approach
Using Multiple Droplets for Digital Microfluidic Biochip   195
  Basudev Saha, Utpal Mandi, and Mukta Majumder

Software Flow for Quantum Computing   206
  Debjyoti Bhattacharjee, Amit Saha, Junde Li, Koustubh Phalak,
  Avimita Chatterjee, Jeremie Pope, Swaroop Ghosh,
  and Anupam Chattopadhyay

Quantum Measurement and Inherent Randomness: A Study on Modified
Hadamard Based Xorshift Pseudorandom Number Generator Algorithm   228
  Rounak Biswas and Utpal Roy

Author Index   241

Pattern Recognition and AI

Resource Description Framework Statement Generation Using Soft Attention Based Hybrid Resnet-Bidirectional Long Short Term Memory Model

Rubaya Khatun$^{(\boxtimes)}$ and Arup Sarkar

Department of Computer and Information Science, Raiganj University, University Road, College Para, Raiganj 733134, West Bengal, India
rkhatun18@gmail.com, arup.sarkar@raiganjuniversity.ac.in

Abstract. Unstructured data processing is a frequent problem with text processing systems. Some of the works are generated RDF statements from unstructured textual data. Extracting the correct feature from the textual data includes some issues with existing works. Efficient feature extraction approach and RDF statement generation technique are necessary to address these issues. This paper introduced an effective hybrid deep learning model, generating RDF statements from a given unstructured textual data. This paper uses data from publicly available datasets namely BBC News and Lonely Planet dataset. The datasets are pre-processed using some approaches such as data normalization, Case folding, Stop word removal, sentence segmentation, and PoS (Parts of Speech) tagging. The pre-processed data is fed into the feature extraction process whereas the corresponding entity and attribute feature values are extracted using the Assimilated N-gram method. From the extracted features, the optimal features are selected using the Walrus optimization (WaOA) algorithm. The Triples are classified using Soft Attention based hybrid ResNet-Bidirectional Long Short Term Memory model (SAtRes_BiLSTM). Finally, the attained triples are then converted and saved in RDF format. The performance of the suggested RDF statement generation is related to diverse emerging methods to prove the efficiency of triple extraction. The proposed method achieved 95%, 91.2%, 90%, and 93.2% respectively in terms of Accuracy, Precision, Recall, and F-Measure.

Keywords: RDF statement generation · unstructured textual data · soft attention · WaOA · ResNet · assimilated N-gram · Bidirectional long short-term memory

1 Introduction

Unstructured data management is a common issue in the Semantic Web and Linked Data industries [1]. RDF is a fundamental data model for displaying any type of online information. Three components make up its foundation: resources, properties, and statements. The resource that needs to be described and that has a URI (Uniform Resource

M. Majumder et al. (Eds.): ICCTE 2023, CCIS 2376, pp. 3–17, 2025.
https://doi.org/10.1007/978-3-031-81935-3_1

Identifier) is the subject. A unique class of resource that describes relationships between other resources is known as a property. URIs are another way to identify them. To create a statement, a specific resource and a named property require an object. A resource or an atomic value, also known as a literal, may make up the object. RDF statements, which have three components, are also referred to as triples (subject, predicate, object) [2]. The information gathered from these triples can be stored as structured data with a graph structure [3]. Two types of unstructured data were used by the models that are namely, non-textual unstructured data such as multimedia files like still photos, MP3 audio files, and videos and textual unstructured data such as word processing files, PowerPoint presentations, combined software memos, email messages, and instant chats [4, 5]. The Semantic Web is a Web of Data, similar to the data that databases contain. The term "Linked Data" can also refer to a group of interconnected datasets on the Web that are supported by tools like RDF and SPARQL (SPARQL protocol and RDF query language for the semantic web). The foundation for publishing and linking data is provided by RDF [6].

The DBpedia project, one of the main sources of structured data on the Web, is an actual example of Linked Open Data [7]. OWL (Web Ontology Language - schema language, or knowledge representation language, of the semantic web), which enables meaning and structure to be added to content in a machine-readable format [8]. To describe the different granularities of temporal data, such as the time point and time interval, some models of temporal data representation based on RDF have been developed. Nowadays, there are several deep learning methods available to process unstructured data [9].

1.1 Motivation

Due to the advancement of computer technology and the growth of datasets, the large dataset is continuously expanding. The main difficulty in extracting RDF triples is the dimension problem, where the total number of features is significantly higher than the total amount of data. Deep learning models are also widely used for effective RDF statement generation due to their increased efficiency. Based on these problems novel approaches are presented in this research paper to obtain enhanced performances.

1.2 Contribution

The major contribution of the proposed study is described as,

- To construct an effective hybrid deep learning model, generate an RDF statement from the given unstructured textual data.
- To remove the unwanted noises, different steps are carried out to make a robust pre-processing stage.
- To improve the RDF statement generation performance, the most particular features are taken out from the pre-processed textual data using a new N-gram method.
- An optimization algorithm is developed in the feature selection stage and thereby reduces the overall processing time, to reduce the feature dimensionality issue.

- To analyse the performance of proposed RDF statement generation by evaluating different parameters and performing comparison over other existing methods.

This research paper is structured as given: Sect. 2 discusses recent related works; Sect. 3 explains the proposed methodology in detail; Sect. 4 describes results and discussions, and Sect. 5 signifies the conclusion for the study.

2 Related Works

Some of the recent research works related to RDF statement generation are surveyed as follows.

Rahman, M.A et al. [10] established a method for extracting RDF triples from raw text data. The author took a collection of news articles and applied different approaches for extracting "subject - verb - object" associations from texts. Initially, the author used syntactic parsing and constructed triples from subject and object syntactic relations. Then contains extracting semantic roles through a frame parser and making triples from agent, predicate, and patient relationships. After testing the aforementioned methods, a manual appraisal was done on a limited quantity of samples with hand labels. The procedure of removing RDF triples can be enhanced both for semantic parsing and for conversion. The author can attempt to extract additional difficult semantic associations such as time, and date and also perform anaphora resolution. The proposed method achieves Recall of 0.46, Precision score of 0.34, and a joint score 0.39 for F-measure as the final result.

In 2021 Orlandi, F et al. [11] presented REF, the first standard for RDF Reification approaches. RDF reification, or writing RDF declarations about RDF assertions, is a data modelling technique. There are several methods for expressing statement-level metadata, often known as "reified" statements, in RDF. This article, examined these metrics using the Standard Reification, Singleton Property, and more modern RDF* reification methods. The author makes available their standard evaluation tools so that the public can grow and compare.

In 2019 Long, Z et al. [12] introduced a contextual features and classification labelling model. Indicators of Compromise (IOCs) are artifacts that can be used to identify processor intrusions and spot cyber-attacks early on. They can be found on a network or in an OS. As a result, they play a significant role in the field of cyber-security. Modern IOCs detection algorithms. This paper suggests utilising an end-to-end neural-based structure labelling model to automatically find IOCs from cyber-security literature without requiring a strong background in cyber-security. The suggested model outperforms conventional sequence labelling models, according to experiments, scoring a normal F1-score of 89.0% on the cyber-security object test set in English and around 81.8% on the Chinese test set.

Sultana, T et al. [13] formed the RDF graph, and a number of significant semantic suggestion rules were derived, along with a suggested compression strategy and an efficient rule mining density method for RDF datasets. In order to improve the mournfulness of rule patterns, find similarities between random pairs of rules, extract the most subtle rules, determine the exact mining threshold, and efficiently learn the rules during the rule mining process from the RDF Knowledge Base, this paper presented grammar-based pattern systems, rules clustering, rules pruning, and Top-k schemes. The proposed rule mining and compression strategy has significantly outperformed the current AMIE+, Rule-based compression, and Triple Bit techniques in terms of compression by approximately 22.10%, 40.5%, and 44%, respectively,

For storing RDF data Zahi, A et al. [14] presented three brand-new non-native method operations. These techniques—known as RDFVP, RDFPC, and RDFSPO—are based on the possessions table, statement table, and vertical partitioning approaches, respectively. The problem of selecting the most appropriate plan for storing RDF data based on dataset characteristics is the main focus of the research. The computational performance achieved by the suggested method is highly competitive.

3   Proposed RDF Statement Generation

The RDF statement generation is needed to process unstructured textual data. The deep learning model for generating RDF statements from the given unstructured textual data is introduced in this research paper. The data required to carry out the proposed approach are collected from BBC News and Lonely Planet datasets. The data are pre-processed using some techniques such as data normalization, Case folding, Stop word removal, sentence segmentation, and PoS (Parts of Speech) tagging. The workflow of the proposed RDF statement generation is depicted in Fig. 1.

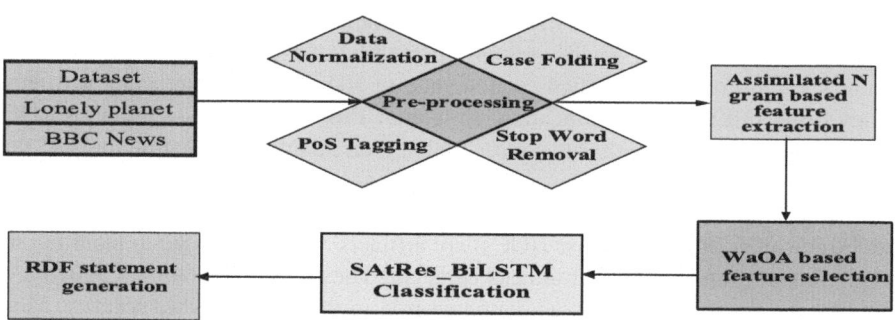

Fig. 1. Workflow of the proposed RDF statement generation

The pre-processed data are fed into the feature extraction process whereas the corresponding entity and attribute feature values are extracted using Assimilated N-gram method. From the extracted features, the optimal features are selected using Walrus optimization (WaOA) algorithm. The Triples are classified using Soft Attention based hybrid ResNet-Bidirectional Long Short Term Memory model (SAtRes_BiLSTM). Finally, the attained triples are then converted and saved in RDF format. The performance of the suggested RDF statement generation is related to diverse emerging methods to prove the efficiency of triple extraction. Also, different performance metrics will be considered for analysing the overall performance.

3.1 Data Acquisition

The proposed model uses two open source datasets namely the BBC news dataset [15] and the Lonely Planet dataset [16], These datasets are input for processing the proposed deep learning model for generating RDF statements from the given unstructured textual data.

3.2 Pre-processing

The pre-processing procedure is to format text in a clear, unified manner so that it can be investigated more effectively. The proposed method uses the following pre-processing approaches.

Data Normalization: By reducing the repetitions by more than two times during the data normalization process, the emotional word "happyyyyy" is used to indicate that "happy" is the correct root word. After that, the word's Levenshtein Distance is calculated to correct any spelling mistakes through updating, deleting, and inserting operations.

Case Folding: In the case folding process identified non-uppercase letters were then converted into uppercase letters.

Stop Word Removal: The process of stop word removal process is to get the words that are used a lot in the text and then remove them because they don't add anything to the understanding of the text.

POS Tagging: POS tagging is also known as grammatical tagging. Based on a word's characteristics and the context in which it appears, POS tags are assigned to it.

3.3 Assimilated N-gram Based Feature Extraction

The feature vector is generated using the Assimilated N-gram, wherein word extraction and TF-IDF are extracted from the N-gram. The term frequency and the inverse term frequency are calculated using the below formula,

$$TF = \frac{freq(t, f)}{|d|} \tag{1}$$

$$IDF = \frac{|D|}{Df(t, D)} \tag{2}$$

where TF represents the term frequency, t and f are respectively term and frequency of the document and D represents the inverse of the document and d represents the document. The ratio between the TF_IDF expressed as,

$$TF\_IDF_{ratio} = \frac{TFIDF \ of \ term \ 't'}{\max(TFIDF \ of \ compund \ of \ 't')} \tag{3}$$

The above formula used to locate important words. TF-IDF based procedure concerns itself with 1 g or 2 g. To normalize the term frequency by using the average frequency, it can be calculated as

$$W_{freq} = \frac{TF}{Averfreq} \tag{4}$$

Word length is intended as the entire distance of the word. Here, $W = w_1, w_2, \ldots\ldots\ldots w_n$ denotes n-gram word, length of the word is represented as,

$$W_{len} = |w_1, w_2, \ldots\ldots\ldots w_n| \tag{5}$$

3.4 Optimal Feature Selection Using WaOA

The significant features from the generated feature vector are selected optimally using the Walrus Optimization (WaOA) Algorithm [17]. The objective of the feature selection is to select the suitable feature that will help to enhance the RDF statement generation accuracy with minimum computation burden. Natural behaviours and social life for the walruses are described as a sharp process of that behaviour of the walrus, three are majorly distinct.

1. Directing others to eat under the direction of the tribe member with the longest tusks
2. Walruses' migration to rocky beaches
3. Combat or flee from predators

From the extracted features, the optimal features are selected using Walrus optimization (WaOA) algorithm. The proposed WaOA consists of three phases,

1. Exploration
2. Migration
3. Exploitation

Exploration: Within the hunting process, the powerful walrus who have the biggest tusks leads the remaining walruses are all set to discover food. The size of the walrus tusk suitable example to express the quality of the goal task values for the nominee results. Accordingly, the best walrus in the group is obtained using which candidate fined the best value for the goal. The procedure of knowing the location of walruses were exactly modelled established on the serving method covered by direction of representative in the group, applying (6) and (7).

$$N_{x,y}^{P1} = N_{x,y} + rand_{x,y} \cdot (BC_y - I_{x,Y} \cdot N_{x,y}) \tag{6}$$

$$N_x = \begin{cases} N_x^{P1}, F_x^{P1} < F_x, \\ N_x, \ else, \end{cases} \tag{7}$$

Here $N_{x,y}$ is the yth decision variable's value suggested by the xth walrus, N_x denotes the xth walrus and F_x denotes the objective value of the xth walrus. N_x^{P1} is the innovative made position for x-th walrus due to the 1st phase, $N_{x,y}^{P1}$ denotes y-th dimension, F_x^{P1} represents objective function rate $rand_{x,y}$ these arbitrary figures from the interval (0, 1). BC is the foremost candidate result it was weight as powerful walrus and $I_{x,y}$ are integers nominated randomly from Eq. (6) and (7).

Migration: The process assumes the WaOA to lead the walruses for searching the hunt place space for finding appropriate zones in the search place. This behavioural process is exactly calculated by applying (8) and (9).

$$N_{x,y}^{P2} = \begin{cases} N_{x,y} + rand_{x,y} \cdot (N_{k,y} - I_{x,y}.N_{x,y}), F_k < F_x; \\ N_{x,y} + rand_{x,y} \cdot (N_{x,y} - N_{k,y}), \ else, \end{cases} \quad (8)$$

$$N_x = \begin{cases} N_x^{P2}, F_x^{P2} < N_x; \\ N_x, \ else, \end{cases} \quad (9)$$

Here F_x^{P2} is an objective function value, N_x^{P2} denotes newly generated point for x-th walrus predicated in 2nd phase. $N_{x,y}^{P2}$ Called the y-th dimension, F_k known as the objective rate of the function. $N_k, K \in \{1, 2, \dots\dots\dots n\}, k \neq x$, it is the position of the particular walrus to emigrate x-th walrus near it, $N_{k,y}$ is its y-th measurement.

Exploitation: Exploitation is an ability in regional identity to investigate space around candidate results. Each walrus is supposed to have a neighbourhood surrounding them, and using (10) and (11), a new random location is first created in this neighbourhood. Then, in accordance with (12), this update location puts back the prior position whether the value of goal purpose is increased.

$$N_{x,y}^{P3} = N_{x,y} + \left(lb_{local,y}^t + \left(ub_{local,y}^t - rand \, .lb_{local,y}^t \right) \right); \quad (10)$$

$$local \ bounds : \begin{cases} lb_{local,y}^t = {lb_y}/{t}, \\ ub_{local,y}^t = {ub_x}/{t}, \end{cases} \quad (11)$$

$$N_x = \begin{cases} N_x^{P3}, F_x^{P3} < F_x; \\ N_x, \ else \end{cases} \quad (12)$$

Here N_x^{P3} represents update created point for the x-th walrus created in 3rd phase. F_x^{P3} is the objective function value, $N_{x,y}^{P3}$ is its y-th dimension, lb_y and ub_x are the subordinate and superior bounds for y-th value, t is the repetition contour, $lb_{local,y}^t$ and $ub_{local,y}^t$ are native lower and native superior bounds permissible for the y-th value.

3.5 SAtRes_BiLSTM Based RDF Statement Generation

The RDF statement generation is devised using the proposed Soft Attention based hybrid ResNet-Bidirectional Long Short Term Memory model (SAtRes_BiLSTM), which is designed by integrating the Soft Attention module, Residual Neural Network, and BiLSTM.

BiLSTM: BiLSTM is able to reserve the location details of text and detention the connected features of text. The social platform text associated with the typical of short text [18]. Problems of less contextual information can be effectively solved by the mechanism

BiLSTM and dubious significance in the short text. In full thought of the context, BiL-STM certifies that every word can attain additional semantic information. From the Fig. 2 $\vec{K}_s$ forward LSTM layer output sequence, $\overleftarrow{K}_s$ backward LSTM layer output sequence, b_s denotes the output sequence a_s denotes the input sequence. b_{s-1} implies the last output, b_{s+1} denotes the first output sequence, a_{s+1} denotes the first input sequence, a_{s-1} implies the last input. $\vec{K}_{s-1}$ and $\vec{K}_{s+1}$ represents the last and first forward LSTM layer sequences. $\overleftarrow{K}_{s-1}$ and $\overleftarrow{K}_{s+1}$ represents the last and first backward LSTM layer sequences respectively. The given sentence of x-words represented through $S = (W_1, W_2, ...W_X)$, the model implants the starting input text among through previous trained inserting matrix. The text feature of the matrix can be characterised as $M \in R^{x \times d}$ here M is the matrix, d is the measurement of word embedding, and x is the vocabulary extent of the given sentences S. BiLSTM text feature represented as $F = (f_1, f_2, ...f_x)$, moreover $f_i = R^{x \times df}$ describe the text feature of ith word. POS information of text is established to further remove the problem. POS feature is denoted as $P = (p_1, p_2, ...p_x)$, where the POS feature of ith word is represented $p_i \in R^{x \times df}$. The text purpose expressed as F and its similar POS feature illustrations P are combined to deeper extraction and produce feature representations $FP = (fp_1, fp_2, ...fp_x)$. The concatenation of the POS feature and Text feature is calculated using the formula (13).

$$fp_i = ConCate\,(f_i, p_i) \tag{13}$$

Soft Attention Module: The plan of attention was derived by the examination that every word provides the content of a sentence inversely especially, not all words contribute regularly to illustration of the sentence definition [19]. The operators read text innately to detect essential, meaningful portions of the sentence. An attention model enhances an extra construction to the linkage, which ever trained as section of the usual model training through back-propagation. As an acceptance, soft attention is used to find where the weights on all positions are generally obtained from Softmax. The result of the attention module is loaded sum description at each location [20]. An attention device provides an attention mark $e_{s,w}$ to all word w and sentence s assumed by (14)

$$e_{s,w} = x(Wh_w, c) \tag{14}$$

where wh denotes the attention mechanism, wh_w denotes the attention mechanism for word w, c denotes the propagation function. x denotes the activation function. Now the probability $a_{s,w}$ for each Wh_w is calculated by (15)

$$a_{s,w} = \frac{\exp(e_{s,w})}{\sum_{k=1}^{W} \exp(e_{s,w})} \tag{15}$$

Residual Network: The residual is the direct connection between two successive layers, the network has short connections. Therefore, it is simple to optimize the deeper networks by applying residual connections; this will increase accuracy in terms of results. Better to reduce terminated features and combine features, so using the residual network. ResNet initiates an avoided residual operation, this is one of the main differences between an ordinary convolutional network and the residual network [21]. The function of the residual network is to avoid the degradation and gradient disappearance caused by the networks and reduce the characteristic parameters. Figure 3 shows the architecture of the residual network. The equation for the residual network is given below (16),

$$Q = WM_x \delta(WM_{x-1}P_{x-1}) + P \qquad (16)$$

Here Q represents output layer x, δ represents the ResNet activation function, WM denotes the mass matrix, P denotes current input layer. WM_x denotes the max matrix of xth position, WM_{x-1} denotes the mass matrix previous position of x and p_{x-1} previous initial position of the xth layer.

The opening and deep residual approaches of the ResNet is calculated in Eq. (17) and (18).

$$P_0 = H_l(P_{l-1}) + P_{l-1} \qquad (17)$$

$$P_Q = H_l(Q_{l-1}) + Q_{l-1} \qquad (18)$$

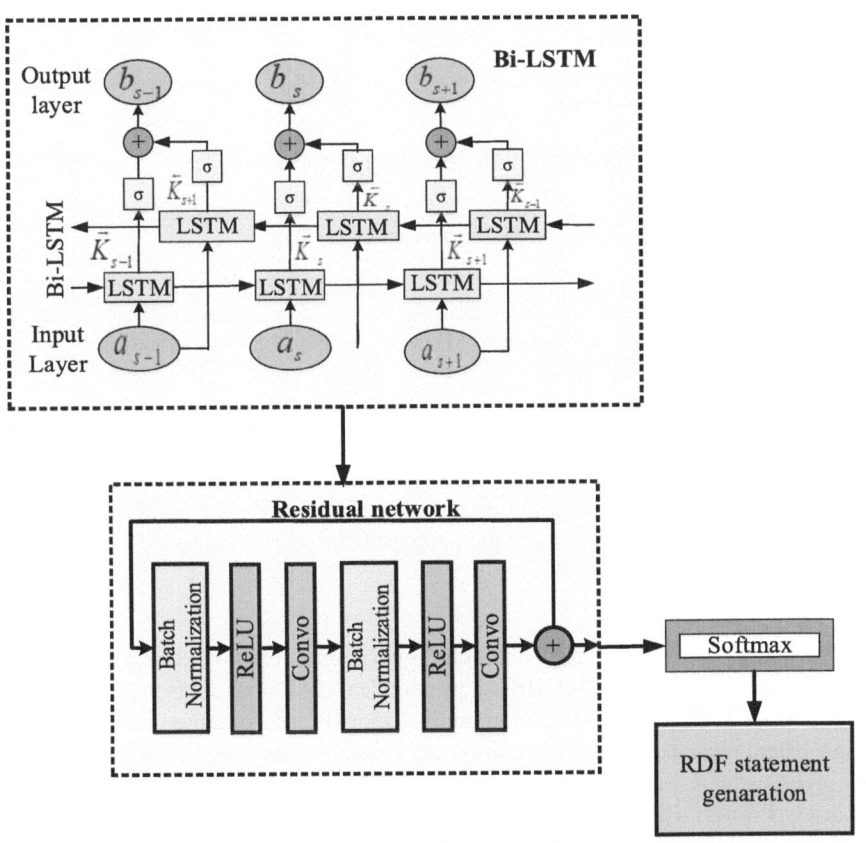

Fig. 2. Architecture of proposed SAtRes_BiLSTM

where P_0 represents the initial feature intensive operation, H_l is the feature combination strategy, P_{l-1} is the previous state of input layer, P_Q denotes the output feature of ResNet, Q_{l-1} denotes previous state of output layer.

Proposed SAtRes_BiLSTM: The Proposed SAtRes_BiLSTM integrates, the BiL-STM, Soft attention module and ResNet. Initially, BiLSTM takes the selected features as its input for extracting the long term dependent features. Then, the soft attention mechanism is utilized for assigning the higher weights to the significant features. After weighting the features, the Residual network is utilized for mapping the weighted features through the activation functions. Finally, using the Softmax classification, the RDF statement generation is devised. The architecture of the Proposed SAtRes_BiLSTM is depicted in Fig. 2.

Thus, using the proposed SAtRes_BiLSTM, the triple is obtained through the Softmax activation function.

3.6 RDF Statement Generation

The outcome of the proposed SAtRes_BiLSTM is the subject, object and predicate for the input text document. The concatenation of subject, predicate and object is performed to generate the RDF statement.

4 Result and Discussion

Table 1 shows the system configuration details for proposed model.

Table 1. System configuration details

Device specification	Details
Processor	Intel(R) Core(TM) i5–3470 CPU @ 3.10 GHz 3.10 GHz
Installed RAM	8 GB
System type	64-bit (Windows 10) operating system, x64-based processor

Softmax is used as the activation function, and dropout factor is set to be 0.2 is the hyperparameter details to implement the proposed classification model.

4.1 Dataset Description

The processing of the suggested statement generation is employed using two various unstructured textual datasets like BBC news and Lonely Planet datasets.

BBC News Article: BBC dataset include 2225 text forms obtained by the BBC news website. There are categorized into five types. Then trimming, aspect of the BBC News space includes 5834 features.

Lonely Planet: There are 96 classes of unstructured documents related to tourism spread across 1801 files. On 300 web pages, the textual description describes popular tourist destinations worldwide.

4.2 Performance Measures

Accuracy, F-Measure, Recall and Precision [22] are utilized for measure the concept of the proposed SAtRes_BiLSTM statement generation process.

4.3 Comparative Analysis

The proposed SAtRes_BiLSTM RDF statement generation technique is matched with several existing approaches. The comparison is performed using corresponding dataset predicated on various performance measures like precision, recall, and accuracy. The detailed analysis of all the performance measures is presented in Table 2.

Table 2: The comparison between existing techniques with suggested RDF statement generation

Model	Dataset	Performance		
		Precision	Recall	Accuracy
Ontology [23]	DBpedia	0.95	0.79	0.87
linear-chain CRF [24]	OIE2016 benchmark framework,	0.32	0.30	0.67
multimodal KG [25]	MSCOCO	0.90	0.60	0.92
ontology learning methods [26]	Brown Corpus	0.74	0.49	0.67
KG Embedding framework [27]	Twitter graph	0.74	0.83	0.77
ASKG [28]	Unruptured	0.76	0.90	0.73
SAtRes_BiLSTM	**BBC news, Lonely Planet dataset**	**0.912**	**0.90**	**0.95**

The accuracy estimated by the proposed SAtRes_BiLSTM is 0.95 using the Lonely Planet and BBC news dataset, which is 87%, 67%, 92%, 67%, 77% and 73% superior outcome compared to the existing methods.

4.4 Discussion

The analysis of the best outcome evaluated by the approaches is portrayed in Table 2. The highest accuracy estimated by the suggested SAtRes_BiLSTM RDF statement generation technique is 0.9897, which is 3.5%, 5.5%, 2.6%, 3.4% and 2.9% improved performance compared to CART, AdaBoost, XGBoost, GBDT, LightGBM existing methods. The highest precision estimated by the proposed SAtRes_BiLSTM RDF statement generation technique is 0.921, which is 6.3%, 14.2%, 3.3%, 6.2%and 4.1% improved performance compared to existing methods. The highest recall estimated by the proposed SAtRes_BiLSTM statement generation technique is 0.90, which is 10%, 23%, 14%, 9% and 3.24% improved performance compared to existing methods. The highest

F-Measure estimated by the proposed SAtRes_BiLSTM statement generation technique is 0.932, which is 11%, 15%, 7%, 10%, 9% improved performance compared to existing methods. Detail about the analysis in given in Table 2 and 3 the Fig. 3 details about all the presentation.

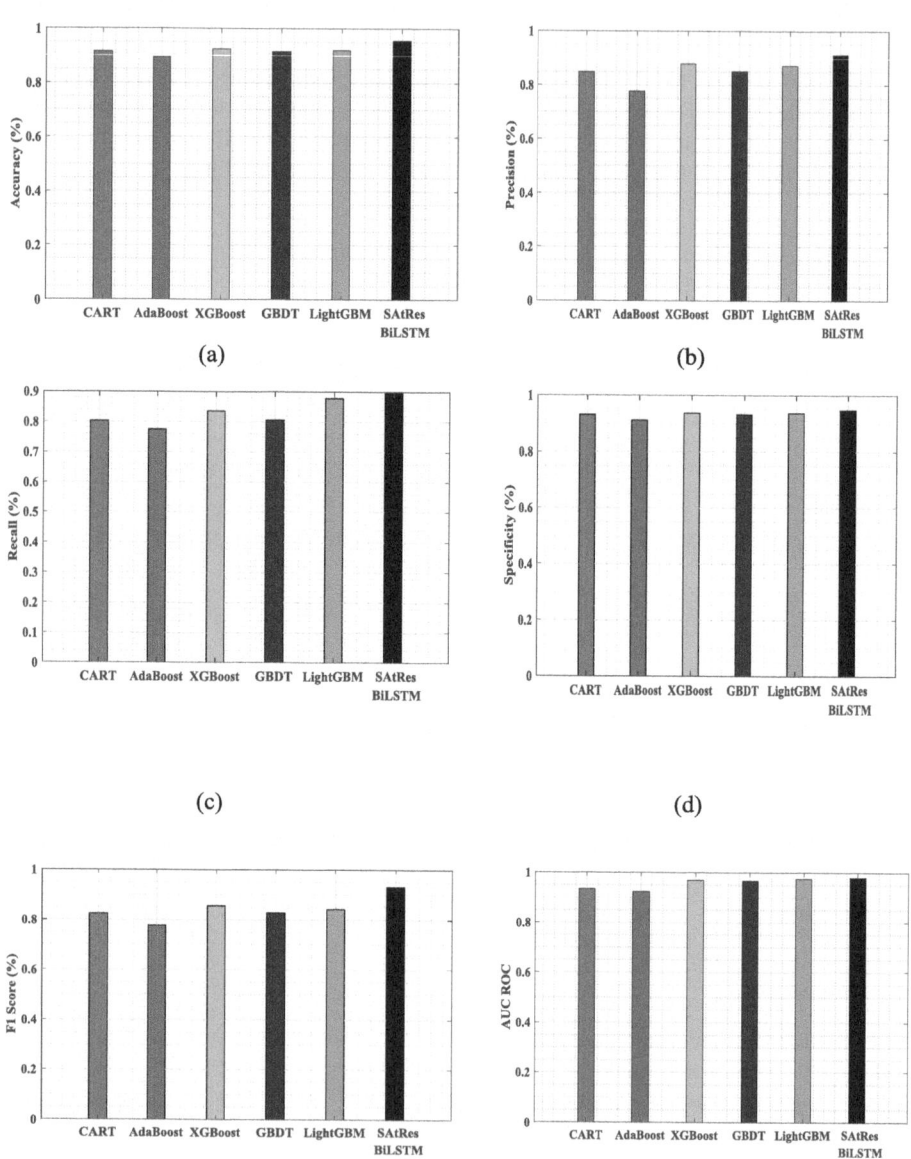

Fig. 3. Comparative Assessment in terms of (a) Accuracy, (b) precision (c) Recall and (d) Specificity, (e) F1 score and (f) AUC ROC

Table 3. Mean and standard deviation analysis of proposed and existing performance [29]

Model	Mean	Standard deviation
CART	0.0347	0.0344
AdaBoost	0.0047	0.0347
XGBoost	0.0753	0.0387
GBDT	0.0465	0.2518
LightGBM	1.3178	0.6843
Proposed (SAtRes_BiLSTM)	2.2456	0.0507

Table 3 gives the analysis about standard deviation and mean results.

5 Conclusion

A novel deep learning model for generate RDF statement from the given unstructured textual data is presented in this research work. The proposed work used two datasets namely BBC news, Lonely Planet dataset. Different pre-processing steps are used in this research to remove the unwanted noises. The most particular features are taken out from the pre-processed textual data using a new N-gram method. The most optimal features are selected using WaOA to reduce the feature dimensionality issue. Then the selected features can be effectively categorized using SAtRes_BiLSTM. Finally, the performances will be compared with the existing models to analyze the model's efficacy. The evaluation of the given model resulted in 95% of accuracy, 91.2% of precision, 90% of recall, 94.8% of specificity, 98% of AUC and 93.2% of F1-score in the given dataset. In the future, a mathematical classification technique could be used to predict the optimal solution using the enhanced optimization strategies and also, more datasets would be considered.

References

1. Zangeneh, P., McCabe, B.: Ontology-based knowledge representation for industrial megaprojects analytics using linked data and the semantic web. Adv. Eng. Inform. **46**, 101164 (2020)
2. Molina, A.I., Toledo, J., Corcho, O., and Fraga, D.C.: Re-construction impact on etadata representation models (2023)
3. Le, N.L., Abel, M.-H., Gouspillou, P.: Improving semantic similarity measure within a recommender system based-on RDF graphs. In International Conference on Information Technology & Systems, pp. 463–474. Springer, Cham (2023)
4. Sarkar, A., Koley, S., Marjit, U., Biswas, U.: A framework for generation of RDF data from HTML. Indian J. Sci. Technol. **8**(16), 1 (2015)
5. Davahli, M.R., et al.: Identification and prediction of human behavior through mining of unstructured textual data. Symmetry **12**(11), 1902 (2020)
6. Ali, W., Saleem, M., Yao, B., Hogan, A., and Ngomo, A.-C.N.: A survey of RDF stores & SPARQL engines for querying knowledge graphs. VLDB J. 1–26 (2022)

7. Rupp, F., Schnabel, B., Eckert, K.: Easy and complex: new perspectives for metadata modeling using RDF-star and Named Graphs. In: Iberoamerican Knowledge Graphs and Semantic Web Conference, pp. 246–262. Springer, Cham (2022)

8. Kiselev, B., Yakutenko, V.: An overview of massive open online course platforms: personalization and semantic web technologies and standards. Procedia Comput. Sci. **169**, 373–379 (2020)

9. Pareti, P., Konstantinidis, G.: A review of shacl: from data validation to schema reasoning for rdf graphs. Reasoning Web International Summer School, 115–144 (2021)

10. Akter, Y.A., Rahman, M.A.: Extracting rdf triples from raw text. In: 2019 1st International Conference on Advances in Science, Engineering and Robotics Technology (ICASERT), pp. 1–4. IEEE (2019)

11. Orlandi, F., Graux, D., O'Sullivan, D.: Benchmarking RDF metadata representations: Reification, singleton property and RDF. In: 2021 IEEE 15th International Conference on Semantic Computing (ICSC), pp. 233–240 (2021)

12. Long, Z., Tan, L., Zhou, S., He, C., Liu, X.: Collecting indicators of compromise from unstructured text of cybersecurity articles using neural-based sequence labelling. In: 2019 International Joint Conference on Neural Networks (IJCNN), pp. 1–8. IEEE (2019)

13. Sultana, T., Lee, Y.-K.: Efficient rule mining and compression for RDF style KB based on Horn rules. J. Supercomput. **78**(14), 16553–16580 (2022)

14. Chaker, I., Zahi, A.: An empirical study on the evaluation of the RDF Storage Systems. (2021)

15. Dogru, H.B., Tilki, S., Jamil, A., Hameed, A.A.: Deep learning-based classification of news texts using doc2vec model. In: 2021 1st International Conference on Artificial Intelligence and Data Analytics (CAIDA), pp. 91–96. IEEE (2021)

16. Cimiano, P.: Ontology Learning and Population from Text: Algorithms, Evaluation and Applications. Springer, vol. 27 (2006)

17. Trojovský, P., Dehghani, M.: A new bio-inspired metaheuristic algorithm for solving optimization problems based on walruses behavior. Sci. Rep. **13**(1), 8775 (2023)

18. Tang, H., Mi, Y., Xue, F., Cao, Y.: An integration model based on graph convolutional network for text classification. IEEE Access **8**, 148865–148876 (2020)

19. Luong, M.-T., Pham, H., Manning, C.D.: Effective approaches to attention-based neural machine translation (2015). arXiv preprint arXiv:1508.04025

20. Xu, K., et al.: Show, attend and tell: neural image caption generation with visual attention. In International Conference on Machine Learning, pp. 2048–2057. PMLR (2015)

21. Cai, W., Liu, B., Wei, Z., Li, M., Kan, J.: TARDB-Net: triple-attention guided residual dense and BiLSTM networks for hyperspectral image classification. Multimed. Tools Appl. **80**, 11291–11312 (2021)

22. Berger, A., Guda, S.: Threshold optimization for F measure of macro-averaged precision and recall. Pattern Recogn. **102**, 107250 (2020)

23. Rosyiq, A., Hayah, A.R., Hidayanto, A.N., Naisuty, M., Suhanto, A., Budi, N.F.A.: Information extraction from Twitter using DBpedia ontology: Indonesia tourism places. In: 2019 International Conference on Informatics, Multimedia, Cyber and Information System (ICIMCIS), pp. 91–96. IEEE (2019)

24. Niklaus, C., Cetto, M., Freitas, A., Handschuh, S.: A canonical context-preserving representation for open IE: extracting semantically typed relational tuples from complex sentences. Knowl.-Based Syst. **268**, 110455 (2023)

25. Norabid, I.A., Fauzi, F.: Rule-based text extraction for multimodal knowledge graph. Int. J. Adv. Comput. Sci. Appl. **13**(5) (2022)

26. Kung, H.-Y., Yu, R.-W., Chen, C.-H., Tsai, C.-W., Lin, C.-Y.: Intelligent pig-raising knowledge question-answering system based on neural network schemes. Agron. J. **113**(2), 906–922 (2021)

27. Abu-Salih, B., et al.: Relational learning analysis of social politics using knowledge graph embedding. Data Min. Knowl. Disc. **35**(4), 1497–1536 (2021)
28. Malik, K.M., Krishnamurthy, M., Alobaidi, M., Hussain, M., Alam, F., Malik, G.: Automated domain-specific healthcare knowledge graph curation framework: Subarachnoid hemorrhage as phenotype. Expert Syst. Appl. **145**, 113120 (2020)
29. Cui, Q., Chen, Q., Liu, P., Liu, D., Wen, Z.: Clinical decision support model for tooth extraction therapy derived from electronic dental records. J. Prosthet. Dent. **126**(1), 83–90 (2021)

Evaluating the Tri-Script Writer Verification System Using a Handcrafted Features and Vision Transformer Learning Approach

Jaya Paul[1(✉)], Kalpita Dutta[1], Anasua Sarkar[1], Kaushik Roy[2], and Nibaran Das[1]

[1] Department of Computer science and Engineering, Jadavpur University,
Kolkata, India
jayapl2005@gmail.com, anasua.sarkar@cse.jdvu.ac.in,
nibaran.das@jadavpuruniversity.in
[2] Department of Computer science, West Bengal State University, Barasat, India

Abstract. A multi-script writer verification system poses a captivating research challenge in the fields of pattern recognition and computer vision. It entails the intricate task of authenticating the identities of writers who produce handwriting samples in various scripts. This scenario introduces complexities demanding robust techniques for accurate recognition across diverse scripts. In West Bengal, the native language is Bangla, with Hindi serving as the second language, while English finds use in official contexts. This study tackles the issue of block-level tri-script (Bangla, Hindi, and English) writer verification systems and presents promising results. The research demonstrates that a combination of handcrafted features outperforms automatically derived features, owing to limitations in the writers within the JUDVLP-TLWVdb dataset. Experimental results indicate that the SMO classifier outperforms other classifiers such as simple logistics and KNN. A novel dataset for writer verification systems using the tri-script approach is introduced, achieving a peak verification accuracy of 91.50% through a combination of Radon Transform, HOG, LBP, and LPQ features. The overall performance of the tri-script approach reaches 91.80%. Furthermore, this study employs the Vision Transformer (ViT) model for writer recognition, showcasing superior performance in comparison to ViT using tri-level block images of the page.

Keywords: Writer verification system · Tri-script · Vision transformer · Text dependent

1 Introduction

Digital technology plays a crucial role in our daily lives, and the ability to recognize and interpret diverse forms of handwriting has become increasingly important. Handwriting-based text recognition, specifically in the context of tri-scripts, presents a challenging area of research [1]. The challenge lies in handling the variations in handwriting style across different scripts. Each script may have its own distinct characteristics and challenges, requiring the system to be adaptable and robust in recognizing and verifying the writer's identity regardless of the script used. Multi-script writer recognition involves identification [2] and verification of handwritten text across different writing systems. Writer verification [3] can be framed as a binary classification problem, where the classifier determines whether a manuscript is authored by a specific writer or not. On the other hand, identification of the writer [4] is considered an n-class problem, which involves the task of identifying the writer of a given handwritten specimen from a set of n potential writers. There are twenty-three (23) official languages in India, each with its own unique script. In total, there are eleven (11) different scripts used to write these languages. Devanagari is the most widely used script across the country. Additionally, English is also commonly used as a script for communication in India, similar to many other countries. For the present work, three scripts were considered: Bangla (used in West Bengal), Devanagari, and English. It is important to emphasize that Bangla and Devanagari scripts are more complex than English, primarily due to the inclusion of compound characters.

The recognition of multi-script handwriting offers a wide range of practical applications. It can enhance the accuracy and efficiency of optical character recognition (OCR) systems [5], enabling the digitization of handwritten documents, historical manuscripts, and multilingual texts. Additionally, it has the potential to revolutionize the field of automated translation, facilitating seamless communication between individuals using different scripts.

The complexity of multi-script writer recognition arises from the inherent variations in handwriting styles, stroke patterns, and structural characteristics across different scripts. Moreover, the presence of cursive writing, ligatures, diacritical marks, and other script-specific features further adds to the complexity of accurate recognition.

To tackle these challenges, researchers and experts in the field have been developing innovative techniques that combine machine learning and deep neural networks [6].

There are two categories of offline writer verification based on the content of writing: text-independent [7] and text-dependent [8]. Text-independent writer verification involves analyzing images of arbitrary texts without relying on fixed text content. In contrast, text-dependent methods require input images with static (fixed) text contents and compare them with registered templates for verification.

Abbas et al. [9] introduced a multi-script writer identification method that incorporates texture features. The authors investigated the effectiveness of using local binary patterns (LBP) column histogram and oBIFs column histogram, along with an SMO classifier. They conducted evaluations on both single-script and multi-script datasets, yielding promising performance results.

This paper introduces an approach for writer verification systems in block-level handwriting scripts, specifically focusing on tri-script datasets. Our approach utilizes various features, including writing orientation analysis and textures, to enhance the verification process. Figure 1 shows our proposed model. In our analysis, we utilized four features: Radon Transformation, Local Binary Pattern (LBP), Histogram of Oriented Gradients (HOG), and Local Phase Quantization (LPQ). To improve the verification rate, we propose a study that combines these features based on their performances at the block level of the document, using the SMO classifier. However, due to the limited number of handwriting samples available from the writers, we are unable to achieve better results using the Vision Transformer (ViT) model [10].

The paper includes the following sections: Sect. 2 presents related works. Section 3 provides a detailed description of the methodology employed in our work, covering aspects such as datasets, data preprocessing techniques, used features, and the experimental framework. Section 4 reports and analyzes the results obtained from the experiments, offering a comprehensive overview of the performance and effectiveness of the proposed model. Finally, in Sect. 5, we summarize the conclusions derived from our proposed model and explore potential directions for future research and enhancement.

2   Related Work

In recent years, transfer learning methods have surpassed numerous other neural network architectures in performance. Sahoo et al. [11] showcased the remarkable effectiveness of transfer learning in large-scale image classification specifically in their proposed approach for offline multi-script handwritten character classification. Roy et al. [12] introduced a Convolutional Neural Network (CNN) based approach for recognizing handwritten multilingual multiscript Indian city names. The experiments encompassed multiple scripts including English, Bangla, and Devanagari. When tested on 106 city names, their method achieved an accuracy of 91.72%.

The Transformer, a revolutionary neural network, has garnered recognition from the academic community for its groundbreaking capabilities, showcasing immense potential in the field of artificial intelligence (AI). Specifically, the Vision Transformer (ViT) has yielded remarkable outcomes in image recognition tasks [13]. Drawing inspiration from this success, ViT has been rightfully employed to tackle other challenging problems in computer vision tasks. For instance, the ViT (Vision Transformer) model, as described in the research paper by Geng et al. [14], has been successfully applied in offline handwritten Chinese character recognition. The LW-ViT (Lightweight ViT) model draws inspiration from MobileViT, with a specific emphasis on parameter and FLOP reduction by minimizing transformer blocks and integrating the MV2 (MobileNetV2) layer into the MobileViT framework. Consequently, the lightweight ViT model achieved an impressive recognition accuracy of 95.8% on the CASIA-HWDB dataset. Dhiaf et al. [15] introduced the MSdocTr-Lite model for page-level multiscript handwriting recognition. Their proposed model addresses the challenge of data scarcity by offering the ability to train on a reasonable amount of data due to its lightweight architecture. Unlike existing segmentation-based approaches, this model can be easily fine-tuned on various scripts without relying on segmentation labels. This is achieved through a straightforward transfer learning process.

3 Proposed Method

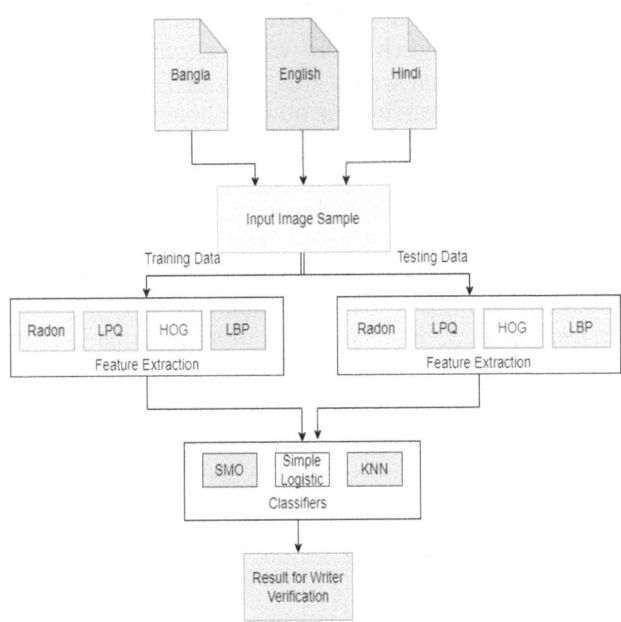

Fig. 1. Block diagram for proposed multi-script writer verification system framework

3.1 Dataset

In this study, we have developed a handwritten tri-script writer verification system dataset called the Jadavpur University Deep Learning in Vision and Language Processing Tri language Writer Verification Database (JUDVLP-TLWVdb dataset). India, being a highly multilingual country, has over a dozen different spoken languages. Hindi and Bengali are two of the most extensively spoken languages in India, alongside the official use of English. To the best of our knowledge, there is currently no publicly available benchmark dataset for a tri-language writer verification system. Therefore, we have created a new offline handwritten tri-language database, as shown in Fig. 2. The dataset comprises samples collected from 31 Indian writers at the document level. Although all writers are native Bengali speakers, they are also fluent in Hindi and proficient in reading and writing English. Bengali documents contain an average of 42 words, English documents consist of approximately 40 words, and Hindi documents typically contain 48 words. Furthermore, the word 'dokan' appears a maximum of 4 times in Bengali, 'shop' occurs a maximum of 3 times in English, and 'dukan' is found up to 3 times in Hindi.

Each writer was instructed to write the same content in multiple languages five times using a standard ball pen with blue or black ink during the data collection phase. The faculty members of the Computer Science and Engineering Department at Jadavpur University assisted in collecting this dataset. For text-dependent analysis, we collected the same writing content in multiple languages at different time intervals. Each writer was provided with A4 size paper and a pen for writing, without any restrictions on the choice of writing equipment. To ensure uniformity and consistency, we designed a data collection form and used multi-script text containing the same meaning for the experiments. In our experiment, we considered various factors such as age, sex, date, time, and vernacular language. Age can change how well people can control their hand movements and create unique handwriting. Gender-related factors, like body differences, may cause handwriting variations. Handwriting can differ depending on someone's mood, alertness, or how they feel, and can also change on different days or times. We included vernacular language to see if our writer verification methods work with local languages in multilingual documents.

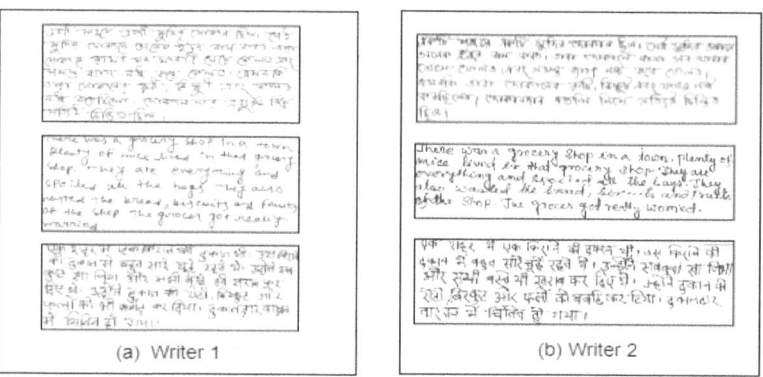

(a) Writer 1   (b) Writer 2

Fig. 2. Sample database images for (a) writer 1, (b) writer 2.

3.2   Data Preprocessing

All the handwritten document text pages are scanned by a scanner (HP Laser-Jet Pro M1136) at 8-bit gray levels with 300 dpi (dot per inch) to produce 2481×3507 dimension digital text images. We label the digital text image data

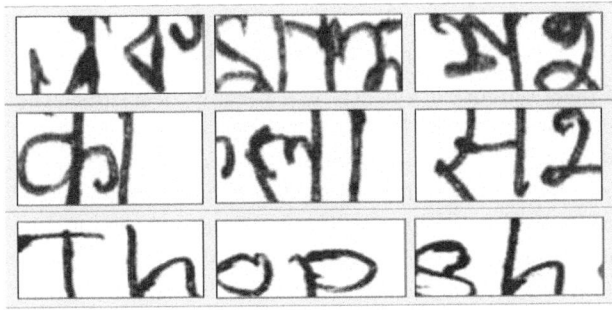

Fig. 3. Block level tri scripts sample data

as per the writer's number and set the number at the preprocessing stage. To convert a digital colour image into a grayscale image, the pixel values are adjusted within the range of 0 to 255 based on their intensities derived from the colour space. We use GIMP to extract the handwritten text content from the gray image. Before using the 'crop' tool on the handwritten text, we correct the skew of the gray images. After the grayscale image is obtained, it can be automatically converted into a binary image using the threshold acquired through Otsu's method [16]. Then we apply the minimum bounding box algorithm [17] on the binary text image to find the minimum-area enclosing rectangle. Figure 3 shows the block level tri-script sample data.

3.3 Hand Crafted Feature

F_{RT}. The Radon Transform [18] is a mathematical technique used in image processing and computer vision to extract information about the linear structures present in an image. It works by projecting the image onto a set of parallel lines and measuring the intensity of the projection at each angle. This feature extracts structure information of provided input images. The dimension of the feature is 180-D for our experiment.

F_{HOG}. The Histogram of Oriented Gradients (HOG) [19] is a widely employed feature descriptor in tasks such as object detection and image recognition. It represents the local gradient or edge directions in an image. HOG works by dividing the image into small 3×3 cells, computing the gradient orientation within each cell, and then grouping the gradient orientations into histograms. The resulting histograms represent the distribution of edge orientations in the image. This feature dimension is 81-D.

F_{LBP}. The texture descriptor Local Binary Patterns (LBP) [20] is commonly employed in image analysis and pattern recognition. It encodes the local texture information of an image by comparing the intensity of a central pixel with its surrounding neighbours. This comparison results in a binary pattern, where each bit represents the relative intensity relationship between the central pixel and its neighbours. The resulting feature dimension is 236-D.

F_{LPQ}. Local Phase Quantization [21] is a texture descriptor that captures the phase information of an image. It represents the local phase variations by quantizing the phase values of the image pixels. LPQ is robust to illumination changes and is commonly used in texture classification and image retrieval tasks. The descriptor is computed by comparing the local phase relationship between pixel pairs in the image, resulting in a compact and informative representation of the image's texture. 256-D.

3.4 $F_{Combine}$

The total combination of features dimension FD_{total} is follows:

$$F_{Combine} = F_{RT} \cup F_{HOG} \cup F_{LBP} \cup F_{LPQ} = 753 \; dimensions \tag{1}$$

3.5   Auto Generated Feature

In this paper, we employ auto-derived features in conjunction with the Vision Transformer (ViT) model. This is depicted in Fig. 8. The ViT model is a deep learning architecture that extends the Transformer model, originally designed for natural language processing tasks, to computer vision tasks. The introduction of the ViT model was presented in [13]. Traditionally, convolutional neural networks (CNNs) have been the prevailing architecture for computer vision tasks, particularly image classification. However, the ViT model takes a distinct approach by leveraging the Transformer architecture, which has exhibited considerable success in natural language processing tasks.

The fundamental concept of ViT revolves around treating an image as a sequence of patches, with each patch representing a small spatial region within the image. After extracting patches from the input image, these patches are flattened into a sequence of 1D vectors. These vectors are then fed into the Transformer model for further processing.

The Transformer architecture incorporates self-attention layers, which allow the model to focus on various patches and capture their interdependencies. Moreover, ViT incorporates positional embeddings to encode the spatial information of the patches, enabling the model to understand the relative positions and arrangements of the image elements. In this paper, we employ the ViT-B model with 16 units for patch embedding. For optimization, we utilize SGD with a fixed learning rate of 0.05 and a momentum of 0.9. The model is trained for 130 epochs, with a batch size of 128. The activation function employed is ReLU, and the training process is parallelized with 32 workers (Fig. 4).

3.6   Experimental Framework

Our experiments were performed using MATLAB R2017b on an i5-8250U core processor with a 1.60 GHz clock and 8 GB RAM using the proposed method

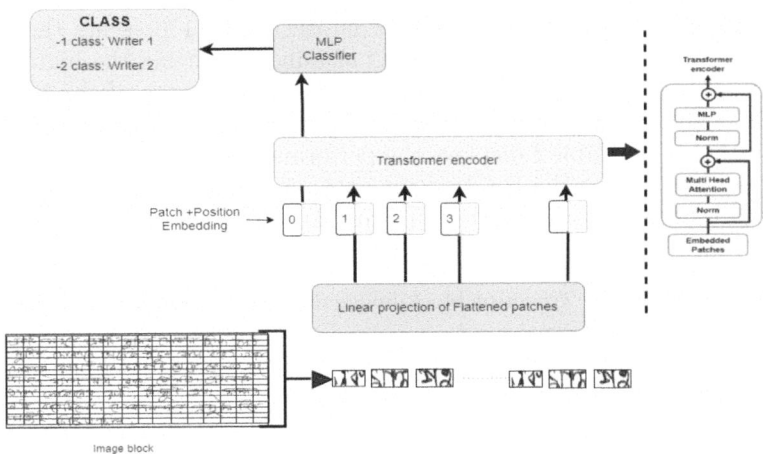

Fig. 4. Architecture of vision Transformer (ViT) model

with a hand-crafted feature set. Additionally, we used Pytorch to run our ViT experiments on a computer with an Intel Core i5-9300H processor clocked at 2.40 GHz, an x64 architecture, 8 GB of RAM, and a 32.51 GB GPU. In our experiments, we utilized version 3.6 of the WEKA software [22] as our simulation tool. We employed three common shallow learners as classifiers: Sequential Minimal Optimization (SMO), K-Nearest Neighbor (KNN), and Simple Logistic. These classifiers were selected based on their widespread usage in various tasks related to handwritten script recognition, including identification and verification, as confirmed by our study. We are using the newly developed JUDVLP-TLWVdb dataset for our experiments. The dataset distribution for the experiment is as follows: a total of 443 pages (148 pages in Bangla + 147 pages in Hindi + 151 pages in English) from 31 writers. The ratio of training to testing data in the JUDVLP-TLWVdb dataset is 3:2. Training and testing data samples are not overlapping. Non overlapping pair writer selection of train and test data distribution is the same. After the pre-processing steps, as explained in Sect. 3.2, the text image is divided into 16 × 16 rows and columns as in the original ViT architecture, the input image is divided into fixed-size patches. A total of 256 blocks are formed on every page. In the training set, on average 2014 blocks level data are considered and on average 1236 blocks data are considered in the test set.

4   Result and Analysis

In this section, we present the results obtained from our experiments. Table 1 showcases a comparison chart that illustrates the performance of our proposed method, which integrates hand-crafted feature sets, in terms of accuracy, precision, recall, and F-Measure. These experiments were conducted using our newly proposed block-level JUDVLP-TLWVdb dataset. Table 1 presents the average performance of 31 writers, with the SMO classifier achieving the highest accuracy of 91.50%. Figure 5 demonstrates that the performance of the block-level JUDVLP-TLWVdb dataset surpasses that of the SimpleLogistic and KNN classifiers. Furthermore, Table 2 displays the performance of a single script which is 89–94%, indicating that the Hindi script outperforms Bangla and English. It is noteworthy that the tri-script overall performance is 91.80%.

Table 1. Tri-script block image dataset performance

Classifier Name	Precision (%)	Recall (%)	F-Measure(%)	Accuracy (%)
SimpleLogistic	86.27	90.53	90.46	90.52
SMO	**91.78**	**91.50**	**91.42**	**91.50**
KNN	74.63	73.14	73.20	73.08

Table 2. Single-script block image dataset performance using SMO classifier

Script Name	Precision (%)	Recall (%)	F-Measure (%)	Accuracy (%)
Bangla	91.55	91.46	91.44	91.45
English	90.09	89.99	89.92	89.97
Hindi	**93.92**	**93.89**	**93.89**	**94.00**

Table 3. Performance based on vision Transformer of block level tri-script writer verification system task

Model Name	Train Loss (%)	Test Loss(%)	Train Accuracy(%)	Testing Accuracy(%)
ViT-B	6.528	1.969	95.46	74.80

In Table 1, it can be observed that the SMO classifier outperforms the SimpleLogistic and KNN classifiers. The workflow of the writer verification methodology utilizing the Vision Transformer (ViT) model is depicted in Table 3. In the JUDVLP-TLWVdb database, we have a total of 31 authors, each with specific training and testing sets. The training and testing ratio is set at 3:2. The verification model is capable of determining whether two multi-script text samples (Bangla, English, and Hindi), namely text1 and text2, belong to the same or different authors. Our model achieves a multi-script verification accuracy of 74.80%.

Table 3 compares the combined handcrafted features model with the deep learning-based ViT model using the same JUDVLP-TLWVdb database. We have provided average training and testing loss of different writers in this Table. Due to the limited number of handwritten samples per writer, the performance does

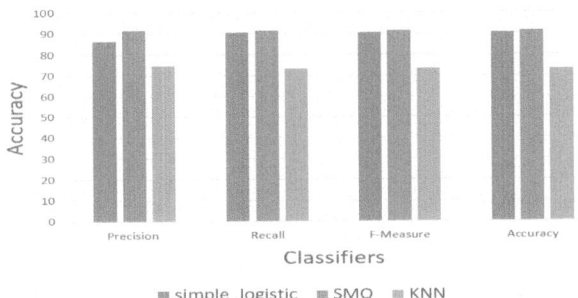

Fig. 5. Comparison of different classifiers performance of Block level dataset

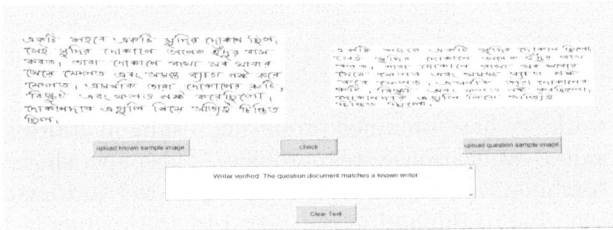

Fig. 6. The known writer document and the question writer document are from the same writer.

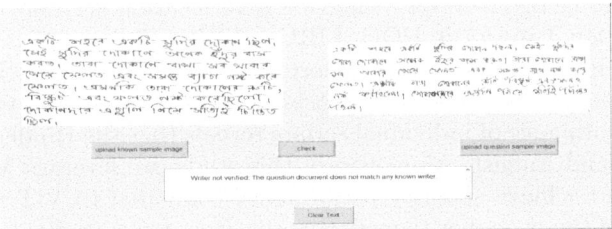

Fig. 7. The known writer document and the question writer document are from different writers.

not improve significantly with auto-derived features compared to handcrafted features. The layout of the writer verification system that we have deployed is depicted in Figs. 6 and 7. We have provided the graph of validation accuracy and validation loss in Fig. 8.

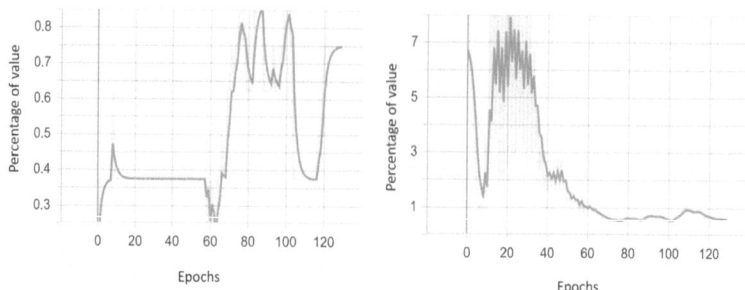

Fig. 8. The graph of validation accuracy and validation loss of writer verification method

5   Conclusion

In our present work, we have obtained promising results in addressing the block-level tri-script writer verification system problem. This paper showcases the superiority of our combination of handcrafted features over automatically derived features, primarily due to the limitations posed by the writers in the JUDVLP-TLWVdb dataset. Our experimental findings also highlight the superior performance of the SMO classifier compared to the other two classifiers, namely simple logistics and KNN. Additionally, we introduce a novel dataset for a writer verification system using tri-script, achieving a verification accuracy of 91.50% by combining Radon Transform, HOG, LBP, and LPQ features.

The script-independent overall performance reaches 91.80%, which is similar to the performance achieved when considering all three scripts together. Analyzing the performance of individual scripts reveals that the Hindi script outperforms Bangla and English. Moreover, in this study, we leverage ViT for writer recognition and achieve superior performance compared to ViT using tri-level block images of the page. For the writer verification task, we divided the original image into 16×16 blocks. For processing, ViT also divided the input image into 16×16 blocks. We came to the conclusion that our proposed system, which uses manual block division, performs better than the ViT model after analysing the results. Based on these promising results, we are planning to further enhance our new dataset at the lines, words, and character levels in the near future.

References

1. Semma, A., Hannad, Y., Siddiqi, I., Lazrak, S., El Youssfi, M., Kettani, E.: Feature learning and encoding for multi-script writer identification. Int. J. Doc. Anal. Recogn. (IJDAR) **25**(2), 79–93 (2022)
2. Bahram, T.: A texture-based approach for offline writer identification. J. King Saud Univ. Comput. Inf. Sci. **34**(8, Part A), 5204–5222 (2022)
3. Aubin, V., Mora, M., Santos, M.: A new approach for writer verification based on segments of handwritten graphemes. Logic J. IGPL **30**(6), 965–978 (2022)

4. Gattal, A., Djeddi, C., Abbas, F., Siddiqi, I., Bouderah, B.: A new method for writer identification based on historical documents. J. Intell. Syst. **32**(1), 20220244 (2023)
5. Rakshit, P., Halder, C., Sk, M.O., Roy, K.: A generalized line segmentation method for multi-script handwritten text documents. Expert Syst. Appl. **212**, 118498 (2023)
6. Malakar, S., Sahoo, S., Chakraborty, A., Sarkar, R., Nasipur, M.: Handwritten Arabic and roman word recognition using holistic approach. Vis. Comput. (2022)
7. Ahmed, B.Q., Hassan, Y.F., Elsayed, A.S.: Offline text-independent writer identification using a codebook with structural features. PLoS ONE **18**(4), 1–31 (2023)
8. Al-Shamaileh, M.Z., Hassanat, A.B., Tarawneh, A.S., Rahman, M.S., Celik, C., Jawthari, M.: New online/offline text-dependent arabic handwriting dataset for writer authentication and identification. In: International Conference on Information and Communication Systems (ICICS), pp. 116–121 (2019)
9. Abbas, F., Gattal, A., Djeddi, C., Siddiqi, I., Bensefia, A., Saoudi, K.: Texture feature column scheme for single- and multi-script writer identification. IET Biometrics **10**(2), 179–193 (2021)
10. Koepf, M., Kleber, F., Sablatnig, R.: Writer identification and writer retrieval using vision transformer for forensic documents. In: Uchida, S., Barney, E., Eglin, V. (eds.) DAS 2022. LNCS, vol. 13237, pp. 352–366. Springer, Cham (2022). https://doi.org/10.1007/978-3-031-06555-2_24
11. Sahoo, S., Kumar, P., Lakshmi, R.: Offline handwritten character classification of the same scriptural family languages by using transfer learning techniques. In: International Conference on Emerging Technologies in Computer Engineering: Machine Learning and Internet of Things (ICETCE), pp. 1–4 (2020)
12. Roy, R.K., Mukherjee, H., Roy, K., Pal, U.: CNN based recognition of handwritten multilingual city names. Multimed. Tools Appl. **81**(8), 11501–11517 (2022)
13. Dosovitskiy, A., et al.: An image is worth 16x16 words: transformers for image recognition at scale. *arXiv preprint*arXiv:2010.11929 (2020)
14. Geng, S., Zhu, Z., Wang, Z., Dan, Y., Li, H.: LW-VIT: the lightweight vision transformer model applied in offline handwritten Chinese character recognition. Electronics **12**(7), 1693 (2023)
15. Dhiaf, M., Rouhou, A.C., Kessentini, Y., Salem, S.B.: MSdocTr-lite: a lite transformer for full page multi-script handwriting recognition. Pattern Recogn. Lett. **169**, 28–34 (2023)
16. Miller, J., Patterson, R., Gantz, D., Saunders, C., Walch, M., Buscaglia, J.: A set of handwriting features for use in automated writer identification. J. Forensic Sci. **62** (2017)
17. Obaidullah, S.M., Halder, C., Santosh, K.C., Das, N., Roy, K.: PHDIndic_11: page-level handwritten document image dataset of 11 official Indic scripts for script identification. Multimed. Tools Appl. **77**, 1643–1678 (2018)
18. Bilan, S., Motornyuk, R., Bilan, S., Galan, O.: User identification using images of the handwritten characters based on cellular automata and radon transform. In: Bilan, S., Elhoseny, M., Hemanth, D.J. (eds.) Biometric Identification Technologies Based on Modern Data Mining Methods, pp. 91–103. Springer, Cham (2021). https://doi.org/10.1007/978-3-030-48378-4_6
19. Hannad, Y., Siddiqi, I., Djeddi, C., El-Kettani, M.E.Y.: Improving Arabic writer identification using score-level fusion of textural descriptors. IET Biometrics **8**(3), 221–229 (2019)

20. Paul, J., Sarkar, A., Das, N., Roy, K.: HOG and LBP based writer verification. In: Bhattacharjee, D., Kole, D.K., Dey, N., Basu, S., Plewczynski, D. (eds.) Proceedings of International Conference on Frontiers in Computing and Systems. AISC, vol. 1255, pp. 3–12. Springer, Singapore (2021). https://doi.org/10.1007/978-981-15-7834-2_1
21. Ahmed Kawther Hussein: Fast learning neural network based on texture for Arabic calligraphy identification. Indones. J. Electr. Eng. Comput. Sci **21**, 1794–1799 (2021)
22. Nascimento, F., Smith, S. L., CostaAbreu, D.: Exploring medieval manuscripts writer predictability: a study on scribe and letter identification. Digit. Stud./le champ numérique. **12**(1) (2022)

Wildlife Detection Using ANN and Other Modern Technology: A Survey of Literatures

Priyodarshini Dhar[1]([✉])[iD] and Rakesh Kumar Mandal[2]

[1] Dynamic Digital Technology, a Motorola Solutions Company,
Kolkata, West Bengal, India
priyodarshini.dhar@gmail.com
[2] Department of Computer Science and Technology, University of North Bengal,
Bagdogra, West Bengal, India
rakeshmandal@nbu.ac.in

Abstract. In recent years, researchers have been trying to build fully automated wildlife detection techniques using modern tools and techniques to prevent human-animal conflict. Habitat destruction due to construction of roads and railways in forest areas has increased human-animal conflict, resulting in loss of both human and animal life. To avoid this conflict and maintain an ecological balance between humans and animals, wildlife detection and early warning systems are used. Manual surveillance from watch towers and traditional systems like electric fencing and trenches to avoid human-animal conflict is often unreliable and requires human interference at all times. With the emergence of new technologies such as Artificial Neural Network, Internet of Things and the availability of high speed internet connection via 4G and low-cost sensors and computers, building reliable automatic wildlife detection and warning systems which require minimum human interference is now possible. In this paper, a comparative study on the existing research papers dealing with fully automatic or semi-automatic wildlife detection techniques using modern technologies has been done. This paper attempts to explain in detail the key points and underlying technology used in the existing research papers in the concerned domain and also find the research gap if any.

Keywords: Wildlife detection techniques · Human-Animal Conflict · Artificial Neural Networks

1 Introduction

A surge in wildlife detection using modern tools and techniques such as Artificial Neural Networks, Deep Neural Networks, Visual Saliency and IoT has been seen over the past few decades. Wildlife detection and alarm generation is necessary for reducing human-wildlife conflict and for preserving wildlife. Vehicle

M. Majumder et al. (Eds.): ICCTE 2023, CCIS 2376, pp. 31–44, 2025.
https://doi.org/10.1007/978-3-031-81935-3_3

collisions leading to animal deaths, elephants and other animals killed on railway tracks due to collision with trains, increased mortality of migratory birds due to human activities and fish kill due to polluted water are some examples of ecological damage caused by humans. 52,847 mammals and 23,346 birds were roadkilled in Flanders, a small region in Belgium from 1960–2020 [25]. In India, in the last 10 years, 186 elephants were killed on railway tracks [29]. Artificial light at night (ALAN), noise pollution, environmental contaminants, wildfires and climate change due to human activities have led to high mortality of migratory birds [15]. Every year an estimated 3,00,000 dolphins and porpoises are killed due to human activities like by catch, vessel strike, oil spills and coastal developments [28]. In North-Eastern India, 6036 individual roadkills of 53 species in a span of 1 year (2016–2017) were recorded [23]. Wildlife detection and early warning systems can be used to avert both wildlife and human loss. Over the past decade, researchers have attempted to build automated wildlife detection and warning systems using traditional Neural Networks, Deep Neural Networks and IoT devices, which are better than traditional techniques such as manual survey, satellite remote sensing in terms of monitoring range and data acquisition lag [6]. Neural networks have generated a huge interest in the last two decades due to the development of more sophisticated neural nets, high speed digital computers and parallel computing. It has gained a lot of popularity in pattern recognition and has shown success in detecting wildlife using image, acoustic, video data sets with high accuracy and less false positives. With high speed networks and low cost IoT devices being available, it has become easier and inexpensive to build wireless sensor networks to detect wild animals and set up alarm systems. In the following section, a brief survey on the various modern wildlife detection techniques has been performed. The following section has been divided into subsections based on the technology and type of data (image, sound) to achieve wildlife detection.

2   Literature Review

In this section research papers on wildlife detection techniques using Deep Neural Networks like AlexNet [12], VGGNet [22], BirdNet [11] and LeNet [13] on different types of datasets such as images, videos and sound and other modern techniques such as visual saliency methods based on histogram contrast algorithm, seismic surveillance using geophones, animal detection and warning systems based on sensors have been studied. For each paper, the main concepts have been summarised and explained the method used in simple terms and stated the research gap if any.

2.1   Wildlife Detection Using Deep Learning Techniques on Image and Video Datasets

Gray et al. [7] applied a customised CNN on aerial images of sea turtles captured during their nesting in coastal areas, to count the number of turtles. Unoccupied aircraft systems (UAS, or drones) were used to collect 1059 aerial images in the coastal areas of Costa Rica during the peak nesting season of olive ridley sea turtles. 467 images were first tagged manually by 3 reviewers in approximately 6 h. 275 of these images were used to train the CNN model and the rest 192 aerial images were used to compare turtle count by manual review and the customised CNN model. For the CNN to enumerate turtles in an image, the input image was divided into 100 * 100 pixel window. The CNN used by the authors consisted of 4 convolutional layers with 64 3 × 3 filters along with max pooling layers and 2 fully connected layers of 1024 neurons. The CNN model achieved an accuracy of 99.83% to detect turtles and 76.5% recall. Recall is described as the number of true positives found by the model divided by the actual number of true positives in the data. However, a low precision of 16.3% was achieved suggesting model overfitting even though a dropout ratio of 0.5 was done on the connections between the neurons of the two connected layers during training. When the authors compared the turtle count done manually and using the CNN model, they found an improvement of 8% in detecting turtles and non-turtles from aerial images using the CNN model. This paper shows that CNN can be used to eliminate manual intervention during animal counting in aerial images with better accuracy. Typical CNNs also have detection limits for dense herds [5]. Bhanupriya et al. [3] used a proposed Convolutional Neural Network (CNN) to detect elephants and cheetahs from images. From the Snapshot Serengeti dataset [24], the authors created an elephant and cheetah image dataset and split the images in 75:25 ratio for training and testing the CNN model. An accuracy of 79.25% while detecting elephants and 86.79% while detecting cheetahs was achieved. Convolutional Neural Network is a feed-forward neural network with an input layer, multiple hidden layers and an output layer. The hidden layer consists of a Convolutional layer, ReLU (Rectified Linear Activation Function), Pooling layer and a fully connected layer. The Convolutional layer uses a filter (or kernel) to scan through the input matrix performing convolution operations to detect patterns and create a feature map. Convolution operation, when done on two functions, expresses how one function affects the shape of another function. The values in the feature map are calculated using the following formula, where I is the input image, f denotes the filter, n and m are the row index and column index of the feature map matrix.

$$M[n, m] = I * f(n, m) = \sum_i \sum_j f(i, j) I[n - i, m - j] \qquad (1)$$

The feature map is then passed onto the ReLU layer, which performs activation on the feature map. It will output the input if it is positive and 0 if the input is negative, hence its non-linearity. It is linear for positive input, but non-linear (0) for negative inputs. The ReLU activation function can be represented by the following equation and Fig. 1.

$$f(x) = max(0, x) \qquad (2)$$

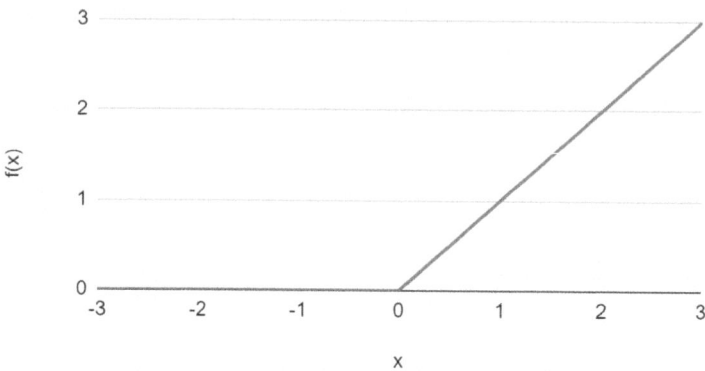

Fig. 1. ReLU Activation Function (Source: Yamashita [31], 2018, p. 7)

The rectified feature map now goes to the Pooling layer. The main function of the pooling function is to reduce the dimensionality of the feature map. The reduced feature map is converted to a 1 dimensional matrix and fed to the fully connected layer which is used for classification. A CNN generally uses the following pattern:

$$[Input] \Rightarrow [Convolution Layer \Rightarrow Pooling Layer] * x \Rightarrow [Fully Connected Layer] * y \Rightarrow [Output] \tag{3}$$

where x and y are integers and "*" indicates repetition. Figure 2 shows the layers in a simple CNN architecture.

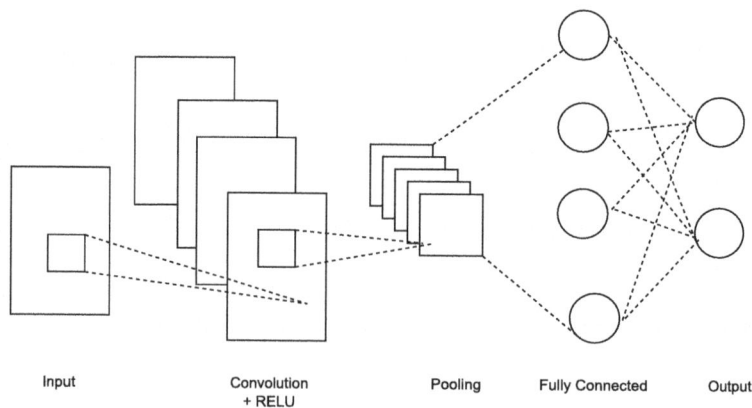

Fig. 2. Layers in Convolutional Neural Network (Source: Phung et al. [18] 2019, p. 1)

Even though CNN is the go-to neural network algorithm for pattern recognition from images, it has some major drawbacks such as overfitting, exploding gradient, and class imbalance [14] especially on small datasets as used in this paper.

Isha et al. [21] showed that Deep Convolutional Neural Networks such as the AlexNet [12] can be used for wildlife detection with greater accuracy than the traditional convolutional neural network. Deep Convolutional Neural Networks have shown great success on image and video pattern recognition [19,20]. The authors used frames from videos to detect elephants using computer vision to prevent human-elephant collisions. The authors used Alex-Net to detect elephants from video frames and then feed the representation to a binary support vector for classification. A mean average precision of 98.621% and a mean average recall of 97.279% was achieved using the above method. Alex-Net uses 5 convolutional layers and 3 fully connected layers. It allows multi-GPU usage, by putting one half of the neurons on different GPUs. This allows more complex and bigger models to be trained in less time compared to the Convolution Neural Network. However, AlexNet tends to show "high error rate due to the small number of layers in the network and the high selection of parameter values" [9]. To avoid this, deeper CNN, such as VGGNet [22], could have been used. AlexNet consists of five convolution layers with max-pooling, two fully connected layers with dropouts and the last fully connected layer is a 1000-way softmax layer. The number of kernel filters and size are different in each of the convolution layers. 96 kernels of size 11×11, 256 kernels of size 5×5, 384 kernels of size 3×3, 384 kernels of size 3×3 and 256 kernels of size 3×3 were used in convolution layers 1,2,3,4 and 5 respectively. The activation function is ReLU for all the layers. Figure 3, shows a simplified diagram of AlexNet architecture.

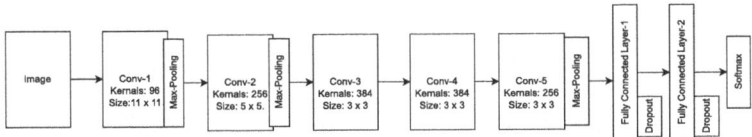

Fig. 3. Simplified diagram of AlexNet architecture (Source: Krizhevsky et al. [12] 2012, p. 5)

In [9], the authors used AlexNet [12] and VGGNet [22] to classify 14 types of animals. The paper showed that VGGNet showed much better accuracy of 91.2% than AlexNet with 67.65% accuracy. Dataset used in this paper is quite small with 150 images for each animal. The author showed that as AlexNet architecture uses an 11×11 convolution filter for its 5 convolutional layers, it causes data loss in between two layers. In contrast, VGGNet uses a 3×3 convolution filter, causing less data loss in between layers. In the next paper the authors used a large image dataset of animals and applied several deep neural networks and showed that the accuracy increases with larger datasets. In [27], the authors showed that with enough data and learning capacity, very deep convolutional neural networks such as AlexNet [12], VGGNet [22], GoogLenet [26], ResNets [8] can be used to automatically detect animal species from camera-trap images. VGGNet16 has 3 fully connected layers, 13 convolutional layers and 5 pooling layers. VGGNet19 has 3 fully connected layers and 16 convolutional layers and 5 pooling layers. For both the varieties of VGGNet, the authors used a very small convolutional filter (3 X 3). GoogLenet [26] is a 22 layer deep neural network. Some major inclusions in GoogLenet are using global average pooling at the end of the network which takes a feature map of 7×7 and averages it to 1×1, using convolutional filters of various sizes (1×1, 3×3, 5×5) in the inception module and using auxiliary classifier for training. Both GoogLenet and VGGNet perform exceptionally well with large data sets. GoogLenet was placed first and VGGNet was placed second in the ImageNet Large Scale Visual Recognition Challenge. Continuing with [27], the authors used the Snapshot Serengeti dataset [24] which contains images of 48 animal species and partitioned it into 4 individual datasets; D1: unbalanced (uneven number of images for each species) dataset, D2: balanced dataset, D3: object in foreground dataset, D4: animal segmented(part of animal visible) dataset. The best result was obtained using deep ConvNet- ResNet-50 on dataset D4 with accuracy top 1: 88.9% and top-5: 98.1%. Top-1 accuracy means the conventional accuracy i.e. how many times the correct answer was selected. Top-5 accuracy is calculated by checking whether among the top 5 answers selected by the model, 1 of them is the correct answer. Delplanque et al. [5] proposed a new CNN model called HerdNet for counting animals like herds of camels, donkeys, sheep and goats in the African landscape of Ennedi reserve. The authors compared this method with state-of-art Faster R-CNN and density based DLA-34. The new model achieved a F1 score of 73.6%. The average confusion rate was however 4% higher than Faster R-CNN.

2.2 Wildlife Detection Using Deep Learning Techniques on Sound Datasets

In [11], the authors developed a deep neural network called BirdNet, derived from ResNets and it consisted of 157 layers and 27 million parameters. The model was trained with hours of audio files of bird acoustics. Data augmentation, data pre-processing and mixup were done. The model aimed to identify bird species from audio clippings which would help researchers, biologists and others. The authors used three datasets of bird sounds from D1: single species recordings (22,960), D2: soundscape annotated data of several recordings (280 h), and D3: soundscape data from bird hotspots (33,670 h). The model was able to achieve a "mean average precision of 0.791" [11] for D1, "a F0.5 score of 0.414" [11] for D2 and "average correlation of 0.251" [11] for D3. In [1], the authors identified chimpanzee calls using Passive Acoustic Monitoring and Recursive Neural Network. The database for Chimpanzee calls are class imbalanced, and so to compensate that the authors applied Spectrogram denoising, Alternative loss functions and Resampling. F1 score: 33% for drumming sound of chimpanzee and 5% for vocalisation sound of chimpanzee. In [20] the authors used deep neural networks to process and detect the sound of whales and compared the results with traditional detection algorithms. The deep learning algorithm produced far less false-positives than the alternative algorithms. The authors examined multiple deep convolutional neural networks such as LeNet [13], BirdNET [11], VGG [22], ResNet [26], a very deep neural network [22] and recurrent neural networks to detect right whale upcalls. The authors trained the neural network with recordings from a single region over a few days. The trained Neural Network was then able to detect/generalise sounds of whales from multiple years across different regions with low false positives.

2.3   Wildlife Detection Using Visual Saliency Detection Method

Visual saliency detection methods simulate how a human perceives a scene and extracts the important information from the scene. In [6], the authors built a wireless multimedia sensor network of cameras and 4G terminal node equipment using ZigBee network protocol to capture images of wild animals as shown in Fig. 4.

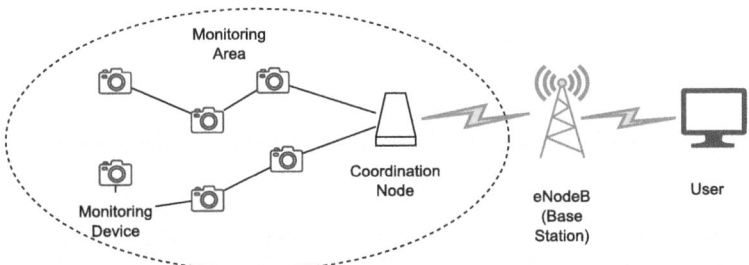

Fig. 4. ZigBee network consisting of multiple monitoring devices for wildlife detection (Source: Feng et al. [6] 2018, p 2)

Infrared sensors placed in the node equipment triggered the camera whenever wild animals entered the monitoring zone. Images captured were first saved in SD card, then sent to coordination nodes and finally transmitted to the data centre. The authors first implemented structure extraction from the images using total variation denoising, which smooths the images and reduces noise in the image. Then the authors applied a histogram contrast algorithm on images captured with wireless multimedia sensor networks. Histogram contrast is used to extract the area of a wild animal from images for providing colour saliency information. The final step involved combining edge detection and saliency information to obtain the final result. The authors obtained an average precision rate of 0.4895, recall rate of 0.7321 and F-measure of 0.53. Chabot et al. [4] synthesised training data of thinly distributed species like polar bears found in the Arctic region and used a convolutional neural network with 50 layers (ResNET-50) to detect

Table 1. Comparative study of handcrafted features based modern techniques for wildlife detection on Image and Seismic waves dataset

Author	Dataset Type	Wildlife	Technique Used	Accuracy achieved/F1 score/mAP	Research Gap
Feng et al. [6]	Image	12 types of wildlife	Saliency visual detection method based on Histogram contrast algorithm.	average precision rate of 0.4895, recall rate of 0.7321, and F-measure of 0.53	This method only detects the presence of wild animals without classifying the animal type.
Wood et al. [30]	Recordings of Seismic Waves	Elephant, Giraffe, Lion	Power spectrum plotted using fast Fourier transform and Hanning window	Accuracy 82% in discriminating different species was achieved	The accuracy will be impacted by the soil type in which a geophone is buried.
Prabu et al. [19]	Recordings of Seismic Waves	Elephant	FastICA to remove noise, FFT for obtaining spectral components, and Dynamic Time Wrapping for matching samples.	93.75% accuracy	Seismic wave pattern changes with soil conditions, lowering the accuracy of detection of wild animals

wildlife automatically from aerial photographs. A large training set is generally required to achieve higher accuracy in CNN. Typically, data augmentation using spatial or spectral transformation is used to increase the training set. The authors used the visual saliency approach for increasing the training set by combining images of polar bears with random background images. An accuracy of 95% was achieved in classifying bear images from aerial photographs.

2.4 Wildlife Detection Using Sensors

In [30], the authors used geophones to detect the footfall of elephant herds or larger animals. Geophones are low cost devices that convert ground movement into voltages. After preprocessing the voltage values, these data can be sent to the server for further analysis. In [16], the authors showed how geophones can be used to record and handle seismic data. In [30], the authors used a single geophone to record footfalls of large animals near a waterhole in Etosha National Park, Namibia, Central Africa. An accuracy rate of 82% was achieved in discriminating different species. To record the seismic activities when elephants walk past it, a vertical geophone (4–5 Hz) was buried at a depth of 15 cm at 1 m distance from the waterhole. A preamplifier was used to increase the recording gain, a sound card converted the analog signal to digital signal and was recorded in a laptop at sampling rate 44.1 kHz. 57 recordings were made which included elephants, giraffe, humans, gemsbok, and lion. 20 recordings per species were

Table 2. Comparative study of Deep Learning Techniques used for wildlife detection on Image/Video dataset-I

Author	Dataset Type	Wildlife	Deep Learning Technique	Accuracy achieved/F1 score/mAP	Research Gap
Gomez et al. [27]	Camera-trap Image	Multiple wildlife	AlexNet, VGGNet, GoogLenet, and ResNets	Top-1: 88.9% Top-5: 98.1%	The Serengeti dataset is unbalanced, I.e there are thousands of images of a handful of animals, and very few images for others. Deep neural networks work well with a large dataset. However in reality, most datasets are small. The paper does not discuss the results for animals having a small dataset.
Banupriya et al. [3]	Camera-trap Image	Elephant, Cheetah	Customised CNN	79.25% (elephants), 86.79% (cheetah) accuracy	It uses a very small and balanced training set, for training the CNN model which should cause the problem of severe overfitting [10] leading to a low accuracy.
Shukla et al. [21]	Images generated from video	Elephant	AlexNet	98.621% - 98.667%	Applying it in real time will be costly and slow. Large volume of data is generated from videos. Processing video to image is time consuming
Inik et al. [9]	Images	Multiple Animals	AlexNet, VGGNet	91.2% with VGGNet and 67.65% with AlexNet	Small and balanced datasets such as used in this paper can cause overfitting problems [10]. Training deeper neural networks with a small training set is difficult due to the vanishing gradient problem.
Gray et al. [7]	Aerial Image	Sea turtles	Customised CNN	99.83% accuracy	A low precision of 16.3% was achieved suggesting model overfitting even though a dropout ratio of 0.5 was done on the connections between the neurons of the two connected layers during training.

extracted from the above recordings. A power spectrum was plotted using a fast Fourier transform and a Hanning window. The authors noticed that the power spectra for elephants were different from other species. The spectrum for

Table 3. Comparative study of Deep Learning Techniques used for wildlife detection on Image/Audio dataset-II

Author	Dataset Type	Wildlife	Deep Learning Technique	Accuracy achieved/F1 score/mAP	Research Gap
Chabot [4]	Aerial Images	Polar Bears	Visual Saliency with ResNet50	95% accuracy in detecting polar bear	The authors manipulated the training set so that the network did not focus on the background colours of the training images, otherwise the model produced false positives. This limitation, called the whole-image classification, is a limitation whereby there is no way to tell the model which parts of the image it should focus on. Training the network with more background images could reduce the false positives. This would require more powerful GPU resources.
Kahl [11]	Audio	Birds	BirdNet	0.791 mean average precision	High-end recording equipment required to detect birds from soundscape data. Noisy training data worsens the performance of BirdNet model [11]. Also this model is not trained with the entire bird population.
Shiu [20]	Audio	Whales	LeNet, BirdNET, VGG, = ResNet and Conv1D+GRU, hybrid RNN	Precision: LeNet with augmentation: 0.903	The data source is dependent on the animal to produce sound. Applying it in real time to detect the presence of the animal will be impossible if no/less sound is produced by the animals

elephants had the most low frequency component and one of the lower high frequency components, leading to a "distinctive low-to-high frequency ratio" [30]. Instead of using Fast Fourier Transform, neural networks such as CNN could be applied to get better results. In [14], the authors describe using a convolutional neural network for seismic event classification on low-power embedded devices using network quantization. In [19], the author designed an "efficient surveillance system to detect elephant intrusion into forest borders using seismic sensors". The author collected seismic data of elephant footfalls from the Hosur and Dharmapuri Forest Division, in Tamil Nadu, India. The raw signals were processed using Fast Independent Component Analysis (FastICA) to remove noise efficiently and Fast Fourier Transform was applied to convert the seismic signal into individual spectral components. The processed signal was matched against previously stored 300 sets of elephant seismic wave patterns using Dynamic Time Wrapping algorithm and on receiving a true positive match, an alert message was generated. The author achieved an elephant detection rate of 93.75%. In [2], Arya et al. proposed an elephant intrusion system algorithm around housing colonies using a two level detection system using infrared sensors and microphones. First, they placed microphones 50 m away from the housing colony which captured sound and sent it to a nearby server using FM transmitter and second, they used Infra-Red sensors at 20 m from a housing colony which got activated if an elephant crossed the IR sensor. The authors used Fast Fourier Transform on the captured signal and used the Mel Frequency spectrum to analyse the sound signal. Later the SVM classifier was used to compare the captured sound with a stored database to detect the sound of elephants. As the proposed system was not implemented in real time, the accuracy of the proposed system is unknown. Similarly in [17], authors Paulchamy and Sadhana shree, proposed an elephant detection system using control sensors like motion sensors, vibration sensors and water level sensors. Vibration sensors were used to detect any vibration velocity for movement of an elephant. The system does not show how it would detect a herd of elephants as the vibration velocity will be different. Motion sensors were used to detect any movement in 120° radius. Again, the motion captured was not compared with any existing database to confirm the presence of elephants and would require manual intervention. The warning system included an Audio Module which produced a loud sound on detection of elephants. All the sensors and warning system were connected to a PIC microcontroller. The system would have been more accurate, had it compared the vibration sensors or motion sensor data with the existing elephant database. In [32], authors Shao and Wang show how ground vibrations can be used to detect human beings falling using machine learning. The same technique could have been used for [17] to improve the detection system. In [32], the authors designed a human model and collected the vibrations of the falls of this model using a low-sensitive accelerometer. The authors proposed a fall detection algorithm based on time of vibration, K-Means Clustering Algorithm and K-Nearest Neighbour Algorithm to classify continuous human movement, human fall and other object fall. Table 1 shows a comparative analysis of the papers that deal with modern techniques to detect wildlife.

Table 2, and Table 3 show the comparative study of deeplearning techniques used for wildlife detection.

3   Conclusion

In this paper, different research papers on wildlife detection techniques based on images, sound, vibrations and videos, each having its pros and cons have been studied. In this study, we saw that wildlife detection using deep learning techniques such as customised CNN, AlexNet, VGGNet, LeNET and other very deep neural networks work effectively on large image and video datasets. RNN and its different variations and BirdNET are applied on audio datasets. In this study we saw how DL techniques can be used to detect and count wildlife like elephants, lions, birds, whales, sea turtles, polar bears without manual intervention. DNN techniques achieve better accuracy than handcrafted feature based techniques according to the survey. However there are few limitations with DNN models. Overfitting problem, vanishing gradient problem, whole-image classification where the model does not understand which part of the image it should focus on during training, are some of the problems discussed in this paper. With small dataset, DNNs tend to produce less accuracy. Training a DNN model required large computing resources. This paper also discusses the solutions which can address the above issues with DNN models. Including dropout ratio of 0.5 in the fully connected layers, increasing the dataset size using data augmentation, training the model with background and foreground image to diminish the whole-image classification problem, increasing computational resources such as GPU to train a model faster and with larger data set, removing noise from audio clips during preprocessing are some of the techniques that researchers can use while training a network. Training a DNN model is a computationally laborious process. It may take several hours to days to train one model. This sets a practical limit on the number of trials one can perform while training a model. From the literature survey, it was found that fully automatic wildlife detection and warning systems are not available. It was also seen that there are very few datasets available for wildlife detection. Most datasets are collected from Africa, Europe and China which are publicly available. India has a rich biodiverse ecosystem with nearly 8% of all mammalian species and 12% of all avian found in the world. However, there are no public image/audio datasets based on Indian wildlife. In future, we can work on creating a public wildlife database for Indian wildlife and also work towards automating the wildlife detection and surveillance systems.

Acknowledgments. The authors wish to thank Dr. Debbrota Paul Chowdhury for his invaluable suggestions on preparing the manuscript.

Disclosure of Interests. The authors have no competing interests to declare that are relevant to the content of this article.

References

1. Anders, F., Kalan, A.K., Kühl, H.S., Fuchs, M.: Compensating class imbalance for acoustic chimpanzee detection with convolutional recurrent neural networks. Eco. Inform. **65**, 101423 (2021). https://doi.org/10.1016/j.ecoinf.2021.101423
2. Arya, S.S., Ramesh, M.V., Divya, P.: Design and simulation of elephant intrusion detection system. In: Proceedings of 2016 6th International Workshop on Computer Science and Engineering, WCSE, pp. 749–753 (2016)
3. Banupriya, N., Saranya, S., Swaminathan, R., Harikumar, S., Palanisamy, S.: Animal detection using deep learning algorithm. J. Crit. Rev **7**(1), 434–439 (2020)
4. Chabot, D., Stapleton, S., Francis, C.M.: Using web images to train a deep neural network to detect sparsely distributed wildlife in large volumes of remotely sensed imagery: a case study of polar bears on sea ice. Eco. Inform. **68**, 101547 (2022). https://doi.org/10.1016/j.ecoinf.2021.101547
5. Delplanque, A., Foucher, S., Théau, J., Bussière, E., Vermeulen, C., Lejeune, P.: From crowd to herd counting: How to precisely detect and count African mammals using aerial imagery and deep learning? ISPRS J. Photogramm. Remote. Sens. **197**, 167–180 (2023)
6. Feng, W., Zhang, J., Hu, C., Wang, Y., Xiang, Q., Yan, H.: A novel saliency detection method for wild animal monitoring images with WMSN. J. Sens. **2018**(1), 3238140 (2018)
7. Gray, P.C., et al.: A convolutional neural network for detecting sea turtles in drone imagery. Methods Ecol. Evol. **10**(3), 345–355 (2019)
8. He, K., Zhang, X., Ren, S., Sun, J.: Deep residual learning for image recognition. In: Proceedings of the IEEE Conference on Computer Vision and Pattern Recognition, pp. 770–778 (2016). arXiv:1512.03385
9. İnik, Ö., Turan, B.: Classification of animals with different deep learning models. J. New Results Sci. **7**(1), 9–16 (2018)
10. Joshi, S., Verma, D.K., Saxena, G., Paraye, A.: Issues in training a convolutional neural network model for image classification. In: Advances in Computing and Data Sciences: Third International Conference, ICACDS 2019, India, pp. 282–293 (2019)
11. Kahl, S., Wood, C.M., Eibl, M., Klinck, H.: Birdnet: a deep learning solution for avian diversity monitoring. Eco. Inform. **61**, 101236 (2021)
12. Krizhevsky, A., Sutskever, I., Hinton, G.E.: Imagenet classification with deep convolutional neural networks. In: Advances in Neural Information Processing Systems, vol. 25 (2012)
13. LeCun, Y., Bengio, Y.: Convolutional networks for images, speech, and time series. Handb. Brain Theory Neural Netw. **3361**(10), 1995 (1995)
14. Meyer, M., et al.: Event-triggered natural hazard monitoring with convolutional neural networks on the edge. In: Proceedings of the 18th International Conference on Information Processing in Sensor Networks, pp. 73–84 (2019)
15. Nemes, C.E., et al.: More than mortality: consequences of human activity on migrating birds extend beyond direct mortality. Ornithol. Appl. **125**(3), duad020 (2023)
16. Onajite, E.: Seismic Data Analysis Techniques in Hydrocarbon Exploration. Elsevier (2013)
17. Paulchamy, B., Sadhanashree, S., Sathya, M., Sudhanthira, R., Vivek, T.: Proficient technique for monitor an elephant intrusion in forest border areas using embedded systems. Asian J. Appl. Sci. Technol. (AJAST) **2**(2), 314–319 (2018)

18. Phung, V.H., Rhee, E.J.: A high-accuracy model average ensemble of convolutional neural networks for classification of cloud image patches on small datasets. Appl. Sci. **9**(21), 4500 (2019)
19. Prabhu, M.: An efficient surveillance system to detect elephant intrusion into forest borders using seismic sensors. Int. J. Adv. Eng. Technol. **7**(1), 166–171 (2016)
20. Shiu, Y., et al.: Deep neural networks for automated detection of marine mammal species. Sci. Rep. **10**(1), 607 (2020)
21. Shukla, P., Dua, I., Raman, B., Mittal, A.: A computer vision framework for detecting and preventing human-elephant collisions. In: Proceedings of the IEEE International Conference on Computer Vision Workshops, pp. 2883–2890 (2017)
22. Simonyan, K., Zisserman, A.: Very deep convolutional networks for large-scale image recognition. arXiv preprint arXiv:1409.1556 (2014)
23. Sur, S., Saikia, P.K., Saikia, M.K.: Speed thrills but kills: a case study on seasonal variation in roadkill mortality on national highway 715 (new) in Kaziranga-Karbi anglong landscape, Assam, India. Nat. Conserv. **47**, 87–104 (2022). https://doi.org/10.3897/natureconservation.47.73036
24. Swanson, A., Kosmala, M., Lintott, C., Simpson, R., Smith, A., Packer, C.: Snapshot Serengeti, high-frequency annotated camera trap images of 40 mammalian species in an African savanna. Sci. Data **2**(1), 1–14 (2015)
25. Swinnen, K.R., et al.: 'Animals under wheels': wildlife roadkill data collection by citizen scientists as a part of their nature recording activities. Nat. Conserv. **47**, 121–153 (2022). https://doi.org/10.3897/natureconservation.47.72970
26. Szegedy, C., et al.: Going deeper with convolutions. In: Proceedings of the IEEE Conference on Computer Vision and Pattern Recognition, pp. 1–9 (2015)
27. Villa, A.G., Salazar, A., Vargas, F.: Towards automatic wild animal monitoring: identification of animal species in camera-trap images using very deep convolutional neural networks. Eco. Inform. **41**, 24–32 (2017). https://doi.org/10.1016/j.ecoinf.2017.07.004
28. Wildlife Conservation Society. https://www.wcs.org/our-work/wildlife/whales-coastal-dolphins. https://doi.org/10.1155/2018/3238140
29. Wildlife Institute of India. https://www.wii.gov.in/capacity_worshop_railway_mortalities. Accessed 8 Nov 2023
30. Wood, J.D., O'Connell-Rodwell, C.E., Klemperer, S.L.: Using seismic sensors to detect elephants and other large mammals: a potential census technique. J. Appl. Ecol. 587–594 (2005)
31. Yamashita, R., Nishio, M., Do, R.K.G., Togashi, K.: Convolutional neural networks: an overview and application in radiology. Insights Imaging **9**, 611–629 (2018). https://doi.org/10.1007/s13244-018-0639-9
32. Zhong, Z., Li, H.: Recognition and prediction of ground vibration signal based on machine learning algorithm. Neural Comput. Appl. **32**(7), 1937–1947 (2020)

Enhancing Learning Outcomes Through the Use of Conducive Learning Spaces

Shabina Modi[1], Deepali Sale[2], Vishal Borate[2], and Yogesh Kisan Mali[3]([✉]) [iD]

[1] Karmaveer Bhaurao Patil College of Engineering, Satara, India
shabina.sayyad@kbpcoes.edu.in
[2] Dr D Y Patil College of Engineering and Innovation, Talegoan, Pune, India
[3] G H Raisoni College of Engineering and Management, Wagholi, Pune, India
yogesh.mali@raisoni.net

Abstract. The learning spaces or the built environment or the physical learning environment, in educational institutions impact the acquisition of 21st century skills and quality education. Proficiency in digital learning is an important competency skill required for 21st century youth. With the developing trend of e-learning, the importance or the role of learning spaces in the improvement of learning outcome is diluting. This study aims to determine the parameters for developing learning spaces to support learning and improve the learning outcomes of the learner. The research is based on the analysis of the existing literature data to find the parameters of learning spaces required for improving cognitive, affective and behavioral aspects of learning outcomes. The research findings put forward five parameters, namely wellness, stimulation, individualization, social interaction and spatial function, required for learning spaces to support learning and improve learning outcomes. The findings have been validated through a pilot survey, conducted in secondary schools in Punjab. This research will help the designers, architects and stakeholders to understand the parameters required to make learning spaces, conducive to learning. This study will further add awareness regarding the importance of learning spaces in improving the learning outcome of students.

Keywords: Learning spaces · learning outcomes · skill development · e-learning · parameters

1 Introduction

In 21st century digital era, the achievement of Sustainable Development Goals-4 of lifelong learning is the main focus of all countries all over the world. Lifelong learning requires not only the development of cognitive skills but also the students' non-cognitive skills. These skills are also the learning outcomes desired from the learner's learning process (Blackmore et al. 2011) [1]. With the growing research in learning outcomes, various means to fulfill the needs of students to live, learn, be social, and be physical have been developed (Anireddy et al. 2022) [2]. Physical learning spaces can support the development of new skills and enhance the learning capabilities of the learner (Strange &

© The Author(s), under exclusive license to Springer Nature Switzerland AG 2025
M. Majumder et al. (Eds.): ICCTE 2023, CCIS 2376, pp. 45–53, 2025.
https://doi.org/10.1007/978-3-031-81935-3_4

Banning, 2015) [4]. With the developing trend of e-learning the importance or the role of learning spaces in the improvement of learning outcome is diluting. People perceive that when the learning can be imparted through e-learning or digital sources, in online mode, then what is the use of developing or investing too much in these learning spaces. This perception has just evolved due to the underrated importance given to learning spaces in the improvement of learning outcome. So, there is need to study and validate the importance of these learning spaces in improving the learning outcomes of the students. According to various researches worldwide, learning spaces moderate the learning outcomes of students (Nair & Doctori, 2019; Sanoff & Walden, 2012; Tanner, 2008; Evan, 2006) [5]. However, countries still lack in the development of learning spaces which can improve learning outcomes of the learners. The physical built environment or the learning spaces in educational institutions are still the same as in the 19th century, and do not meet the need of competency-based learning required to develop 21st-century skills (Baafi, 2020) [6].

The study's objectives are to understand the relationship between the learning outcomes and the learning spaces and, secondly, to determine the parameters of learning spaces which can support learning and improve the learning outcomes of the students. The study will help the architects, designers and school leaders better understand the parameters required for learning spaces to improve learning outcomes.

2 Literature Study

Spaces can change the students' behavior (Imms & Kvan, 2021) [7]. There are three components of spaces, namely physical, social and cultural. Physical component can affect social (Martin, 2003) as well as cultural component (Haghighi & Jusan, 2012) [8]. Hence, understanding the relationship between the physical spaces and the behavior is essential (Leung & Fung, 2005) [9], for creating an environment favorable to students' learning behavior in educational institutions (Talvert & Mor-Avi, 2019) [10]. Spaces have the potential to accommodate, satisfy or enable a variety of desired experiences during learning (Imms & Kvan, 2021) [11]. Learning spaces also affect cognitive, social and emotional behaviour and influence the learner's personality development (Sanoff & Walden, 2012) [12]. Learning spaces support pedagogy and enhance teacher-student relationships (Barrett et al. 2015) [13] to develop and improve the skills or the learning outcomes (Rao, 2020) [14]. The impact of learning, which deals with the acquisition of affective, behavioral and cognitive learning outcomes (UNESCO, 2013), is not for the short term. Instead, it lasts almost throughout the learner's life (Owoseni et al. 2020) [15].

Learning spaces need changes to adapt to the new mode of learning or pedagogy (Barrett et al. 2019) and face the challenges of 21st-century learning. Most students and teachers want improvement and change in the traditional classroom learning spaces (Closs, Mahat, & Imms, 2021) [16]. The learning spaces can stimulate students' engagement in the learning process, influence their behaviour and attitude towards learning and develop their skills and mental ability (Owoseni et al. 2020) [17]. Nepal (2016) determined an indirect relationship between spaces for learning and outcomes. Satisfaction of the students (Fig. 1) with three factors of learning spaces, namely, environmental factors, design factors, and facilities available, will enhance the students' perceived performance or learning outcomes (Munir et al. 2018) [18]. According to Hunley & Schaller (2009), the learning spaces can be assessed for learning through students' engagement and students' satisfaction.

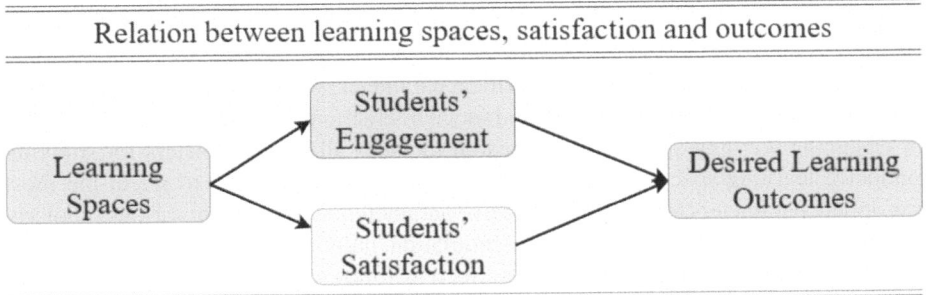

Fig. 1. Relation between learning spaces, satisfaction and outcomes

Various researchers have studied the parameters of learning spaces that support the student's learning outcomes. Closs, Mahat & Imms (2021) studied the influence of physical, pedagogical and psychological aspects of learning spaces on the learning experience and found out that the outcomes are affected by individualization and social interaction. Imms & Kvan (2021) [19], in their research on the impact of learning space on cognitive, affective & regulative aspects of learning outcomes, bring out four parameters: wellness, individualization, social interaction and stimulation. Owoseni et al. (2020) [20] found the parameters for learning spaces conducive to learning outcomes as Spatial function, social interaction, wellness and stimulation. Blackmore et al. (2012) find individualization, wellness, spatial function and stimulation as the main parameters of learning spaces. An empirical study concluded that learning environments positively impact learners' motivation and cognition (Singh et al. 2021). Barrett et al. (2015) [21] determined the impact of classroom design features on students' academic achievement and found naturalness, individualization and stimulation as the main principles of designing a classroom.

3   Experimental Methodology

A qualitative study has been conducted for the research. Data for the study is collected from research papers and books published between 2007 to 2022 worldwide. Most of the studies have been conducted in the United States of America, the United Kingdom and Australia. Only few studies have been conducted in developing countries. Now, there is need of studies in developing country like India too, with the growing awareness towards outcome-based learning [22]. Moreover, most of the studies are limited to academic achievement and have yet to consider two other aspects of learning outcomes, namely, the affective and behavioral. So, data has been collected, compiled and analyzed from 17 relevant studies. The findings of the analysis have been further validated using pilot survey. The teachers were the respondents from different secondary schools in Punjab, a state in North India [23].

4   Data Analysis

An analysis of the recent empirical studies (Fig. 2) conducted by various researchers has been done to determine the parameters for conducive learning spaces [24]. Most researchers have found wellness as the critical parameter; some also consider stimulation and individualization, whereas others find social interaction and spatial function essential for designing learning spaces. This research includes all five parameters for studying conducive learning spaces [25]. Face validity was also conducted to select the most appropriate parameters for the study. The five parameters selected from various scholarly studies and the face validity is wellness, stimulation, individualization, social interaction and spatial function. They have been described as follows:

Parameters for conducive Learning spaces	Llorens-Gámez et al. (2022)	Imms & Kvan (2021)	López-Chao et al. (2020)	Ovesoni et al. (2020)	Munir et al (2018)	Gilavand et al. (2016)	Barrett et al. (2015)	Ramli et al. (2013)	Davies et al. (2013)	Blackmore et al. (2012)	Ghaziani (2012)	Sanoff & Walden (2012)	Barrett & Zhang (2009)	Uline (2009)	Tanner (2008)	Pamela et al. (2007)	Woolner et al. (2007)
1 Wellness	✓	✓	✓	✓	✓	✓	✓	✓	✓	✓	✓	✓	✓	✓	✓	✓	✓
2 Stimulation	✓	✓	✓	✓	✓	✓	✓	✓		✓			✓	✓		✓	✓
3 Individualization		✓				✓	✓	✓				✓	✓	✓	✓		
4 Social Interaction					✓							✓	✓	✓	✓		✓
5 Spatial function	✓	✓	✓	✓	✓					✓		✓			✓		✓

Fig. 2. Analysis of researches for parameters of conducive learning spaces

4.1   Wellness

Wellness relates to all people's mental and physical benefits and supports their knowledge, skills, motivation, and actions (Ardoin et al. 2022). Therefore, linking to nature is a strategy often used to promote health and wellness among people (Braus et al. 2022; Uline, 2009). Barrett & Zhang (2009) defined the naturalness design principle that deals with optimum delighting levels, optimum acoustics, optimum learning temperature and optimum air quality levels and link to nature as it directly or indirectly improves cognitive skills (Barrett et al. 2019; Tanner, 2015). The fundamental environmental variables could affect the ability of a child to concentrate and learn in a classroom. A noise outside the class distracts and negatively affects mental wellness [26].

4.2 Stimulation

An appropriate stimulation level impacts the learning outcomes considerably (Barrett et al. 2015). The classroom's physical learning spaces impact the student's motivation towards learning and learning activities, behavior, and attendance at schools. Various facilities within the learning spaces motivate and stimulate the students differently (Baafi, 2020). Visual stimulation within a classroom relates to colour (Pamela et al. 2007), complexity (Uline, 2009) and appearance (Barrett & Zhang, 2009); and can influence the attention, mood, behavior and achievement of the students. A study found more distraction in more visually complex learning spaces among students and behavioral problems in children living [27].

4.3 Individualization

Individualization includes the elements of ownership and flexibility, which satisfies the child's needs (Barrett et al. 2015). Flexibility and choice for the arrangement of space must be provided to create a sense of belonging among the students (Kariippanon et al. 2019). The low density of classroom occupancy positively affects individualization and academic scores (Barrett et al. 2019). Connectivity between different spaces to create ease for the students to move from one place to another also helps create a sense of ownership of the place or the school (Uline, 2009; Tanner, 2008). Ergonomic scales of spaces similar to the body of pupils lead to a strong feeling of individualization among the students [28].

4.4 Social Interaction

The physical environment cannot be there without the social environment (Uline, 2009), and thus positive social interaction can be encouraged through central activity pockets (Barrett & Zhang, 2009). Seating arrangements can also influence student-teacher interaction and student-student relationships and thus impact student behaviour (Haghighi & Jusan, 2012). The social interaction between students increases with transparency within the spaces (Uline, 2009) and shaded outdoor spaces (Ghaziani, 2012), positively affecting the student's behavior [29].

4.5 Spatial Function

Experiential learning focuses on doing, active engagement, reflecting, processing, and applying the experience in the real world (Braus et al. 2022). Spatial function profoundly affects student engagement and behaviour and thus impact learning outcomes [30]. The learning experience and behaviour can only be improved by the students' engagement also showed that function of learning spaces affect student engagement, teachers' pedagogy and learning outcomes [31]. Spatial function must support active engagement in class interaction and discussions, to support students to develop social skills, regulate their behaviour, control their emotions and create a sense of relatedness to peers and teachers [32].

5 Findings and Discussion

The parameters of wellness and stimulation directly impact all three aspects of learning outcomes, whereas individualization and social interaction directly impact the affective and behavioral learning outcomes; and indirectly impact the cognitive learning outcome (Fig. 3). Spatial function directly impacts the affective and behavioral learning outcome, and indirectly impacts the cognitive learning outcomes.

The achievement of all three aspects of learning outcomes, namely cognitive, affective and behavioral, will lead to competency and outcome-based learning for quality education [33].

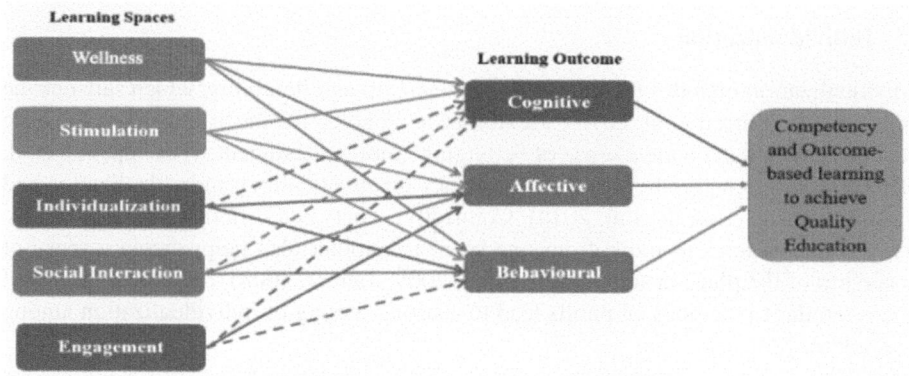

Fig. 3. Analysis of the relationship between parameters of learning spaces and learning outcome

If the parameters for learning spaces are considered while designing the learning spaces, then the learning outcomes of the learners can be improved. It will help develop skills required for student-centric, 21st-century skills. The improvement in the learning outcomes will lead to the development of skills and competencies for the holistic development of the learner; and the achievement of quality education.

6 Conclusion and Future Recommendations

The research concludes that the parameters, namely wellness, stimulation, individualization, social interaction, and spatial function, for learning spaces should be fulfilled to improve the learning outcomes of the learner and equip the learners with competency-based skills. These parameters will increase the satisfaction level of the learner as well as the students' engagement in learning, which is the need for outcome-based learning. Implementing the parameters in designing educational institutes will help develop the required skills for the learners to be competent in the real world. Moreover, this research is novel in determining the parameters for designing learning spaces to improve learning outcomes. Although the digital learning has its importance in the 21st century but the importance of learning spaces can't be neglected if we want to achieve the desired learning outcomes. So, the solution is to involve technology in the learning spaces to achieve the benefits from both of them, in enhancing learning and improving the learning outcomes of the students.

Future research is recommended to determine new parameters of learning spaces. The validation of these parameters for designing learning spaces can be done in other various countries too. It is essential to validate the relationship between the design of learning spaces and learning outcomes through a quantitative study. Moreover, the validation of this research is done only in a state of India, so further research can validate it all throughout the country to generalize the result.

References

1. More, P.B., Jadhav, A.N., Khatik, I., Singh, S., Borate, V.K., Mali, Y.K.: Sign language recognition using hand gestures. In: 2024 3rd International Conference for Advancement in Technology (ICONAT), pp. 1–5. GOA (2024). https://doi.org/10.1109/ICONAT61936.2024.10774685

2. Mali, Y., Pawar, M.E., More, A., Shinde, S., Borate, V., Shirbhate, R.: Improved pin entry method to prevent shoulder surfing attacks. In: 2023 14th International Conference on Computing Communication and Networking Technologies (ICCCNT), pp. 1–6. Delhi (2023). https://doi.org/10.1109/ICCCNT56998.2023.10306875

3. Mali, Y.K., Mohanpurkar, A.: Advanced pin entry method by resisting shoulder surfing attacks. In: 2015 International Conference on Information Processing (ICIP), pp. 37–42. Pune (2015). https://doi.org/10.1109/INFOP.2015.7489347

4. Nalawade, S.A., Pattnaik, R., Kadam, S., Lodha, P.P., Mali, Y.K., Borate, V.K.: Smart contract system with block-chain capability for improving supply chain management. In: 2024 3rd International Conference for Advancement in Technology (ICONAT), pp. 1–7. GOA (2024). https://doi.org/10.1109/ICONAT61936.2024.10774955

5. Patil, S.P., Zurange, S.Y., Shinde, A.A., Jadhav, M.M., Mali, Y.K., Borate, V.: Upgrading energy productivity in urban city through neural support vector machine learning for smart grids. In: 2024 15th International Conference on Computing Communication and Networking Technologies (ICCCNT), pp. 1–5. Kamand (2024). https://doi.org/10.1109/ICCCNT61001.2024.10724069

6. Modi, S., Modi, M., Alone, V., Mohite, A., Borate, V.K., Mali, Y.K.: Smart shopping trolley using Arduino UNO. In: 2024 15th International Conference on Computing Communication and Networking Technologies (ICCCNT) pp. 1–6. Kamand (2024). https://doi.org/10.1109/ICCCNT61001.2024.10725524

7. Mehta, U., Chougule, S., Mulla, R., Alone, V., Borate, V.K., Mali, Y.K.: Instant messenger forensic system. In: 2024 15th International Conference on Computing Communication and Networking Technologies (ICCCNT), pp. 1–6. Kamand (2024). https://doi.org/10.1109/ICCCNT61001.2024.10724367

8. Shimpi, P., Balinge, B., Golait, T., Parthasarathi, S., Arunima, C.J., Mali, Y.: Job crafter - the one-stop placement portal. In: 2024 15th International Conference on Computing Communication and Networking Technologies (ICCCNT) pp. 1–8. Kamand (2024). https://doi.org/10.1109/ICCCNT61001.2024.10725010

9. Ingale, V., Wankar, B., Jadhav, K., Adedoja, T., Borate, V.K., Mali, Y.K.: Healthcare is being revolutionized by AI-powered solutions and technological integration for easily accessible and efficient medical care. In: 2024 15th International Conference on Computing Communication and Networking Technologies (ICCCNT), pp. 1–6. Kamand (2024). https://doi.org/10.1109/ICCCNT61001.2024.10725646

10. Mulani, U., Nandgaonkar, V., Mulla, R., Sonavane, S., Borate, V.K., Mali, Y.K.: Smart contract system with blockchain capability for improved supply chain management traceability and transparency. In: 2024 15th International Conference on Computing Communication and Networking Technologies (ICCCNT), pp. 1–7. Kamand (2024). https://doi.org/10.1109/ICC CNT61001.2024.10723871

11. Sonawane, S., Mulani, U., Gaikwad, D.S., Gaur, A., Borate, V.K., Mali, Y.K.: Blockchain and Web3.0 based NFT marketplace. In: 2024 15th International Conference on Computing Communication and Networking Technologies (ICCCNT), pp. 1–6. Kamand (2024). https://doi.org/10.1109/ICCCNT61001.2024.10724420

12. Mandale, P., Modi, S., Jadhav, M.M., Khawate, S.S., Borate, V.K., Mali, Y.K.: Investigation of different techniques on digital actual frameworks toward distributed denial of services attack. In: 2024 15th International Conference on Computing Communication and Networking Technologies (ICCCNT), pp. 1–6. Kamand (2024). https://doi.org/10.1109/ICCCNT61001.2024.10725776

13. Sengupta, D., Nalawade, S.A., Sharma, L., Kakade, M.S.J.V., Borate, K., Mali, Y.K.: Enhancing file security using hybrid cryptography. In: 2024 15th International Conference on Computing Communication and Networking Technologies (ICCCNT), pp. 1–8. Kamand (2024). https://doi.org/10.1109/ICCCNT61001.2024.10724120

14. More, A., Khane, S., Jadhav, D., Sahoo, H., Mali, Y.K.: Auto-shield: Iot based OBD application for car health monitoring. In: 2024 15th International Conference on Computing Communication and Networking Technologies (ICCCNT), pp. 1–10. Kamand (2024). https://doi.org/10.1109/ICCCNT61001.2024.10726186

15. Wanaskar, U.H., Dangore, M., Raut, D., Shirbhate, R., Borate, V.K., Mali, Y.K.: A method for re-identifying subjects in video surveillance using deep neural network fusion. In: 2024 15th International Conference on Computing Communication and Networking Technologies (ICCCNT) pp. 1–4. Kamand (2024). https://doi.org/10.1109/ICCCNT61001.2024.10726255

16. More, A., Ramishte, O.L., Shaikh, S.K., Shinde, S., Mali, Y.K.: Chain-checkmate: chess game using blockchain. In: 2024 15th International Conference on Computing Communication and Networking Technologies (ICCCNT), pp. 1–7. Kamand (2024). https://doi.org/10.1109/ICC CNT61001.2024.10725572

17. Palkar, J.D., Jain, C.H., Kashinath, K.P., Vaidya, A.O., Borate, V.K., Mali, Y.K.: Machine learning approach for human brain counselling. In: 2024 15th International Conference on Computing Communication and Networking Technologies (ICCCNT), pp. 1–8. Kamand (2024). https://doi.org/10.1109/ICCCNT61001.2024.10723852

18. Dangore, M., Modi, S., Nalawade, S., Mehta, U., Borate, V.K., Mali, Y.K.: Revolutionizing sport education with AI. In: 2024 15th International Conference on Computing Communication and Networking Technologies (ICCCNT), pp. 1–8. Kamand (2024). https://doi.org/10.1109/ICCCNT61001.2024.10724009

19. Dangore, M., Bhaturkar, D., Bhale, K.M., Jadhav, H.M., Borate, V.K., Mali, Y.K.: Applying random forest for IoT systems in industrial environments. In: 2024 15th International Conference on Computing Communication and Networking Technologies (ICCCNT), pp. 1–7. Kamand, India (2024). https://doi.org/10.1109/ICCCNT61001.2024.10725751

20. More, A., Shinde, S.R., Patil, P.M., Kane, D.S., Mali, Y.K., Borate, V.K.: Advancements in early detection of lung cancer using YOLOv7. In: 2024 5th International Conference on Smart Electronics and Communication (ICOSEC), pp. 1739–1746. Trichy (2024). https://doi.org/10.1109/ICOSEC61587.2024.10722534

21. Vaidya, A.O., Dangore, M., Borate, V.K., Raut, N., Mali, Y.K., Chaudhari, A.: Deep fake detection for preventing audio and video frauds using advanced deep learning techniques. In: 2024 IEEE Recent Advances in Intelligent Computational Systems (RAICS), Kothamangalam, pp. 1–6. Kerala (2024)

22. Sawardekar, S., Mulla, R., Sonawane, S., Shinde, A., Borate, V., Mali, Y.K.: Application of modern tools in Web 3.0 and blockchain to innovate healthcare system. In: Rawat, S., Kumar, A., Raman, A., Kumar, S., Pathak, P. (eds.) Proceedings of Third International Conference on Computational Electronics for Wireless Communications. ICCWC 2023. LNNS, vol. 962. Springer, Singapore (2025). https://doi.org/10.1007/978-981-97-1946-4_2

23. Modi, S., Mali, Y., Kotwal, R., Kisan Borate, V., Khairnar, P., Pathan, A.: Hand gesture recognition and real-time voice translation for the deaf and dumb. In: Jain, S., Mihinduku-lasooriya, N., Janev, V., Shimizu, C.M. (eds.) Semantic Intelligence. ISIC 2023. LNEE, vol. 1258, pp. 435–449. Springer, Singapore (2024). https://doi.org/10.1007/978-981-97-7356-5_35

24. Bhongade, A., Dargad, S., Dixit, A., Mali, Y.K., Kumari, B., Shende, A.: Cyber threats in social metaverse and mitigation techniques. In: Somani, A.K., Mundra, A., Gupta, R.K., Bhat-tacharya, S., Mazumdar, A.P. (eds.) Smart Systems: Innovations in Computing. SSIC 2023. Smart Innovation, Systems and Technologies, vol. 392, pp. 455–467. Springer, Singapore (2024). https://doi.org/10.1007/978-981-97-3690-4_34

25. Mali, Y.: Upadhyay, T.: Fraud detection in online content mining relies on the random forest algorithm. SWB **1**(3), 13–20 (2023)

26. Kale, H., Aswar, K., Yadav, Y.M.K.: Attendance marking using face detection. Int. J. Adv. Res. Sci. Commun. Technol. 417–424

27. Inamdar, F., Ojha, D., Ojha, C.J., Mali, D.Y.: Job title predictor system. Int. J. Adv. Res. Sci. Commun. Technol. 457–463 (2024)

28. Jagdale, S., Takale, P., Lonari, P., Khandre, S., Mali, Y.: Crime awareness and registration system. Int. J. Sci. Res. Sci. Technol. **5**(8) (2020)

29. Modi, S., Mali, Y., Sharma, L., Khairnar, P., Gaikwad, D.S., Borate, V.: A protection app-roach for coal miners safety helmet using IoT. In: Jain, S., Mihindukulasooriya, N., Janev, V., Shimizu, C.M. (eds) Semantic Intelligence. ISIC 2023. LNEE, vol. 1258, pp. 377–388. Springer, Singapore (2024). https://doi.org/10.1007/978-981-97-7356-5_30

30. Mali, Y.K., Sharma, L., Mahajan, K., Kazi, F., Kar, P., Bhogle, A.: Application of CNN algorithm on X-Ray images in COVID-19 disease prediction. In: 2023 IEEE International Carnahan Conference on Security Technology (ICCST), pp. 1–6. Pune (2023). https://doi.org/10.1109/ICCST59048.2023.10726852

31. Modi, S.: Automated attendance monitoring system for cattle through CCTV. REDVET, **25**(1), 1025–1034 (2024)

32. Mali, Y., Chapte, V.: Grid based authentication system. Int. J. Adv. Res. Comput. Sci. Manage. Stud. **2**(10), 93–99 (2014)

33. Asreddy, R., Shingade, A., Vyavhare, N., Rokde, A., Mali, Y.: A survey on secured data transmission using RSA algorithm and steganography. Int. J. Sci. Res. Comput. Sci. Eng. Inf. Technol. (IJSRCSEIT) **4**(8), 159–162 (2019). ISSN 2456–3307

34. Pathak, J., Sakore, N., Kapare, R., Kulkarni, A., Mali, Y: Mobile rescue Robot. Int. J. Sci. Res. Comput. Sci. Eng. Inf. Technol. (IJSRCSEIT), **4**(8), 10–12 (2019). ISSN 2456–3307

Smart Diagnosis Using Symptoms for Seeking a Specialist Doctor

Bidyut Das$^{(\boxtimes)}$ and Rishu Kumar

Haldia Institute of Technology, Haldia Purba Medinipur 721657, West Bengal, India
bidyut2002in@gmail.com

Abstract. Technological advancement in the healthcare industry has been helping people to suggest hospitals and doctors to travel for treatment, where to admit, and which hospitals are the simplest for treating the desired disease. This paper intends to use machine learning (ML) algorithms for smart diagnosis to classify a disease using the patient's symptoms. People can use this model to predict their illness before seeking medical attention from a specialist doctor. This research focuses on existing techniques applied to human disease diagnosis in the medical field. Three different models are applied to a dataset for this experiment. Experimental results and analysis show that it is possible to construct a self-diagnosis system in the healthcare system to prevent wrong-directional treatment.

Keywords: Disease prediction · Machine learning · Human disease · Symptoms · Healthcare

1 Introduction

Every person at some point in life falls ill, some more often than others [1]. People then face some symptoms depending on the disease they are diagnosed with; medical professionals use these symptoms to identify the disease and treat the patient accordingly [2]. Disease classification based on symptoms is a preliminary task in the field of healthcare. It involves identifying the underlying disease or condition that a patient may have based on the symptoms they are experiencing. Accurate and timely disease classification can aid in early diagnosis, treatment planning, and improved patient outcomes. Traditional methods of disease classification relied heavily on manual examination and expert knowledge, which can be time-consuming and subject to human error. With the advent of machine learning and data analysis techniques, automated disease classification systems have gained prominence, offering efficient and reliable solutions [3]. As we know, the world has been facing a pandemic of COVID-19, also known as Corona Virus, since 2019. It has affected millions of people across the globe and has led to many deaths [4]. There is a sense of fear in the people. Individuals require a stage where they can set up particular manifestations, which can then predict the illness and allow the sufferer to breathe in a whisper of assistance.

M. Majumder et al. (Eds.): ICCTE 2023, CCIS 2376, pp. 54–63, 2025.
https://doi.org/10.1007/978-3-031-81935-3_5

The problem addressed in this research is the classification of diseases based on symptoms. The goal is to develop machine-learning models that can accurately predict the disease or condition using the symptoms reported by the patient. The research aims to overcome the limitations of manual diagnosis and provide a scalable and efficient solution for disease classification.

Section 2 describes related previous literature. Section 3 includes challenges and limitations in this field of research. Methods are included in Sect. 4. Section 5 and Sect. 6 present the dataset, challenges of collecting data, and data preprocessing techniques. Evaluation measures are shown in Sect. 7. Section 8 depicts the results and discussion. Lastly, Sect. 9 includes the conclusion of this paper. The prime contributions are:

- Prepare a reliable tabular dataset with symptoms and corresponding disease labels collected from various sources.
- Utilize ML models for disease classification based on symptoms and evaluate the performance.
- Compare and analyze the results of different ML models for disease prediction where ensemble learning performs better than the others.

2   Previous Related Literature

Several previous studies have focused on disease classification using symptoms and explored different approaches to enhance the accuracy and effectiveness of classification models. This section includes a few previous literature available on disease prediction using symptoms. Since healthcare research and machine learning are continually evolving, it is essential to focus on recent publications to stay updated with the latest developments.

Uddin et al. [5] studied the comparative performances of supervised ML algorithms for disease prediction. They applied many ML techniques to the same data of disease prediction for comparison. Tarawneh et al. [6] proposed a hybrid approach using data mining techniques for heart disease prediction. Their heart disease prediction system combined all algorithms into a single method. Their result shows that the combined model, which used all of the individual techniques, had superior accuracy. Keniya et al. [7] framed a system for disease prediction using multiple ML techniques. Over 230 diseases were included in the dataset. The diagnosis system identified the disease of individuals using their symptoms. Out of all methods, the weighted KNN method produced the best results. The weighted KNN's prediction accuracy was 93.5%. Deepthi et al. [8] described the disease prediction from symptoms by applying ML algorithms such as Decision Tree, Random Forest, and Naive Bayes utilized on their dataset to predict the disease. Grampurohit et al. [9] developed a classifier using machine learning to solve health-related problems by assisting physicians in predicting diseases. Their dataset contained 4920 patients' records with 41 diseases. They also presented a comparative result analysis of ML algorithms used for this experiment. Hamsagayathri et al. [10] presented a survey paper on ML-based disease prediction using symptoms. The main focus of this study was the collection of ML algorithms and approaches for disease diagnosis and decision-making. Maram et al. [11] proposed an algorithm for disease prediction using symptoms with the help of Big data analysis. Their dataset considered 400 symptoms with 147 diseases. They analyzed the performance of their proposed algorithm with ML techniques. Joshe et al. [12] collected 101 patients' data from hospitals with COPD[1] in two classes. They identified the most appropriate ML algorithm, which is efficient and accurate for COPD prediction. Their findings suggested that the Decision Tree and the Logistic Regression performed better than all other ML algorithms. Ahsan et al. [13] reviewed the articles on ML-based Disease Diagnosis (MLBDD) published between 2012 and 2021. Researchers are mainly interested in some diseases, including Diabetes, Heart disease, Kidney disease, Breast cancer, Parkinson's, and Alzheimer's diseases, which have the possibility of using machine learning/deep learning-based techniques. They also addressed some other ML-based methods for disease diagnosis. Ccolak et al. [14] studied disease prognosis on weighted symptom data. The dataset included 133 symptoms with 42 disease classes. There are 306 patients' data containing diverse types of cases. Various supervised ML methods, including SVM, KNN, and Random Forest, were employed in their study. They demonstrated that the XGBoost algorithm produced the best accuracy. Hema et al. [15] employed three ML methods (DT, NB, and RF) for their study. They displayed results using a GUI interface with these three machine-learning approaches, and feature extraction was carried out depending on disease symptoms.

3   Challenges and Limitations

Despite the progress in symptom-based disease classification, several challenges and limitations exist. Some of these include as follows.

Data Quality and Availability: Obtaining high-quality and comprehensive symptom datasets can be challenging. Incomplete or inaccurate symptom data can impact the performance of classification models. Additionally, obtaining labeled data for a wide range of diseases may be difficult, leading to imbalanced datasets and bias in classification results.

Defining Symptom Data: Symptoms reported by patients can be subjective and may vary in their descriptions or severity. Defining and standardizing symptom data across distinct individuals or healthcare settings can be challenging. Further research is needed to develop standardized symptom encoding systems or ontologies to improve consistency and comparability.

Class Imbalance: Some diseases may be rare, resulting in imbalanced class distributions in the dataset. Imbalanced datasets can lead to biased models that

---

[1] https://www.who.int/news-room/fact-sheets/detail/chronic-obstructive-pulmonary-disease-(copd).

favor the majority class. Addressing class imbalance and ensuring that rare diseases are represented adequately in the dataset is crucial for accurate disease classification.

Overfitting and Generalization: Models trained on symptom data should be able to generalize well to unseen cases. It might be a challenge to avoid overfitting, which occurs when a model gives good accuracy on training data but not works well on test data. Cross-validation, regularization, and model selection techniques can help mitigate overfitting and improve model generalization.

Explainability: Interpreting and explaining the decisions taken by symptom-based classification models is important in the healthcare domain. Black-box models, such as deep learning models, lack interpretability. Developing techniques to include previous or self-lab testing reports for the model's predictions can increase trust and facilitate adoption in clinical practice.

Addressing these challenges and limitations is crucial for the advancement and adoption of symptom-based disease classification models. Further research and development efforts should focus on data quality, interpretability, generalization, and addressing the specific challenges related to symptom-based classification in healthcare.

4 Methods

Various techniques and algorithms have been employed for symptom-based disease classification. Traditional ML techniques, such as support vector machine, random forest, and naive Bayes, have been widely used. These algorithms leverage the patterns and relationships between symptoms and diseases to classify new instances [16].

Decision Tree (DT): It is a non-parametric hierarchical ML algorithm for classification. It represents a series of decisions and their possible consequences. The decision tree can predict the target variable value by learning decision rules derived from the features.

K-Nearest Neighbors (KNN): It is a supervised learning classifier that groups individual data points based on closeness to classify or predict data [17].

Naive Bayes (NB): It is a probabilistic ML technique that classifies the input data points using the Bayes Theorem internally [18].

Support Vector Machine (SVM): It is a supervised ML algorithm commonly applied for classification problems [19]. In the SVM approach, every data point is presented as a point in N-dimensional space (N denotes the number of features), with each feature representing the value of a specific coordinate. The classification task is performed by locating the hyper-plane which distinguishes the two classes.

Random Forest (RF): It is an ensemble learning (Bagging) approach that performs the classification using several decision trees [20]. In a random forest, all the individual decision trees are weak classifier, and the combined output of these weak decision trees are a strong classifier.

The above ML algorithms are used for predicting disease and comparing the results of the individual algorithms on our dataset in Sect. 8.

5 Dataset Description

Data collection is crucial in disease classification. It involves gathering relevant data representing symptoms and disease labels. The collection process should collect a representative and diverse dataset that covers a wide range of diseases and associated symptoms. It should ensure the data is reliable and properly labeled with accurate disease information. The following are key considerations for data collection in disease classification:

Symptom Data: Identify the symptoms or clinical features relevant to the diseases of interest. It can include subjective symptoms reported by patients, objective measurements from tests, or diagnostic criteria defined by medical professionals.

Disease Labels: Clearly define the target disease labels for classification. These labels should represent the specific diseases or conditions that the classification model aims to predict.

We have used a dataset from Kaggle[2]. It has two CSV files. The first is used for model training, and the second is for model testing. The two CSV files are combined to create a single dataset with 4962 entries (rows). We have removed the duplicate entries (rows) and considered 305 records (rows) in the dataset for this experiment. The Kaggle dataset has been collected from many sources, such as medical records, surveys, or online health platforms. It consists of 133 columns. The first 132 columns denote the symptoms, and the end column indicates the disease class. This dataset has 42 different disease classes. We have tested our models using 10-fold cross-validation.

6 Data Pre-processing

Data preprocessing is a prime step in disease classification to ensure the quality and suitability of the data for analysis. The following are common data preprocessing steps applied to prepare the data for disease classification:

Cleaning the Data: Before feeding the data to the model for training, we cleaned it. The refined data can improve the performance of machine-learning models.

Feature Selection: All features may not contribute equally to disease classification. The most informative features with a strong correlation with the target

---

[2] https://www.kaggle.com/datasets/kaushil268/disease-prediction-using-machine-learning.

variable (disease labels) have been identified for classification. We have removed uninformative or redundant features to reduce noise and dimensionality.

Normalization: Data transformation techniques have been applied to normalize the data distribution, handle skewed data, or scale the features to a similar range. Common transformations include logarithmic or power transformations, z-score normalization, or min-max scaling. We have used the logarithmic transformation to normalize the dataset.

Encoding Categorical Data: Categorical variables have been encoded into numerical values for ML algorithms. Encoding techniques include one-hot encoding, label encoding, or ordinal encoding is applied depending on the nature of the categorical variables. We have applied one-hot encoding for encoding categorical values.

Handling Imbalanced Data: Imbalanced data means some disease classes have significantly fewer samples than others. Oversampling, under-sampling, or class weights can be employed to address this issue and ensure a balanced representation of different disease classes. We have used oversampling to balance the dataset.

7   Evaluation Metrics

Evaluation metrics are used to examine the performance of disease classification models and measure their accuracy in predicting disease labels based on symptom data. The choice of evaluation metrics depends on the particular requirements and characteristics of the classification task. These metrics provide information about the capacity of the model to classify diseases correctly, detect true positives and negatives, and handle imbalanced data. The choice of evaluation metrics should align with the objectives and consider the specific characteristics of the disease classification task [21]. The following metrics are used to evaluate our ML models for disease classification.

Accuracy: The accuracy is measured by the proportion of instances correctly classified out of the total number of instances. It is the most straightforward evaluation metric but may not be sufficient when dealing with imbalanced datasets.

Confusion Matrix: The classifier's predictions are shown clearly using a confusion matrix, which displays the numbers of true positives, true negatives, false positives, and false negatives. It enables the calculation of various evaluation metrics along with precision, recall, and F1 score.

Cross-Validation: The k-fold cross-validation is used to evaluate the model's performance more accurately. It divided the dataset into several subsets (folds). The model is trained and tested multiple times on various folds using 10-fold cross-validation. The accuracy of the model is calculated as an average of all results.

8   Results and Discussion

A comparative analysis is conducted to evaluate the results of different ML models for disease classification. Multiple models, such as Decision tree, Gaussian NB, KNN, SVM, and random forest are applied, and their performance is compared on our dataset.

The dataset is split into training (70%) and test data (30%). The models are trained on a training dataset, and the performances are evaluated on the test dataset. The evaluation results are shown in Table 1. Figure 1 depicts the comparative accuracy of ML models for disease prediction.

Table 1. Performance of ML models for disease prediction

Model	Precision	Recall	F1-Score
Decision Tree	0.80	0.72	0.75
K-Nearest Neighbour	0.96	0.96	0.96
Gaussian Naive Bayes	1.0	1.0	1.0
Support Vector Machine	0.95	0.93	0.94
Random Forest	1.0	1.0	1.0

The results reveal that Gaussian Naive Bayes and Random Forest outperform other models across all evaluation metrics. It indicates its superiority in refers to precision, recall, and F1 score. The model's strong generalization capabilities are crucial in real-world applications, where robust performance on unseen data is essential.

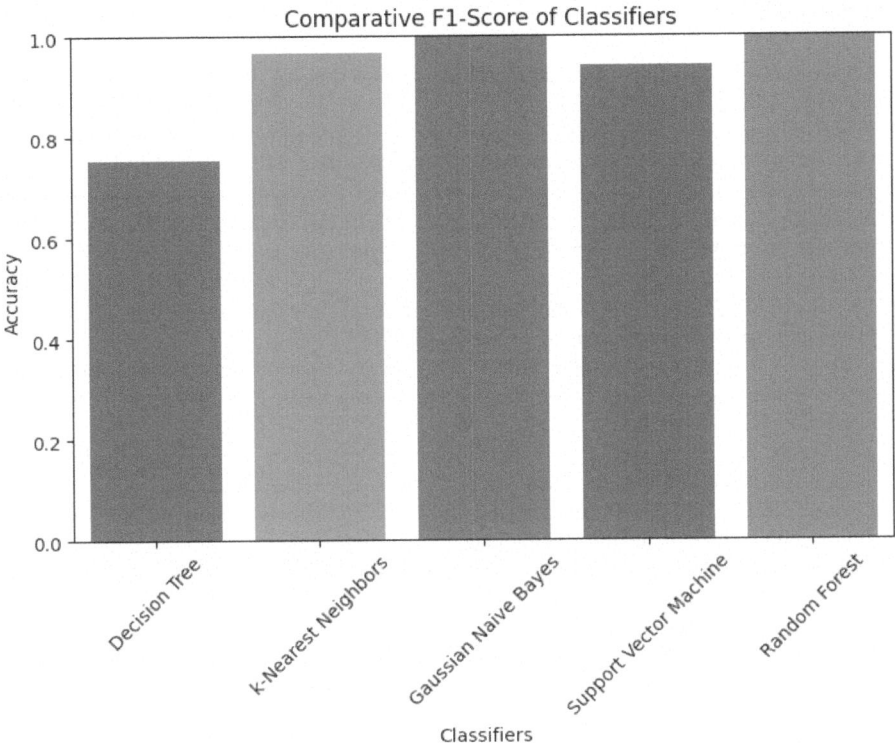

Fig. 1. Comparative result of ML models for disease prediction

9   Conclusion

Research on disease classification based on symptoms is an emerging machine-learning application in healthcare. We have seen the value of health and lives during the pandemic. There is nothing more valuable than a healthy life. Serving the people and solving their health-related problems is our motto throughout the development of this task. We have developed this model for classifying and analyzing diseases using ML algorithms for early diagnosis and treatment of diseases depending on the symptoms provided by the patient. Random Forest, or Gaussian Naive Bayes, is a promising classification approach for classifying diseases. The disease classification project based on symptoms holds significant potential for improving healthcare outcomes. Machine learning models can be utilized for accurate and automated disease classification, facilitating early diagnosis and effective treatment. Continued research and development in this area can contribute to advancements in healthcare and benefit patients worldwide.

References

1. Caplan, A.L.: The concepts of health, illness, and disease. Companion Encycl. Hist. Med. **1**, 233–248 (1993)
2. Knottnerus, J.A.: The effects of disease verification and referral on the relationship between symptoms and diseases. Med. Decis. Making **7**(3), 139–148 (1987)
3. Ferdous, M., Debnath, J., Chakraborty, N.R.: Machine learning algorithms in healthcare: a literature survey. In: 2020 11th International Conference on Computing, Communication and Networking Technologies (ICCCNT), pp. 1–6. IEEE (2020)
4. Pal, R., Sekh, A.A., Kar, S., Prasad, D.K.: Neural network based country wise risk prediction of COVID-19. Appl. Sci. **10**(18), 6448 (2020)
5. Uddin, S., Khan, A., Hossain, M.E., Moni, M.A.: Comparing different supervised machine learning algorithms for disease prediction. BMC Med. Inform. Decis. Making **19**(1), 1–16 (2019)
6. Tarawneh, M., Embarak, O.: Hybrid approach for heart disease prediction using data mining techniques. In: Barolli, L., Xhafa, F., Khan, Z.A., Odhabi, H. (eds.) EIDWT 2019. LNDECT, vol. 29, pp. 447–454. Springer, Cham (2019). https://doi.org/10.1007/978-3-030-12839-5_41
7. Keniya, R., et al.: Disease prediction from various symptoms using machine learning (2020). Available at SSRN 3661426
8. Deepthi, Y., Kalyan, K.P., Vyas, M., Radhika, K., Babu, D.K., Krishna Rao, N.V.: Disease prediction based on symptoms using machine learning. In: Sikander, A., Acharjee, D., Chanda, C.K., Mondal, P.K., Verma, P. (eds.) Energy Systems, Drives and Automations. LNEE, vol. 664, pp. 561–569. Springer, Singapore (2020). https://doi.org/10.1007/978-981-15-5089-8_55
9. Grampurohit, S., Sagarnal, C.: Disease prediction using machine learning algorithms. In: 2020 International Conference for Emerging Technology (INCET), pp. 1–7. IEEE (2020)
10. Hamsagayathri, P., Vigneshwaran, S.: Symptoms based disease prediction using machine learning techniques. In: 2021 Third International Conference on Intelligent Communication Technologies and Virtual Mobile Networks (ICICV), pp. 747–752. IEEE (2021)
11. Maram, B., Kumar, K.S., Gampala, V.: Symptoms based disease prediction using bigdata analytics. Turk. J. Physiotherapy Rehabil. **32**(3), 3228–3234 (2021)
12. Joshe, M.D., Emon, N.H., Islam, M., Ria, N.J., Masum, A.K.M., Noori, S.R.H.: Symptoms analysis based chronic obstructive pulmonary disease prediction in Bangladesh using machine learning approach. In: 2021 12th International Conference on Computing Communication and Networking Technologies (ICCCNT), pp. 1–5. IEEE (2021)
13. Ahsan, M.M., Luna, S.A., Siddique, Z.: Machine-learning-based disease diagnosis: a comprehensive review. Healthcare **10**(3), 541 (2022)
14. Çolak, M., Sivri, T.T., Akman, N.P., Berkol, A., Ekici, Y.: A study of disease prediction on weighted symptom data using deep learning and machine learning algorithms. In: 2022 International Conference on Theoretical and Applied Computer Science and Engineering (ICTASCE), pp. 116–119. IEEE (2022)
15. Hema, P., Darbha, A., Sunny, N., Naganjani, R.V.: Disease prediction using symptoms based on machine learning algorithms and natural language processing. In: 2023 International Conference on Artificial Intelligence and Knowledge Discovery in Concurrent Engineering (ICECONF), pp. 1–7. IEEE (2023)

16. Costa, V.G., Pedreira, C.E.: Recent advances in decision trees: an updated survey. Artif. Intell. Rev. **56**(5), 4765–4800 (2023)
17. Peterson, L.E.: K-nearest neighbor. Scholarpedia **4**(2), 1883 (2009)
18. Webb, G.I., Keogh, E., Miikkulainen, R.: Naïve bayes. Encycl. Mach. Learn. **15**(1), 713–714 (2010)
19. Noble, W.S.: What is a support vector machine? Nat. Biotechnol. **24**(12), 1565–1567 (2006)
20. Rigatti, S.J.: Random forest. J. Insur. Med. **47**(1), 31–39 (2017)
21. Zhou, J., Gandomi, A.H., Chen, F., Holzinger, A.: Evaluating the quality of machine learning explanations: a survey on methods and metrics. Electronics **10**(5), 593 (2021)

A Comprehensive Review and Future Prospects of Lie Detection Using Machine Learning

Debanil Chanda$^{(\boxtimes)}$ (iD) and R. K. Mandal (iD)

Department of Computer Science and Technology, University of North Bengal, Siliguri 734013, West Bengal, India
dchanda6@gmail.com, rakeshmandal@bu.ac.in

Abstract. Lie detection plays a vital role in various fields, including law enforcement, security and forensic psychology. With the advancement of machine learning techniques, there has been a significant shift towards using computational models for lie detection. This research article presents a comprehensive review of lie detection methods using machine learning approaches. It explores the diverse range of machine learning algorithm employed in deception detection, their performance metrics, and the datasets used for training and evaluation. Additionally, the article discusses the challenges, limitations and future perspectives in the field of machine learning based lie detection. The aim is to provide researchers, practitioners and policymakers with an in-depth understanding of the latest and greatest advancement in this rapidly evolving field.

Keywords: Lie detection · Machine learning · Deception detection · Physiological signals · Facial expression · Natural language processing

1 Introduction

1.1 Background and Significance

Lie, the act of deliberately misrepresentation of false information, has been a comprehensive aspect of human communication throughout history [1]. As society evolves, the significance of precise deception detection grows in diverse contexts, including law enforcement, security, intelligence gathering and social interactions [2]. Lie detection, the process of identifying deceptive behaviour or dishonesty, has thus becomes a crucial field of research with far-reaching implications. The prevalence of deception in human communications has fueled a longstanding search for effective lie detection methods [3]. Throughout history, numerous traditional techniques, such as direct questioning methods and non-verbal behaviour analysis, have been used to identify deceptive individuals [1, 4]. However, these methods often lack scientific rigor, suffer from subjective interpretation and produce inconsistent results, making them unreliable for critical applications [5]. In recent decade, the field of lie detection has witnessed significant advancements driven by the rapid progress of machine learning techniques [6]. Machine learning, a subset of artificial intelligence, empowers computers to learn patterns and make data-driven

M. Majumder et al. (Eds.): ICCTE 2023, CCIS 2376, pp. 64–77, 2025.
https://doi.org/10.1007/978-3-031-81935-3_6

conclusions without explicit programming [7–9]. By harnessing the power of large-scale data analysis and complex algorithms, machine learning has demonstrated the potential to transform lie detection into a more objective, efficient and accurate process [6]. The growing importance of lie detection in various fields, including law enforcement, national security, forensic psychology and corporate investigations, enhance the significance of this research article. Accurate and reliable lie detection method can help law enforcement agencies in solving crimes, intelligence agencies in identifying potential threats and legal proceedings in justice [2, 10]. Moreover, the implications of deception go beyond traditional domains, extending to areas like human-computer interaction, customer service and relationship management [11, 12]. Ensuring truthfulness in digital communications and interactions is crucial for building trust and maintaining integrity in the digital age [13]. Despite the potential benefits, the application of machine learning in lie detection is not without challenges. The need to handle sensitive personal data raises ethical concerns regarding privacy, data protection and potential biases [5]. It is imperative to strike a balance between leveraging advanced technologies for deception detection and safeguarding individuals' rights and privacy [14, 15].

The importance of this research article stems from its comprehensive review of lie detection using machine learning techniques. By critically assessing the various methodologies, algorithms and datasets, this article seeks to provide researchers, practitioners and policymakers with an up-to-date understanding of the latest and greatest advancement in the field. The knowledge presented in this article can help researchers identify the most suitable machine learning methods for their lie detection studies and encourage interdisciplinary collaboration between experts in machine learning and lie detection. Furthermore, the insights gained from this review can guide the development of robust, transparent and ethical lie detection systems that can withstand scrutiny and confidently applied in real-world scenarios. This research article's background and significance lie in its exploration of how machine learning has transformed the landscape of lie detection. By addressing the challenges, potential biases and ethical considerations, the article aims to foster advancements in the field and contribute to the development of more reliable and responsible lie detection technologies.

1.2 Objective of the Study

The main goal of this research article is to provide a comprehensive review of lie detection using machine learning techniques. The study aims to explore the diverse range of machine learning algorithms and methodologies employed in deception detection, providing a critical analysis of their strengths, limitations and potential applications. Specifically, the objectives of the study are as follows:

Review Lie Detection Techniques: The article will examine various lie detection techniques enabled by machine learning, including but not limited to physiological signals analysis, facial expression recognition, speech and voice analysis, and natural language processing [15–18]. Each technique's underlying principles, algorithms and relevant studies will be reviewed.

Explore Multimodal Fusion Approaches: The article will investigate the integration of multimodal fusion techniques, where multiple sources of data, such as physiological signals, facial expressions and speech are combined to enhance the accuracy and robustness of lie detection systems [19, 20].

Evaluate Machine Learning Algorithms: The study will comprehensively evaluate the performance of machine learning algorithms in lie detection, such as Support Vector Machine (SVM), Random Forest (RF), Artificial Neural Network (ANN), Convolutional Neural Networks (CNN) and Recurrent Neural Network (RNN) will be discussed in terms of their applicability and efficacy for lie detection [7–9].

Discuss Performance Evaluation and Datasets: The study will analyze the commonly used performance evaluation metrics in machine learning based lie detection and assess the publicly available datasets used for training and evaluating lie detection models [7–9].

Address Ethical Considerations: Ethical considerations play a vital role in deploying machine learning based lie detection systems responsibly. The article will address privacy concerns, data protection, potential biases and the importance of fairness and transparency in the development and application of such technologies [14, 15].

Identify Challenges and limitations: By critically evaluating existing methodologies, the study aims to identify the challenges and limitations faced in lie detection using machine learning, including issues related to overfitting, generalization and the interpretability of complex models [7–9].

Provide Future Perspectives: The article will outline future research directions and potential advancements in the field of machine learning based lie detection. Topics such as ensemble methods, incremental learning and explainable AI will be explored [21–23].

Overall, the study seeks to provide researchers, practitioners and policymakers with an up-to-date understanding of the latest and greatest advancement in lie detection using machine learning. By achieving these objectives, the research article aims to foster responsible and effective application of machine learning techniques in lie detection, with implications spanning law enforcement, security, legal proceeding and various other domains.

2   Lie Detection Techniques

This section of the research article provides an in-depth exploration of various lie detection techniques that have been empowered by machine learning methodologies. These techniques include physiological signals analysis, facial expression recognition, speech and voice analysis, as well as natural language processing. Each technique offers unique insights into the realm of deception detection.

2.1   Physiological Signals Analysis

Physiological signals have been recognized as potential indicators of stress and emotional arousal linked to deception [24]. Machine learning algorithms can analyze these signals to detect patterns associated with deceptive behavior. By training models on physiological data collected during controlled experiments or real-world scenarios, researchers have developed algorithms capable of discriminating between truth-tellers and liars with enhanced accuracy [25].

2.2 Facial Expression Recognition

Human faces convey a wealth of information, including emotions and potential deception cues [26]. Machine learning techniques, particularly computer vision algorithms, have enabled the analysis of micro expressions, subtle facial movements that occur involuntarily and can reveal concealed emotions [15]. These algorithms detect and classify facial features to identify patterns indicative of deceptive behavior. The integration of deep learning approaches has significantly improved the accuracy of facial expression analysis for lie detection [27].

2.3 Speech and Voice Analysis

The human voice carries a plethora of cues that can be indicative of deception, including changes in pitch, tone, and speech rate. Machine learning models, including both traditional methods like Support Vector Machines (SVM) and advanced deep learning architectures like Recurrent Neural Networks (RNN), can be trained to identify patterns in speech associated with deception [28, 29]. By analyzing voice characteristics such as pitch variation and hesitation patterns, these models offer a computational framework for discerning truth from falsehood.

2.4 Natural Language Processing (NLP)

Language is a powerful medium for conveying information, including deception. Natural Language Processing (NLP) techniques enable the analysis of linguistic patterns, sentiment, and discourse markers to detect deceptive content within text [18, 30]. Machine learning algorithms, such as sentiment analysis models and recurrent neural networks (RNN), are applied to text data to identify linguistic cues that may indicate deceit [18, 31]. This approach is particularly valuable in scenarios where textual communication is prevalent, such as digital interactions and written statements.

2.5 Multimodal Fusion

Recognizing that deception detection cues are often multifaceted, researchers have explored the integration of multiple data sources through multimodal fusion techniques. By combining information from physiological signals, facial expressions, speech, and linguistic patterns, machine learning algorithms can leverage the complementary nature of these cues to enhance overall accuracy. Multimodal approaches address the limitations of relying solely on a single data source and create a more robust foundation for lie detection systems [19, 20].

These techniques not only enhance accuracy but also provide valuable insights into the underlying physiological, facial, vocal, and linguistic cues associated with deception. The following sections of this article will delve deeper into the machine learning algorithms employed in these techniques and evaluate their efficacy in lie detection.

3 Machine Learning Algorithms for Lie Detection

Machine learning algorithms play a pivotal role in transforming raw data from various lie detection techniques into actionable insights. These algorithms are capable of identifying intricate patterns, discriminating between truthful and deceptive behaviors, and enhancing the accuracy of deception detection systems. This section explores the diverse spectrum of machine learning algorithms utilized in lie detection, ranging from traditional methods to advanced deep learning architectures.

3.1 Support Vector Machine (SVM)

Support Vector Machines (SVM) are a class of supervised learning algorithms primarily used for classification tasks including lie detection. They aim to find a hyperplane that maximizes the margin between different classes while minimizing the classification error [9]. Mathematically, for a binary classification problem, the decision boundary of SVM can be represented as given in Eq. (1).

$$w^T x = 0 \tag{1}$$

where, w and x are vectors and $w^T x$ is the dot product of the vectors.

3.2 Random Forest (RF)

Random Forest is a powerful ensemble learning algorithm for lie detection that combines multiple decision trees to improve classification accuracy [32]. The prediction of a random forest model can be expressed as given in Eq. (2).

$$\hat{y}RF = mode(\hat{y}_1, \hat{y}_2, \ldots\ldots, \hat{y}_n) \tag{2}$$

where, $\hat{y}RF$ is the final prediction of the random forest model and $\hat{y}_i$ represents the prediction of the ith decision tree.

3.3 Artificial Neural Network (ANN)

ANNs belong to a category of machine learning algorithms inspired by the neural structure of the human brain. ANNs are composed of interconnected nodes or neurons arranged in layers, including input, hidden and output layers. ANNs stand as powerful tools for lie detection, offering the ability to model complex relationships and unravelling complex patterns [9, 33]. The output of a neuron in an ANN can be represented as given in Eq. (3).

$$a_j = \sigma \left(\sum_{i=1}^{n} \left(w_{ij} x_i + b_j \right) \right) \tag{3}$$

where, a_j is the output of neuron j, w_{ij} is the weight between input i and neuron j, x_i is the input value, b_j is the bias of neuron j and σ is the activation function.

3.4 Convolutional Neural Network (CNN)

In lie detection, when facial expressions use as input data, Convolutional Neural Networks (CNN) play vital role to analyze the data from images and videos [34]. CNN mainly consists of three layers:

Convolutional layer can be represented as given in Eq. (4).

$$c_{i,j} = \sum_m \sum_n x_{i-m,j-n} \times k_{m,n} \tag{4}$$

where, $c_{i,j}$ is the output at position (i, j), $x_{i-m,j-n}$ represents input values at position $(i\text{-}m, j\text{-}n)$ and $k_{m,n}$ is the filter kernel.

Pooling layer can be represented as given in Eq. (5).

$$y_{i,j} = max_m max_n x_{i-m,j-n} \tag{5}$$

where, $y_{i,j}$ is the pooled output and $x_{i-m,j-n}$ is the input at position $(i\text{-}m, j\text{-}n)$.

Fully connected layer can be represented as given in Eq. (6).

$$y = f(wx + b) \tag{6}$$

Here, y represents the output vector, f denotes the activation function, w stands for the weight matrix, x represents the input vector and b signifies the bias vector.

3.5 Recurrent Neural Network (RNN)

RNNs are designed to handle sequential data, processing inputs not as instances but as sequences [35]. In lie detection, where cues are dynamic, RNNs play a vital role. The hidden state update in a basic RNN can be defined as given in Eq. (7).

$$h_t = \sigma(w_{hx} + w_{hh}h_{t-1} + b_h) \tag{7}$$

where, h_t is the hidden state at time step t, w_{hx} is the weight matrix between input x_t and hidden state h_t, w_{hh} is the weight matrix associated with the previous hidden state h_{t-1}, b_h is the bias term and σ is the activation function.

4 Performance Evaluation and Datasets

Performance evaluation is a critical aspect of assessing the efficacy of machine learning algorithm in lie detection. Rigorous evaluation metrics and suitable datasets are essential for validating the accuracy, precision and generalizability of deception detection models. This section delves into the common evaluation metrics employed and highlights the significance of appropriate datasets in the advancement of lie detection research.

Evaluating the performance of lie detection models is crucial to assess their accuracy and generalizability. A comprehensive experiment conducted on "Real Life Trial Deception Detection Dataset" [36, 37]. The dataset consists of deceptive and truthful video clips, deceptive and truthful transcriptions and gesture annotations (facial expressions, head movement and hand gesture). Here, the gesture annotations are used as a dataset for the experiment. The dataset consists of 121 samples, each containing 39 features that encapsulate facial expression, head movement and hand gesture.

4.1 Evaluation Metrics

The following machine learning algorithms were considered for the study: SVM, RF, multi-layer perceptron (MLP), CNN and RNN. The following metrics are employed to quantify the effectiveness of the models [38]:

Accuracy. Accuracy measures the portion of correct predictions among all predictions made by the model, calculated as given in Eq. (8).

$$Accuracy = \frac{Number\ of\ correct\ predictions}{Total\ number\ of\ predictions} \times 100\% \tag{8}$$

Precision and Recall. Precision measures the proportion of true positive predictions among all the predicted positives as given in Eq. (9).

$$Precision = \frac{True\ positives}{True\ positives + False\ positives} \tag{9}$$

Recall assesses the fraction of true positives relative to the total actual positives as given in Eq. (10).

$$Recall = \frac{True\ positives}{True\ positives + False\ negatives} \tag{10}$$

F1-Score. The F1-Score offers an equilibrium-based evaluation of a model's performance as given in Eq. (11).

$$F1 - Score = 2 \times \frac{Precision \times Recall}{Precision + Recall} \tag{11}$$

Roc. This is a plot depicting the true positive rate against the false positive rate.

Table 1 represents the performance of different lie detection model on the above said dataset and Fig. 1 depicts the graphical representation of the results shown in Table 1.

These results highlight the efficacy of various lie detection models. The MLP exhibited the highest performance, achieving an accuracy of 97% and balanced F1-Score of 97%.

Table 1. Comparative Analysis

Model	Accuracy	Precision	Recall	F1-Score	ROC
SVM	0.87	0.84	0.91	0.88	0.84
RF	0.92	0.94	0.90	0.93	0.92
MLP	0.97	0.98	0.96	0.98	0.94
CNN	0.80	0.82	0.80	0.81	0.81
RNN	0.76	0.77	0.76	0.77	0.76

These results demonstrate the capability of Multi-layer Perceptron in effectively capturing and exploiting the complex patterns of facial expression, head movement and hand gesture. The performance variation across models emphasizes the importance of selecting the appropriate architecture for a given lie detection scenario.

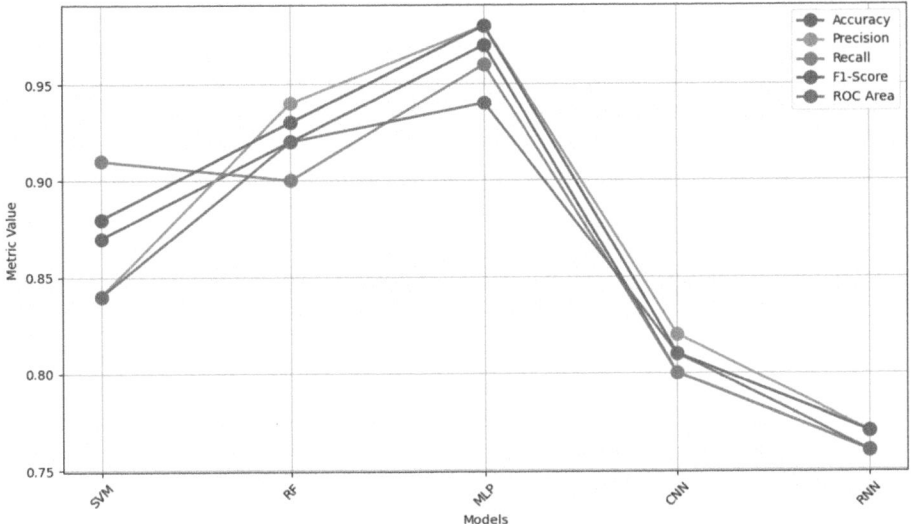

Fig. 1. A graphical comparison of various evaluation metrics as shown in the Table 1

4.2 Publicly Available Datasets

Access to high quality and diverse datasets is crucial for training, validating and comparing lie detection models. Several publicly available datasets are commonly used in the lie detection community.

Real-life Trial Deception Detection Dataset. The dataset consists of real-life trial data. The dataset also contains the statements made by the speakers in the video [36, 37].

Open-Domain Deception Dataset. The dataset contains short open domain truth and lies form 512 users. It has a total of 7168 one-line statements out of which 3598 are truth and 3569 are lies. It also includes the demographic information of each user [39].

Multilingual Deception Dataset. This dataset consists of short truthful deceptive essays [40].

Bag of Lies Dataset. This is a multimodal dataset consists of video, audio, EEG and eye gaze data from 35 unique subjects [41].

These datasets provide a foundation for lie detection research, enabling researchers to develop and evaluate machine learning models under various scenarios and conditions.

4.3 Challenges in Dataset Collection

Collecting high quality datasets for lie detection presents several challenges that researchers need to address. Acquiring authentic data where individuals are naturally engaged in deceptive behaviours can be exceptionally challenging [42, 43]. Collecting data that involves personal interactions or sensitive information raises privacy and ethical concerns. Lie detection datasets often exhibit imbalanced class distributions, where the number of truthful instances significantly outweighs deceptive instances or vice versa.

People employ various strategies while lying, capturing the diversity of lie strategies and ensuring that the dataset covers a broad range of deceptive behaviors is a complex task. For lie detection in real-time scenarios (e.g., interviews, negotiations), collecting data with accurate temporal alignment is challenging. The timing of responses and physiological signals needs to be synchronized accurately to train effective models.

Addressing these challenges requires careful planning, ethical considerations and innovative solutions.

5 Ethical Consideration and Privacy Issues

The integration of machine learning into lie detection brings forth a series of ethical considerations and privacy concerns that must be carefully addressed to ensure responsible and accountable deployment of these technologies. This section delves into the ethical implications associated with machine learning-based lie detection, emphasizing the importance of transparency, fairness, and individual rights.

5.1 Privacy and Data Protection

One of the foremost ethical considerations in lie detection is the protection of individuals' privacy and personal data [44, 45]. The collection of physiological signals, facial expressions, speech, and textual information for analysis raises questions about the ownership, storage, and potential misuse of sensitive information. Adhering to robust data protection protocols, anonymization, and secure storage mechanisms is essential to prevent unauthorized access and protect individuals' confidentiality.

5.2 Bias and Fairness

Machine learning algorithms can unintentionally adopt biases present in their training data, resulting in unfair and discriminatory results [46]. Biases rooted in cultural, demographic, or socioeconomic factors could result in disparate treatment of different groups. Ensuring fairness requires careful consideration during dataset curation, feature selection, and algorithm design, as well as the continuous monitoring and mitigation of bias in lie detection models [47].

5.3 Interpretability and Transparency

The "black-box" nature of some advanced machine learning models raises concerns about their interpretability. Understanding how decisions are reached is pivotal for building trust and accountability [48]. Efforts should be made to develop interpretable models that provide explanations for their predictions, enabling stakeholders to comprehend and validate the rationale behind deception classification outcomes.

6 Challenges and Limitations

While machine learning has offered remarkable advancements in the field of lie detection, numerous challenges and limitations underscore the complexity of this endeavor. This section delves into the multifaceted hurdles that researchers and practitioners encounter when applying machine learning to deception detection, ranging from technical intricacies to ethical considerations.

6.1 Lack of Standardization

The absence of standardized protocols for collecting and labeling deceptive behavior data poses a significant challenge [49]. Variations in experimental designs, data collection procedures, and ground truth definitions hinder the comparability and reproducibility of results across different studies. A lack of standardized benchmarks makes it arduous to benchmark the performance of various machine learning techniques consistently.

6.2 Overfitting and Generalization

Overfitting, wherein a model captures random noise in the training data instead of the underlying patterns, can result in limited performance when applied to new, unseen data. Balancing model complexity and data scarcity is particularly challenging, and efforts must be made to ensure that the trained models are capable of extrapolating beyond the training set [50].

6.3 Interpretability of Complex Models

Deep learning architectures, while powerful, often lack interpretability due to their intricate architectures [27]. This lack of transparency raises concerns about the trustworthiness of model decisions and hampers the ability to diagnose errors. Striking a balance between model performance and interpretability is a formidable challenge in the development of machine learning-based lie detection systems.

7 Future Perspectives

This section explores potential directions and innovations that could shape the future of deception detection, leveraging advancements in machine learning to address existing challenges and open new horizons.

7.1 Ensemble Methods and Hybrid Approaches

Ensemble methods, which combine the predictions of multiple machine learning models, hold potential for enhancing the robustness and reliability of lie detection [21]. Hybrid approaches that fuse information from different lie detection techniques, such as physiological signals analysis, facial expression recognition, and linguistic analysis, could lead to more comprehensive and accurate deception detection systems.

7.2 Incremental Learning

The ability of machine learning models to adapt to new information in real-time is crucial for lie detection in dynamic scenarios [22]. Incremental learning techniques could facilitate the continuous refinement and adaptation of models, enabling them to better capture evolving patterns of deception and counteract adversarial strategies.

7.3 Explainable AI

The pursuit of more transparent and interpretable machine learning models remains a priority. Advancements in explainable AI can bridge the gap between model complexity and human comprehensibility, enabling stakeholders to understand the rationales behind model predictions and promoting trust in deception detection systems [23].

8 Conclusion

This research article provides a comprehensive review of lie detection techniques using machine learning, emphasizing their strength, limitations and potential applications. It addresses the challenges associated with data collection, evaluation and ethical considerations.

Furthermore, it highlights the future directions of research in the field of machine learning-based lie detection, with a focus on explainability, fairness and real-time implementation. By promoting a deeper understanding of machine learning approaches for lie detection, this article aims to contribute to the development of reliable and robust lie detection systems.

References

1. Otasowie, O.: Application of machine learning in deception detection. In: Advances in Intelligent Systems and Computing, pp. 61–76, Springer, Cham (2020). https://doi.org/10.1007/978-3-030-52243-8_6
2. Caso, L., Palena, N., Carlessi, E., Vrij, A.: Police accuracy in truth/lie detection when judging baseline interviews. Psychiatry, Psychol. Law. **26**, 841–850 (2019). https://doi.org/10.1080/13218719.2019.1642258
3. Brennen, T., Magnussen, S.: Research on non-verbal signs of lies and deceit: a blind alley. Front. Psychol. **11** (2020). https://doi.org/10.3389/fpsyg.2020.613410
4. Bond, C.F., Levine, T.R., Hartwig, M.: New findings in non-verbal lie detection. Detecting Deception. 37–58 (2014). https://doi.org/10.1002/9781118510001.ch2
5. Constâncio, A.S., Tsunoda, D.F., Silva, H. de F.N., Silveira, J.M. da, Carvalho, D.R.: Deception detection with machine learning: a systematic review and statistical analysis. PLOS ONE. **18**, e0281323 (2023). https://doi.org/10.1371/journal.pone.0281323
6. Lai, V., Tan, C.: On Human predictions with explanations and predictions of machine learning models. In: Proceedings of the Conference on Fairness, Accountability, and Transparency. ACM, New York, NY, USA (2019)
7. Géron, A.: Hands-On Machine Learning with Scikit-Learn and TensorFlow: Concepts, Tools, and Techniques to Build Intelligent Systems. "O'Reilly Media, Inc." (2017)

8. Brownlee, J.: Machine learning mastery with python: understand your data, create accurate models, and work projects end-to-end. Mach. Learn. Mastery (2016)
9. Srinivasaraghavan, A.: Machine Learning (2014)
10. Frank, M.G., Feeley, T.H.: To catch a liar: challenges for research in lie detection training. J. Appl. Commun. Res. **31**, 58–75 (2003). https://doi.org/10.1080/00909880305377
11. Abootalebi, V., Moradi, M.H., Khalilzadeh, M.A.: A new approach for EEG feature extraction in P300-based lie detection. Comput. Methods Programs Biomed. **94**, 48–57 (2009). https://doi.org/10.1016/j.cmpb.2008.10.001
12. Chen, J.I.-Z., Lai, K.-L.: Deep convolution neural network model for credit-card fraud detection and alert. June 2021. **3**, 101–112 (2021). https://doi.org/10.36548/jaicn.2021.2.003
13. Alharbi, A., Dong, H., Yi, X., Tari, Z., Khalil, I.: Social media identity deception detection. ACM Comput. Surv. **54**, 1–35 (2021). https://doi.org/10.1145/3446372
14. Haut, K.G., Sen, T., Lomakin, D., Hoque, E.: A mental trespass? Unveiling truth, exposing thoughts, and threatening civil liberties with noninvasive ai lie detection. IEEE Trans. Technol. Soc. **3**, 132–142 (2022). https://doi.org/10.1109/tts.2022.3172556
15. Oravec, J.A.: The emergence of "truth machines"?: Artificial intelligence approaches to lie detection. Ethics Inf. Technol. **24** (2022). https://doi.org/10.1007/s10676-022-09621-6
16. Vicianova, M.: Historical techniques of lie detection. Eur. J. Psychol. **11**, 522–534 (2015). https://doi.org/10.5964/ejop.v11i3.919
17. Mbaziira, A., Jones, J.: A text-based deception detection model for cybercrime. In: International Conference on Technology Management, pp. 1–8 (2016)
18. Oshikawa, R., Qian, J., Wang, W.Y.: A Survey on Natural Language Processing for Fake News Detection. https://arxiv.org/abs/1811.00770
19. Abouelenien, M., Perez-Rosas, V., Mihalcea, R., Burzo, M.: Detecting deceptive behavior via integration of discriminative features from multiple modalities. IEEE Trans. Inf. Forensics Secur. **12**, 1042–1055 (2017). https://doi.org/10.1109/tifs.2016.2639344
20. Chebbi, S., Jebara, S.B.: Deception detection using multimodal fusion approaches. Multimed. Tools Appl. **82**, 13073–13102 (2021). https://doi.org/10.1007/s11042-021-11148-9
21. Ahmad, I., Yousaf, M., Yousaf, S., Ahmad, M.O.: Fake news detection using machine learning ensemble methods. Complexity **2020**, 1–11 (2020). https://doi.org/10.1155/2020/8885861
22. Silva, R.M., Pires, R.P., Almeida, T.A.: Incremental learning for fake news detection. J. Inf. Data Manage. **13**(6) (2023). https://doi.org/10.5753/jidm.2022.2542
23. Schemmer, M., Hemmer, P., Nitsche, M., Kühl, N., Vössing, M.: A meta-analysis of the utility of explainable artificial intelligence in human-AI decision-making. In: Proceedings of the 2022 AAAI/ACM Conference on AI, Ethics, and Society. ACM, New York, NY, USA (2022). https://doi.org/10.1145/3514094.3534128
24. Bhutta, M.R., Hong, M.J., Kim, Y.-H., Hong, K.-S.: Single-trial lie detection using a combined fNIRS-polygraph system. Frontiers in Psychology. 6, (2015). https://doi.org/10.3389/fpsyg.2015.00709
25. Khan, W., Crockett, K., O'Shea, J., Hussain, A., Khan, B.M.: Deception in the eyes of deceiver: a computer vision and machine learning based automated deception detection. Expert Syst. Appl. **169**, 114341 (2021)
26. Owayjan, M., Kashour, A., Al Haddad, N., Fadel, M., Al Souki, G.: The design and development of a Lie Detection System using facial micro-expressions. In: 2012 2nd International Conference on Advances in Computational Tools for Engineering Applications (ACTEA). IEEE (2012). https://doi.org/10.1109/ictea.2012.6462897
27. Mehendale, N.: Facial emotion recognition using convolutional neural networks (FERC). SN Appl. Sci. **2** (2020). https://doi.org/10.1007/s42452-020-2234-1
28. Nasri, H., Ouarda, W., Alimi, A.M.: ReLiDSS: novel lie detection system from speech signal. In: 2016 IEEE/ACS 13th International Conference of Computer Systems and Applications (AICCSA). IEEE (2016). https://doi.org/10.1109/AICCSA.2016.7945789

29. Talaat, F.M., Deep Neural Network for Lie Detection Based on Voice Stress. www.res earchgate.net/profile/Fatma-M-Talaat/publication/366606445_Deep_Neural_Network_for_ Lie_Detection_Based_on_Voice_Stress/links/63aae158097c7832ca6df388/Deep-Neural-Network-for-Lie-Detection-Based-on-Voice-Stress.pdf

30. Rubin, V.L., Chen, Y., Conroy, N.K.: Deception detection for news: three types of fakes. Proc. Assoc. Inf. Sci. Technol. **52**, 1–4 (2015). https://doi.org/10.1002/pra2.2015.145052010083

31. Jain, N., Kumar, A., Singh, S., Singh, C., Tripathi, S.: Deceptive reviews detection using deep learning techniques. In: Natural Language Processing and Information Systems. pp. 79–91. Springer, Cham (2019). https://doi.org/10.1007/978-3-030-23281-8_7

32. Breiman, L.: Random Forests. Mach. Learn. **45**, 5–32. https://doi.org/10.1023/A:101093340 4324

33. Fausett, L.V.: Fundamentals of Neural Networks: Architectures, Algorithms and Applications. Pearson Education India (2006)

34. Lecun, Y., Bottou, L., Bengio, Y., Haffner, P.: Gradient-based learning applied to document recognition. Proc. IEEE **86**, 2278–2324 (1998). https://doi.org/10.1109/5.726791

35. Goodfellow, I, Yoshua, B., Aaron, C.: Deep Learning. MIT Press, Cambridge (2016)

36. Pérez-Rosas, V., Abouelenien, M., Mihalcea, R., Burzo, M.: Deception detection using real-life trial data. In: Proceedings of the 2015 ACM on International Conference on Multimodal Interaction. ACM, New York, NY, USA (2015). https://doi.org/10.1145/2818346.2820758

37. Sen, M.U., Perez-Rosas, V., Yanikoglu, B., Abouelenien, M., Burzo, M., Mihalcea, R.: Multimodal deception detection using real-life trial data. IEEE Trans. Affect. Comput. **13**, 306–319 (2022). https://doi.org/10.1109/taffc.2020.3015684

38. Hastie, T., Friedman, J., Tibshirani, R.: The Elements of Statistical Learning. Springer, New York (2001). https://doi.org/10.1007/978-0-387-21606-5

39. Pérez-Rosas, V., Mihalcea, R.: Experiments in open domain deception detection. In: Proceedings of the 2015 Conference on Empirical Methods in Natural Language Processing. Association for Computational Linguistics, Stroudsburg, PA, USA (2015)

40. Pérez-Rosas, V., Mihalcea, R.: Cross-cultural deception detection. In: Proceedings of the 52nd Annual Meeting of the Association for Computational Linguistics (Volume 2: Short Papers). Association for Computational Linguistics, Stroudsburg, PA, USA (2014)

41. Gupta, V., Agarwal, M., Arora, M., Chakraborty, T., Singh, R., Vatsa, M.: Bag-of-lies: a multimodal dataset for deception detection. In: 2019 IEEE/CVF Conference on Computer Vision and Pattern Recognition Workshops (CVPRW). IEEE (2019). https://doi.org/10.1109/CVPRW.2019.00016

42. Dinges, L., Fiedler, M.-A., Al-Hamadi, A., Hempel, T., Abdelrahman, A., Weimann, J., Bershadskyy, D.: Automated deception detection from videos: using end-to-end learning based high-level features and classification approaches. https://arxiv.org/abs/2307.06625

43. Alaskar, H., Sbaï, Z., Khan, W., Hussain, A., Alrawais, A.: Intelligent techniques for deception detection: a survey and critical study. Soft. Comput. **27**, 3581–3600 (2022). https://doi.org/10.1007/s00500-022-07603-w

44. Boyne, S.M.: Data Protection In The United States. Am. J. Comp. Law. **66**, 299–343 (2018). https://doi.org/10.1093/ajcl/avy016

45. Ienca, M., Malgieri, G.: Mental data protection and the GDPR. J. Law Biosci. **9** (2022). https://doi.org/10.1093/jlb/lsac006

46. Mehrabi, N., Morstatter, F., Saxena, N., Lerman, K., Galstyan, A.: A survey on bias and fairness in machine learning. ACM Comput. Surv. **54**, 1–35 (2021). https://doi.org/10.1145/3457607

47. Weld, G., Ayton, E., Althoff, T., Glenski, M.: Leveraging community and author context to explain the performance and bias of text-based deception detection models. https://arxiv.org/abs/2104.13490

48. Strobel, M.: Aspects of transparency in machine learning. In Proceedings of the 18th International Conference on Autonomous Agents and MultiAgent Systems, pp. 2449–2451 (2019)
49. Vrij, A.: Verbal lie detection tools. Detecting Deception. 1–35 (2014). https://doi.org/10.1002/9781118510001.ch1
50. Yang, J.-T., Liu, G.-M., Huang, S.C.-H.: Emotion transformation feature: novel feature for deception detection in videos. In: 2020 IEEE International Conference on Image Processing (ICIP). IEEE (2020). https://doi.org/10.1109/ICIP40778.2020.9190846

Searching Optimizers for Deep Learning Based Hyperspectral Image Classification

Anish Sarkar[1], Utpal Nandi[1(✉)], Bachchu Paul[1], Sudipta Kr. Ghosal[2],
Moirangthem Marjit Singh[3], Jyotsna Kumar Mandal[4],
and Nuruzzaman Faruqui[5]

[1] Department of Computer Science, Vidyasagar University, Midnapore, West Bengal,
India
nandi.3utpal@gmail.com

[2] Department of Computer Science and Technology, Behala Government Polytechnic,
Kolkata, West Bengal, India

[3] Department of Computer Science and Engineering, North Eastern Regional
Institute of Science and Technology, Nirjuli, Arunachal Pradesh, India

[4] Department of Computer Science and Engineering, University of Kalyani, Nadia,
West Bengal, India

[5] Department of Software Engineering, Daffodil International University, Dhaka,
Bangladesh
faruqui.swe@diu.edu.bd

Abstract. In spite of its capacity to capture fine-grained features, hyperspectral images (HSIs), which provide extensive spectral and spatial information, have become more important in a variety of applications. Unlike traditional images, they give a multidimensional representation that allows for improved discrimination and analysis. This work analyses the effect of optimizers on the performance of the deep learning (DL) based models for hyperspectral image classification (HSIC) in terms of overall accuracy (OA), actual accuracy (AA), and $Kappa$. The research emphasizes the critical function of optimizers in deep learning model training for hyperspectral image processing, impacting convergence dynamics and generalization proficiency. The chosen testbed for testing the influence of optimizers is HybridSN, which is well-known for its ability to combine 2D and 3D convolutions to successfully extract both spectral and spatial data. Additionally, HybridSN is the ideal choice for investigating optimizer impacts on hyperspectral image classification outcomes because to its straightforward architecture and constant performance across varied datasets such as Indian Pines (IP), Pavia University (PU), and Salinas (SA). This study juxtaposes numerous optimizers for classification tasks on different datasets, emphasizing DiffMoment as the best performer for the IP dataset, while AdamP, RAdam, and DiffMoment shine on the PU dataset, and all optimizers do well except SGD on the SA dataset.

Keywords: Hyperspectral image (HSI) · classification · hybrid convolution network · optimizers · deep learning

© The Author(s), under exclusive license to Springer Nature Switzerland AG 2025
M. Majumder et al. (Eds.): ICCTE 2023, CCIS 2376, pp. 78–92, 2025.
https://doi.org/10.1007/978-3-031-81935-3_7

1   Introduction

A hyperspectral image is one that captures information over a large number of tiny and continuous spectral bands in the electromagnetic spectrum. According to prior research [1], the analysis of HSI entails the extraction of critical data from spectral bands collected by equipment at particular distances, all without necessitating direct interaction with the item under investigation. Using HSI technology, data may be retrieved from a variety of distinct spectral bands encompassing the whole electromagnetic spectrum. Visible light (from 0.4 m to 0.7 m) and short-wave infrared (from 0.7 m to 2.4 m) are included within these bands. Furthermore, as mentioned in another source [2] the assessment of the brightness characteristics of objects in the mid to far infrared region is only possible through the use of HSI. However, mainstream methods face various obstacles in extracting valuable information from individual pixels in both multispectral and RGB images, which is why a straightforward substitution with HSI is not viable. HSI is useful in a variety of applications, including municipal planning [14], natural resource exploitation [15], ecosystem viability assessment [16], and food processing [3]. HSI has lately found several uses in the defense industry, including the identification of landmines and the mapping of coastal zones. Furthermore, exact spectral data gathering has been performed on satellites, airplanes, and warships utilizing HSI [4].

Image classification, as described in [17], is a fundamental challenge in the realm of computer vision that entails classifying images into preset categories. The assignment is recognizing and classifying items, scenarios, or patterns within images. Similar with other fields of research like natural language processing [18,19], speech processing [29–31], image compression [32,33], etc., the Deep Learning (DL) based models have been frequently used for HSI classification successfully throughout the years. The DL model predicts the class of an image using convolutional and fully connected layers. A succession of 3D or 2D convolution layers is used in the majority of DL-based algorithms, followed by pooling and activation layers, and finally flattening and softmax activation layers.

A number of studies have also investigated the potential of Convolutional Neural Networks (CNNs) for HSIC. Liu et al. [5] developed a fusion network using 2D convolutions in their paper. This method amalgamates a shallow and deep feature extractor to acquire spectral data that was supplemented with both broad global and precise local spatial context. Notably, the researchers used an attention module based on Squeeze and Excitation (SE) approaches to increase the effectiveness of both spectral and spatial aspects. Y. Xu et al. [6] devised an HSIC-specific robust self-ensembling network (RSEN). This unique method combines a foundational network with an ensemble network, making use of both supervised and unsupervised losses from annotated and unannotated data. The use of unannotated data in this unique technique helps to improve network training. As part of their attempts to promote self-ensembling learning, the researchers also developed a consistency filter. Z. Gong et al. [7] introduced a deep manifold embedding technique (DMEM) tailored specifically for HSIC. By representing individual classes as nonlinear manifolds and leveraging geodesic distances, DMEM effectively encapsulated the intrinsic data structures. The manifold was partitioned into sub-classes through hierarchical clustering. Preserving geodesic distances among lower-dimensional features, DMEM accounted for sub-class distributions and inter-class associations. D. Hong et al. [8] used CNNs and graph convolution networks (GCNs) to classify hyperspectral images. Their "miniGCN" approach enhanced large-scale GCN training, improving classification and allowing out-of-sample data inference without re-training. Fusion approaches (additive, multiplicative, and concatenation) for combining CNNs and GCNs were investigated and shown to outperform individual models.

Other studies have used 3D CNNs for hyperspectral image classification in addition to 2D CNNs. Ahmad et al. [9] investigated this possibility by using a unique HSIC technique based on 3D CNNs. They input the hyperspectral data into their approach in the form of 3D overlapping patches, which aided in the construction of 3D feature maps. Notably, they used a 3D kernel function to successfully apply over numerous neighboring spectral bands. Sun et al. [10] improved the M3D-CNN model into M-3DCNN-Attention, combating overfitting by using the mixup technique for virtual samples, diversifying the dataset, and adding convolutional block attention modules (CBAMs) to 3D convolutional and ReLU layers, thereby improving discriminative qualities in hyperspectral image classification (HSIC). In the field of HSI classification, H. Zhang et al. [11] suggested a new technique based on autonomous architectural search. Their method attempted to accurately categorize hyperspectral images by utilizing a specialized 3D asymmetric decomposition search space capable of handling both spectral and geographical data. This pixel-level categorization approach not only lowered compute requirements but also reached competitive performance levels. Significantly, the suggested technique outperformed current strategies in terms of inference speed.

The 2D convolutional network is well-known for its computational efficiency in acquiring spatial data, which is a significant benefit. However, a fundamental disadvantage of the 2D CNN is its poor performance as an extractor for spectral features. This issue is notable since spectral feature extraction is critical for HSIC. In contrast, while the 3D CNN shines as an effective spectral feature extractor, its processing requirements are significantly larger. Another disadvantage is the need for a large number of training samples for deep 3D CNNs, which is worsened by the lack of publicly available hyperspectral image datasets. To address the shortcomings both of the 3D CNNs and 2D CNNs, Roy et al. [11] evolved HybridSN, a unique HSIC network. Their novel method merged 3D CNNs with 2D CNNs in an unusual way: three successive layers of 3D convolution were followed by a single layer of 2D convolution. Surprisingly, their suggested model had no pooling layers. In response to the same problem, Yang et al. [12] proposed Synergistic CNN (SyCNN), an alternate hybrid approach. This method deftly combined 2D and 3D CNNs, allowing for the successive extraction of spectral and spatial characteristics.

The objective of this survey research paper is to explore the influence of different optimizers on the classification performance of deep learning-based HSIC models. Through a focused examination of optimization techniques, this study aims to analyze their impact on overall accuracy (OA), average accuracy (AA), and Cohen's Kappa coefficient ($Kappa$) in the context of hyperspectral image classification. By assessing these metrics, this research intends to provide valuable insights into optimizing deep learning models for accurate HSIC.

The succeeding sections of the article are arranged in the following order: Sect. 2 emphasizes selecting an appropriate model based on its ability to extract spectral-spatial features. Following this, Sect. 3 introduces the optimizers chosen for evaluation, and Sect. 4 illustrates and analyses comparative results while also introducing the datasets chosen for evaluation. Finally, Sect. 5 summarises the study's observations and gives concluding insights into the role of optimizers in hyperspectral image classification.

2   Deep Learning Based Model Selection

In this study review, HybridSN [11] was chosen as the model for examining the influence of various optimizers on the performance of the HSIC model. HybridSN, a recently established model, combines the benefits of 2D and 3D convolution. HybridSN combines the powers of 2D CNNs in collecting local spatial elements inside HSIs with the capabilities of 3D convolution in grabbing non-local spectral characteristics. HybridSN has exhibited cutting-edge performance across a range of HSIC tasks by feeding the input hsi cube through three layers of 3D convolution layers and then a single layer of 2D convolution layer. The overall structure of the HybridSN has shown in Fig. 1.

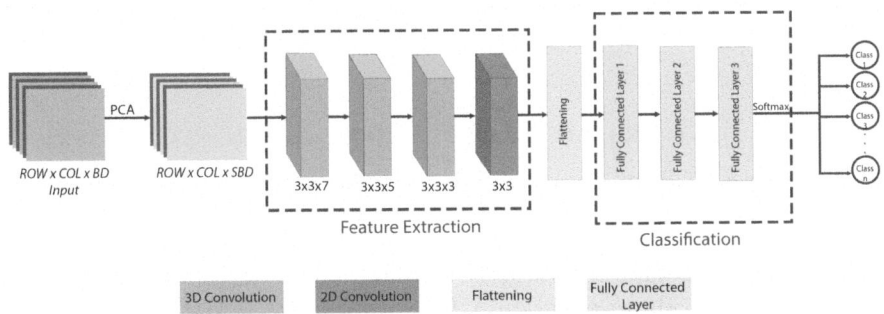

Fig. 1. The architecture of HybridSN model

Among the various models considered for our study, including 2DCNN, RSEN, DMEM, 3D CNN, M3D CNN, CBAM, FSKNet, and SyCNN, the decision to focus on HybridSN stems from the following benefits: 1. Its simplicity facilitates ease of training and interpretation; 2. Its documented efficacy across diverse hyperspectral image classification tasks strengthens its selection; 3. The novelty of HybridSN implies a limited body of existing research, providing an opportunity to contribute unique insights into the interplay between optimizers and HybridSN's performance; 4. HybridSN's versatility allows experimentation with different architectures. The synthesis of 3D Convolution and 2D Convolution elevates HybridSN to the level of an excellent feature extractor, greatly expanding its potential.

This research project is expected to contribute to the progress of the cutting edge of hyperspectral image categorization. An improved understanding of improving this model's performance is sought by a thorough study of several optimizers on HybridSN. Such discoveries have the potential to accelerate the improvement of hyperspectral image categorization systems for future breakthroughs.

3   Overview of Different Optimizers Used in Deep Learning Model

It is already mentioned in the previous section, that this article aims to compare the effects of different optimizers on the HSIC models. In total, seven very popular optimizers are chosen for this evaluation, including SGD [22], Adam [23], AdamP [24], AdamW [25], RAdam [26], DiffGrad [27], and DiffMoment [28]. This section goes into depth about the optimizers that were chosen for evaluation.

Stochastic Gradient Descent (SGD) [22] is a machine learning optimization approach that is frequently utilized. It changes model parameters iteratively to minimize a loss function. It computes the gradient of the loss concerning a subset of training data (a mini-batch) in each iteration and modifies the parameters to lower the loss in the opposite direction of the gradient. The SGD optimizer's step-up rule is:

$$\sigma_{it+1} = \sigma_{it} - \alpha \cdot \nabla LOSS(\sigma_{it}; inp_{i:i+n}, tout_{i:i+n}) \tag{1}$$

In the above equation, σ_{it} represents the value of the parameter at the it^{th} iteration, while α representing the learning rate.The loss function is defined by $LOSS$ in the above equation. $inp_{i:i+n}$ and $tout_{i:t+n}$ represents the small batches of the input data and targeted output value from the index i to $i + n$, respectively. In the above equation $\nabla LOSS(\sigma_{it}; inp_{i:i+n}, tout_{i:i+n})$ represents the loss gradient corresponding to the σ_{it}, which is calculated on the small batches of the input data.

The Adam [23] optimizer is a well-known DL optimization technique. It combines the benefits of both the Adaptive Gradient Algorithm (AdaGrad) and the Root Mean Square Propagation (RMSProp) algorithms. Adam manually modifies the learning rates for each parameter depending on the historical gradient and squared gradient values. The step-up rule for the Adam optimization function is:

$$\sigma_{it+1} = \sigma_{it} - \alpha \cdot \frac{\widehat{F}_{it+1}}{\sqrt{\widehat{S}_{it+1}} + \epsilon} \qquad (2)$$

In the above equation, σ_{it} denotes the value of the parameter at the it^{th} iteration, while the α representing the learning rate. $\widehat{F}_{it+1}$ and $\widehat{S}_{it+1}$ defines the estimated bias corrected at the first and second moment. To prevent the division by zero a small constant ϵ is added.

The AdamP [24] optimizer is an Adam optimizer modification that adds a weight decay penalty term directly into its updating process. This advancement makes it easier to control weight loss. The following equation captures the mathematical description of the AdamP optimizer's update step:

$$\sigma_{it+1} = \sigma_{it} - \alpha \cdot \frac{\widehat{F}_{it+1}}{\sqrt{\widehat{S}_{it+1}} + \epsilon} - \lambda \cdot \alpha \cdot \rho \cdot \frac{\|\sigma_{it}\|_2^\rho}{\|gl_{it}\|_2^{\rho-1}} \qquad (3)$$

Where, σ_{it+1} represents the parameter value at it^{th} iteration, while λ and α representing the weight decay coefficient and the learning rate, respectively. The fist and second moment bias corrected moment is defined by the $\widehat{F}_{it+1}$ and $\widehat{S}_{it+1}$, respectively. gl_{it} denotes the gradient loss with respect to the iteration step it. A hyperparameter ρ is used to control the power of the Lp-norm in the penalty term, and a small constant ϵ is utilized to prevent the division with zero.

The AdamW [25] optimizer is an Adam optimizer modification that directly includes weight decay regularisation in its updating process. This reduces overfitting and enhances the optimization process. The following equation summarises the mathematical formulation of the AdamW optimizer's update step:

$$\sigma_{it+1} = \sigma_{it} - \alpha \cdot \frac{\widehat{F}_{it+1}}{\sqrt{\widehat{S}_{it+1}} + \epsilon} - \lambda \cdot \alpha \cdot \sigma_{it} \qquad (4)$$

In the above equation, σ_{it} represents the value of parameter at it^{th} iteration step, α defines the learning rate. The bias corrected estimates are defined by $\widehat{F}_{it+1}$ and $\widehat{S}_{it+1}$. A small constant ϵ is used to eliminated the division by zero, and a weight decay coefficient is defined by λ.

The Rectified Adam (RAdam) [26] optimizer is an improvement to the regular Adam optimizer that tackles its convergence concerns. RAdam dynamically changes the learning rate throughout training to improve stability and performance. To counteract the adaptive momentum, the optimizer employs a rectification term. The following equation expresses RAdam's update step mathematically:

$$\sigma_{it+1} = \sigma_{it} - \frac{\alpha}{\sqrt{\widehat{S}_{it+1}} + \epsilon} \cdot \widehat{F}_{it+1} + (\rho - 1) \cdot (gl_{it}^2 - \widehat{S}_{it+1}) \qquad (5)$$

Where, σ_{it} defines the value of parameter at it^{th} iteration step, and the learning rate is represented by α. $\widehat{F}_{it+1}$ and $\widehat{S}_{it+1}$ defines the first and second moment bias corrected estimates, respectively. In the above equation ρ defines a hyperparameter and a small constant represented by ϵ.

The diffGrad [27] optimizer is an adaptive learning rate approach that modifies the step size for each parameter depending on the gradient difference between the current and the near past. The diffGrad optimizer's step up rule is supplied by the following equation:

$$\sigma_{it+1,i} = \sigma_{it,i} - \frac{\alpha_{it} \times \xi_{it,i} \times \widehat{F}_{it,i}}{\sqrt{\widehat{S}_{it,i} + \epsilon}} \tag{6}$$

In the above equation, $\sigma_{it,i}$ defines the value of ith parameter at it^{th} iteration. $\widehat{F}_{it,i}$ and $\widehat{S}_{it,i}$ represents the first and second moment bias corrected estimates. A small positive constant ϵ is added to overcome division by zero, and $\xi_{it,i}$ represents the diffGrad friction coefficient.

The DiffMoment [28] optimizer is a deep neural network training optimization algorithm that aims to optimize the step size based on the shifting information between the first and second moment estimations of gradients or momentums. This method is intended to solve limitations in existing optimizers and perhaps improve the training performance of neural networks, notably CNNs, by using different activation functions. The step up rule for the DiffMoment optimizer is follows:

$$\sigma_{it+1,i} = \sigma_{it} - \frac{\alpha \cdot \left(\widehat{F}_{it,i} \cdot \frac{1}{abs(F_{it,i} - S_{it,i})}\right)}{\left(\sqrt{\widehat{S}_{it,i}} + \epsilon\right)} \tag{7}$$

Where, the $\sigma_{it+1,i}$ refers to the value of $i^t h$ parameter at it^{th} iteration. The learning rate defined by α, while the first and second order bias corrected estimates defined by $\widehat{F}_{it,i}$ and $\widehat{S}_{it,i}$. To eliminate the situation of divisible by zero, a small positive integer ϵ is added.

4 Comparative Results and Analysis

This section compares the effects of several optimizers on the performance of the HSIC model chosen for assessment. Each experiment uses the same parameters for training and validation ratios, patch size, number of assessment samples, and the usage of ReLU activation functions as the benchmark to guarantee a consistent comparison. The experimental setting entails selecting identical samples for each experiment. Furthermore, throughout the cross-validation procedure, a same amount of samples are chosen for each training session. However, because these samples are often picked at random, their results may alter when applied over other duration.

The occurrence of overlap between training and test groups is a common difficulty in modern research. While training and validation samples are chosen at random, testing uses the full dataset, resulting in a biased yet trustworthy result. To address this risk, the current study used random sample selection while guaranteeing that no crossovers or overlaps occurred.

4.1 Selected Dataset for Evaluation

A thorough study is carried out utilizing three commonly used datasets: Indian Pines (IP) [20], University of Pavia (PU) [13], and Salinas Scene (SA) [21]. The goal is to evaluate the performance of several optimizers on an HSIC model. Given their extensive use in remote sensing applications, these datasets are ideal candidates for assessment studies aiming at improving our knowledge of the effect of optimizers on the model. Tables 1 provide detailed information on the dateset.

The AVIRIS sensor was used to acquire the Indian Pines (IP) [20] dataset at Indiana's northern test site. The dataset comprises 224 spectral bands spanning the wavelength range of 400 nm to 2500 nm after 24 useless bands were removed. The image has a resolution of 20 meters per pixel (MPP) and a size of 145×145 pixels. It contains descriptions and ground truth maps for each of the 16 plant classes. The RGB source image and ground truth representation of the IP dataset are depicted in Fig. 2.

The University of Pavia (PU) [13] dataset comprises hyperspectral images acquired by the ROSIS sensor in an agricultural region near Pavia, Italy. The dataset has a pixel resolution of 2.5 meters per pixel (MPP) and covers an area of (610×340) pixels. It has 103 spectral bands with wavelengths ranging from 430 to 860 nm and nine unique land cover classes: asphalt, fields, gravel, trees, metal sheets, bare ground, bitumen, masonry, and shadows. The ground truth information for the PU dataset are depicted in Fig. 3.

Table 1. Overview of three popular HSI datasets

Particulars	IP	PU	SA
1: Year	1992	2001	2001
2: Source	AVIRIS	ROSIS-03	AVIRIS
3: Spatial	145×145	610×340	512×217
4: Spectral	220	115	224
5: Wavelength	400–2500	430–860	-
6: Samples	21025	207400	54129
7: Classes	16	9	16
8: Sensor	Aerial	Aerial	Aerial
9: Resolution	20 m	1.3 m	3.7 m

As explained in [21], the Salinas Scene dataset (SA) comprises of hyperspectral images taken by the AVIRIS sensor over the Salinas Valley in California.

Each pixel of the Earth measures around 1.3 m by 1.3 m and has 145×145 pixels and 224 spectral bands. The image, which included vegetation, bare soils, and vines, was composed entirely of at-sensor radiance data covering 512 lines by 217 samples. The Salinas dataset's ground truth consisted of 16 unique groups. Figure 4 depicts the ground truth for the SA dataset.

4.2   The Comparative Study

All experiments were carried out using a Dell G15 5530 laptop, which was powered by an Intel Core i5 13450HX CPU, an Nvidia RTX 3050 6 GB GPU, and 16 GB of DDR5 4800 MHz RAM. Using 15 spectral bands across all datasets ensured rigorous homogeneity throughout the studies.

This thorough methodology ensured that the outcomes were evaluated and compared on an equal playing field. When the model was trained and tested with different optimizers, a patch size of 25×25 was used for all experiments. The datasets were used for testing and training in a 70:30 split ratio, respectively. To provide a fair evaluation, a total of 100 epochs are specified as a norm for all experiments. The experiments is performed for all three datasets including IP, PU, SA using SGD [22], Adam [23], AdamP [24], AdamW [25], RAdam [26], DiffGrad [27], and DiffMoment [28] optimizers.

From Table 2, almost all the optimizers including Adam, AdamP, AdamW, DiffGrad, and DiffMomen perform similarly and showing and show a consistence performance for the classification of the IP dataset. But when the experiment was performed with the SGD optimizer, then it showed very poor performance. If the in-depth analysis of the classification result for IP dataset is performed, then it will be visible that the model shows the highest performance when it is tested with the DiffMoment optimizer. The prediction masked for IP dataset using different optimizers has been shown in Fig. 2

It can be visible from Table 3 that, all the optimizers including Adam, AdamP, AdamW, RAdam, DiffGrad, and DiffMoment perform well and shows consistency in the classification performance for the PU dataset. When the model is tested with the SGD optimizer, it shows comparatively poor performance and shows performance drops for some classes. The detailed analysis shows that the model gives its best performance for the classification of the PU dataset when it is tested with the AdamP, RAdam, or DiffMoment optimizer. The prediction masked for PU dataset using different optimizers has been shown in Fig. 3.

Table 3 shows that the model gives almost the best performance when it is tested with all of the optimizers including SGD, Adam, AdamP, AdamW, RAdam, DiffGrad, and DiffMoment for SA dataset. But it can be visible from Table 3 that the model shows minor drops in its performance when it is tested with the SGD optimizer. The prediction masked for SA dataset using different optimizers has been shown in Fig. 4.

The confusion matrices of best performance for all three datasets, using SGD [22], Adam [23], AdamP [24], AdamW [25], RAdam [26], DiffGrad [27], and DiffMoment [28] has been shown in Fig. 5.

Table 2. The comparison of class-wise accuracy and *OA*, *AA*, and *Kappa* using different optimizers for IP dataset (The best performances are marked in bold)

Classes	SGD [22]	Adam [23]	AdamP [24]	AdamW [25]	RAdam [26]	DiffGrad [27]	DiffMoment [28]
1: Alfalfa	0	93.10	93.10	100	100	96.43	96.43
2: Corn-notill	0	100	99.76	99.19	99.08	99.30	100
3: Corn-mintill	0	99.59	100	100	100	100	100
4: Corn	0	95.95	97.26	97.92	93.38	99.55	97.26
5: Grass-pasture	0	99.66	99.65	100	100	99.53	100
6: Grass-trees	0	100	99.54	100	100	100	100
7: Grass-pasture-mowed	0	100	100	100	100	100	100
8: Hay-windrowed	0	99.65	99.65	100	99.65	100	99.65
9: Oats	0	100	100	100	100	100	100
10: Soybean-notill	0	99.83	100	100	100	99.83	100
11: Soybean-mintill	23.97	99.46	99.26	99.26	99.39	99.46	99.53
12: Soybean-clean	0	100	99.72	99.72	99.72	99.72	100
13: Wheat	0	100	100	100	100	98.40	100
14: Woods	0	99.87	99.21	100	99.74	99.74	99.87
15: Building-Grass-Trees-Drives	0	100	100	100	99.57	100	100
16: Stone-Steel-Towers	0	98.25	93.22	98.21	97.82	98.25	98.25
OA	23.97	99.63	99.45	99.63	99.45	99.56	**99.76**
AA	6.25	98.37	99.53	99.39	99.57	99.67	**99.86**
Kappa	0	99.57	99.37	99.57	99.37	99.50	**99.72**

Table 3. Comparison of *OA*, *AA* and *Kappa* using different optimizers for PU and SA datasets (The best performances are marked in bold)

	University of Pavia (PU)			Salinas Scene (SA)		
	OA	*AA*	*Kappa*	*OA*	*AA*	*Kappa*
SGD [22]	87.85	74.87	83.76	98.18	99.03	97.97
Adam [23]	99.96	99.94	99.94	100	100	100
AdamP [24]	**99.98**	**99.96**	**99.97**	100	100	100
AdamW [25]	99.92	99.87	99.90	100	100	100
RAdam [26]	**99.98**	**99.96**	**99.97**	100	100	100
DiffGrad [27]	99.94	99.88	99.92	100	100	100
DiffMoment [28]	**99.98**	**99.97**	**99.97**	100	100	100

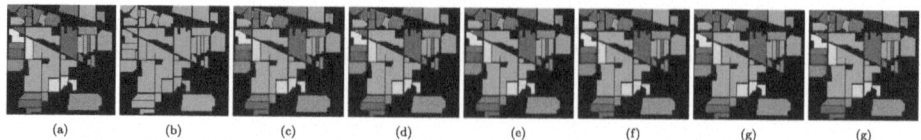

Fig. 2. Predicted feature maps for IP dataset using different optimizer, while considering ReLU as activation function: (a) Ground Truth, (b) Using SGD, (c) Using Adam, (d) Using AdamP, (e) AdamW, (f) Using RAdam, (g) Using DiffGrad, and (h) Using DiffMoment

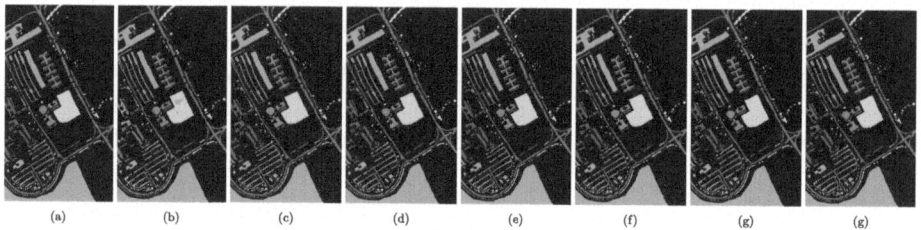

Fig. 3. Predicted feature maps for PU dataset using different optimizer, while considering ReLU as activation function: (a) Ground Truth, (b) Using SGD, (c) Using Adam, (d) Using AdamP, (e) AdamW, (f) Using RAdam, (g) Using DiffGrad, and (h) Using DiffMoment

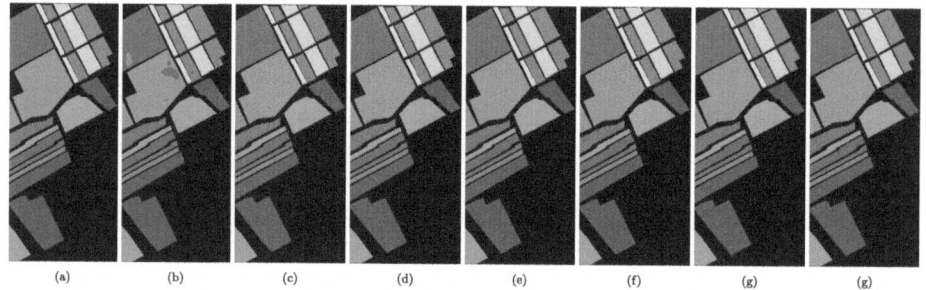

Fig. 4. Predicted feature maps for SA dataset using different optimizer, while considering ReLU as activation function: (a) Ground Truth, (b) Using SGD, (c) Using Adam, (d) Using AdamP, (e) AdamW, (f) Using RAdam, (g) Using DiffGrad, and (h) Using DiffMoment

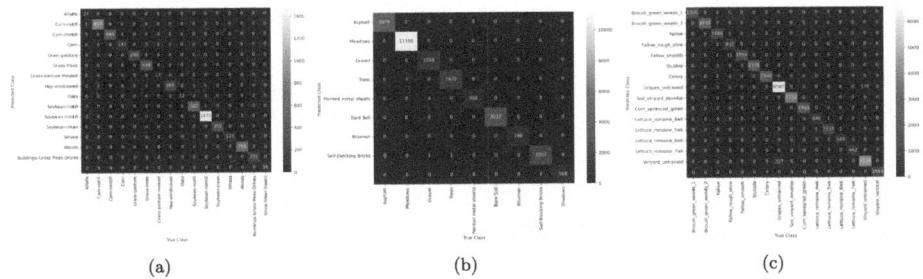

(a)   (b)   (c)

Fig. 5. Confusion matrices for performances for three datasets using best performing optimizers, while considering ReLU as activation function: (a) IP using DiffMoment, (b) PU using DiffMoment, (c) SA using DiffMoment,

5   Conclusion

In conclusion, this study conducted a thorough investigation of the effects of various optimizers on the performance of the chosen HybridSN model. SGD, Adam, AdamP, AdamW, RAdam, DiffGrad, and DiffMoment were among the optimization methods used in the experiment. To establish the influence of these optimization strategies on classification performance, we looked into the subtle interplay between them and the HybridSN architecture.

Following an in-depth analysis and experimentation, a startling discovery was made. Various optimizers were evaluated for classification tasks on different datasets in the tests. Notably, for the IP dataset, the DiffMoment optimizer showed the best performance, whereas SGD performed poorly. AdamP, RAdam, and DiffMoment optimizers performed well on the PU dataset, whereas SGD showed the worst performance. Finally, except SGD, all optimizers performed well in the SA dataset, with no substantial variance among them. The HybridSN model demonstrated its most spectacular and durable performance when subjected to the DiffMoment optimizer. This finding emphasizes the importance of optimizer selection in determining the HybridSN model's performance for HSIC tasks. DiffMoment's effectiveness demonstrates the critical relevance of specialized optimization tactics in realizing the entire potential of sophisticated models like HybridSN.

As we conclude our study, we must consider the larger ramifications of our findings. The link between optimization tactics and model performance is a fluid field that is constantly developing as machine learning techniques progress. As future endeavors push the boundaries of optimization techniques and model designs, this work will serve as a crucial reference point in navigating the complex link between optimization and model performance.

References

1. Ahmad, M., et al.: Spatial prior fuzziness pool-based interactive classification of hyperspectral images. Remote Sens. **11**(9), 1136 (2019). https://doi.org/10.3390/rs11091136

2. Hong, D., et al.: Interpretable hyperspectral artificial intelligence: when nonconvex modeling meets hyperspectral remote sensing. IEEE Geosci. Remote Sens. Mag. **9**(2), 52–87 (2021). https://doi.org/10.3390/app10196862

3. Ayaz, H., Ahmad, M., Mazzara, M., Sohaib, A.: Hyperspectral imaging for minced meat classification using nonlinear deep features. Appl. Sci. **10**(21), 7783 (2020). https://doi.org/10.3390/app10217783

4. Xing, F., Yao, H., Liu, Y., Dai, X., Brown, R. L., Bhatnagar, D.: Recent developments and applications of hyperspectral imaging for rapid detection of mycotoxins and mycotoxigenic fungi in food products. Crit. Rev. Food Sci. Nutr. **59**(1), 173–180 (2019). https://doi.org/10.1080/10408398.2017.1363709

5. Li, X., Ding, M., Pižurica, A.: Deep feature fusion via two-stream convolutional neural network for hyperspectral image classification. IEEE Trans. Geosci. Remote Sens. **58**(4), 2615–2629 (2019). https://doi.org/10.1109/TGRS.2019.2952758

6. Xu, Y., Du, B., Zhang, L.: Robust self-ensembling network for hyperspectral image classification. IEEE Trans. Neural Netw. Learn. Syst. https://doi.org/10.1109/TNNLS.2022.3198142

7. Gong, Z., Hu, W., Du, X., Zhong, P., Hu, P.: Deep manifold embedding for hyperspectral image classification. IEEE Trans. Cybern. **PP**, 1–14 (2021). https://doi.org/10.1109/TCYB.2021.3069790

8. Hong, D., Gao, L., Yao, J., Zhang, B., Plaza, A., Chanussot, J.: Graph convolutional networks for hyperspectral image classification. IEEE Trans. Geosci. Remote Sens. **59**(7), 5966–5978 (2020). https://doi.org/10.1109/TGRS.2020.3015157

9. Ahmad, M., Khan, A.M., Mazzara, M., Distefano, S., Ali, M., Sarfraz, M.S.: A fast and compact 3-D CNN for hyperspectral image classification. IEEE Geosci. Remote Sens. Lett. **19**, 1–5 (2020). https://doi.org/10.1109/LGRS.2020.3043710

10. Sun, K., Wang, A., Sun, X., Zhang, T.: Hyperspectral image classification method based on M-3DCNN-Attention. J. Appl. Remote Sens. **16**(2), 026507 (2022). https://doi.org/10.1117/1.JRS.16.026507

11. Zhang, H., Gong, C., Bai, Y., Bai, Z., Li, Y.: 3-D-ANAS: 3-D asymmetric neural architecture search for fast hyperspectral image classification. IEEE Trans. Geosci. Remote Sens. **60**, 1–19 (2021). https://doi.org/10.1109/TGRS.2021.3079123

12. Yang, X., et al.: Synergistic 2D/3D convolutional neural network for hyperspectral image classification. Remote Sens. **12**(12), 2033 (2020). https://doi.org/10.3390/rs12122033

13. Huang, X., Zhang, L.: A comparative study of spatial approaches for urban mapping using hyperspectral ROSIS images over Pavia City, northern Italy. Int. J. Remote Sens. **30**(12), 3205–3221 (2009). https://doi.org/10.1080/01431160802559046

14. Heldens, W., Esch, T., Heiden, U., Müller, A., Dech, S.: Exploring the demands on hyperspectral data products for urban planning: a case study in the Munich region. In: 9th International Proceedings on Proceedings, pp. 1–2 (2009)

15. Yokoya, N., Chan, J.C.W., Segl, K.: Potential of resolution-enhanced hyperspectral data for mineral mapping using simulated EnMAP and Sentinel-2 images. Remote Sens. **8**(3), 172 (2016). https://doi.org/10.3390/rs8030172

16. Dai, X., et al.: Ecological vulnerability assessment of a China's representative mining city based on hyperspectral remote sensing. Ecol. Indicators **145**, 109663 (2022). https://doi.org/10.1016/j.ecolind.2022.109663

17. Sun, Y., Xue, B., Zhang, M., Yen, G.G.: Evolving deep convolutional neural networks for image classification. IEEE Trans. Evol. Comput. **24**(2), 394–407 (2020). https://doi.org/10.1109/TEVC.2019.2916183

18. Nandi, U., Ghorai, A., Singh, M.M., Changdar, C., Bhakta, S., Pal, R.K.: Indian sign language alphabet recognition system using CNN with diffGrad optimizer and stochastic pooling. Multimed. Tools Appl. **82**(7), 9627–9648 (2023). https://doi.org/10.1007/s11042-021-11595-4

19. Ghorai, A., Nandi, U., Changdar, C., Si, T., Singh, M.M., Mondal, J.K.: Indian sign language recognition system using network deconvolution and spatial transformer network. Neural Comput. Appl. **2**(5), 99–110 (2023). https://doi.org/10.1007/s00521-023-08860-y

20. Baumgardner, M.F., Biehl, L.L., Landgrebe, D.A.: 220 Band AVIRIS Hyperspectral Image Data Set: June 12, 1992 Indian Pine Test Site 3, September 2015. . https://doi.org/10.4231/R7RX991C. https://purr.purdue.edu/publications/1947/1

21. Plaza, A., Martinez, P., Plaza, J., Perez, R.: Dimensionality reduction and classification of hyperspectral image data using sequences of extended morphological transformations. IEEE Trans. Geosci. Remote Sens. **43**(3), 466–479 (2005). https://doi.org/10.1109/TGRS.2004.841417

22. Kiefer, J., Wolfowitz, J.: Stochastic estimation of the maximum of a regression function. Ann. Math. Stat. 462–466 (1952)

23. Kingma, D.P., Ba, J.: Adam: a method for stochastic optimization. arXiv preprint arXiv:1412.6980 (2014). https://doi.org/10.48550/arXiv.1412.6980

24. Heo, B., et al.: AdamP: slowing down the slowdown for momentum optimizers on scale-invariant weights. In: International Conference on Learning Representations (ICLR) (2021)

25. Loshchilov, I., Hutter, F.: Decoupled weight decay regularization. arXiv preprint arXiv:1711.05101 (2017)

26. Liu, L., et al.: On the variance of the adaptive learning rate and beyond. arXiv preprint arXiv:1908.03265

27. Dubey, S.R., Chakraborty, S., Roy, S.K., Mukherjee, S., Singh, S.K., Chaudhuri, B.B.: diffGrad: an optimization method for convolutional neural networks. IEEE Trans. Neural Netw. Learn. Syst. **31**(11), 4500–4511 (2019). https://doi.org/10.1109/TNNLS.2019.2955777

28. Bhakta, S., Nandi, U., Si, T., Ghosal, S.K., Changdar, C., Pal, R.K.: DiffMoment: an adaptive optimization technique for convolutional neural network. Appl. Intell. **2022**, 1–15. https://doi.org/10.1007/s10489-022-04382-7

29. Paul, B., Dey, T., Das Adhikary, D., Guchhai, S., Bera, S.: A novel approach of audio-visual color recognition using KNN. In: Das, A.K., Nayak, J., Naik, B., Dutta, S., Pelusi, D. (eds.) Computational Intelligence in Pattern Recognition. AISC, vol. 1349, pp. 231–244. Springer, Singapore (2022). https://doi.org/10.1007/978-981-16-2543-5_20

30. Paul, B., Phadikar, S., Bera, S.: Indian regional spoken language identification using deep learning approach. In: Giri, D., Buyya, R., Ponnusamy, S., De, D., Adamatzky, A., Abawajy, J.H. (eds.) Proceedings of the Sixth International Conference on Mathematics and Computing. AISC, vol. 1262, pp. 263–274. Springer, Singapore (2021). https://doi.org/10.1007/978-981-15-8061-1_21

31. Paul, B., Phadikar, S.: A novel pre-processing technique of amplitude interpolation for enhancing the classification accuracy of Bengali phonemes. Multimed. Tools Appl. **82**, 7735–7755 (2023). https://doi.org/10.1007/s11042-022-13594-5
32. Nandi, U., Mandal, J.K.: Adaptive region based huffman compression technique with selective code interchanging. In: Meghanathan, N., Nagamalai, D., Chaki, N. (eds.) Advances in Computing and Information Technology. Advances in Intelligent Systems and Computing, vol. 176. Springer, Heidelberg (2012). https://doi.org/10.1007/978-3-642-31513-8_75
33. Nandi, U., Mandal, J.K.: Efficiency and capability of fractal image compression with adaptive quardtree partitioning. Int. J. Multimed. Appl. (IJMA) **5**(4) (2013). https://doi.org/10.5121/ijma.2013.5404

Efficient Crop Recommendation System: A Machine Learning Based Approach

Swagatika Tripathy[1], Dibya Ranjan Das Adhikary[2($\boxtimes$)], and Premansu Sekhara Rath[1]

[1] Department of Computer Science and Engineering, GIET University, Gunupur, India
[2] Department of Computer Science and Engineering, Siksha 'O' Anusandhan Deemed to be University, Bhubaneswar, India
dibyadasadhikary@soa.ac.in

Abstract. Internationally, agricultural research has increased the optimization of economic profit and is a very large and crucial field to acquire additional benefits. But it can be improved by utilizing various technological tools, resources, and processes. Nowadays, machine learning techniques helps farmers in making informed decision about which crop to cultivate in their land. This research offers empirical proof that we are able to categorize agricultural area's dataset according to soil parameters utilizing different machine learning classification techniques. Additionally, in order to suggest the appropriate crop for higher yield, we looked into the most effective algorithm with strong prediction accuracy.

Keywords: Machine learning · classification algorithms · soil testing · Confusion Matrix · Precession Agriculture

1 Introduction

A country's economy cannot function without its agricultural sector. It serves as the foundation of the economic structure in our nation. The selection of the right crop for cultivation is one of the major issues faced by farmers. The choice of crop is influenced by a number of variables, including temperature, soil type, climate conditions, market prices, etc. The efficiency, cost-effectiveness, and resource usage of conventional agricultural practices and techniques are major problems. Better methods are required in order to raise farmers' standards of living as well. Agriculture has changed over time as a result of globalization by incorporating the newest methods and technologies to raise living standards. In agriculture Sector, Precession Agriculture (PA) is a developing technology among other technologies and methods. Site-specific farming is the core emphasis of PA [1].

In PA, crop recommendation is the most important field. Crop recommendations are based on a variety of parameters, and by using PA techniques one can identify the parameters and improve crop selection. A type of software called recommendation system helps users to find the products that best suit their needs, preferences, or tastes. A problem that has gained popularity and is common among e-commerce, social networking, and content-based websites uses knowledge discovery methodologies to provide

M. Majumder et al. (Eds.): ICCTE 2023, CCIS 2376, pp. 93–101, 2025.
https://doi.org/10.1007/978-3-031-81935-3_8

personalized suggestions for material, commodities, or solutions during an interaction. The agricultural and food industries can benefit from its tremendous power. Machine learning has been incorporated into the agriculture sector to develop efficient, cost-effective solutions to the issues that farmers confront. Researchers can use computer models to conduct preliminary testing to see how a variety might do when examined with altered sub-atmosphere, soil compositions, climatic layouts, and other parameters. Modern agricultural researchers are conducting larger-scale tests of their concepts and creating noticeably more precise, ongoing projections.

In order to function intelligently, machine learning employs sophisticated algorithms and a set of predetermined rules. It makes use of historical data to identify patterns, and depending on the analysis it generates, executes the intended task in accordance with the stated rules and algorithms. Throughout the entire cycle of planting, growing, and harvesting, machine learning is pervasive. The soil type, kind of land, and the amount of macronutrients in the soil are the main factors influencing crop output. The goal of this effort is to forecast the crops that can be grown in the soil and classify soil samples based on the macronutrients present. In this the inputs that has been used to determine the crops to recommend include the soil's electrical conductivity, pH, soil type, soil texture, nitrogen (N), phosphorus (P), and potassium (K) levels, temperature and rainfall. The goal of this study was to gain an understanding of the field, explore how various machine learning classifiers interacted with the dataset, and select the most accurate and predictive method.

2 Literature Review

Several research have explored the crop recommendation systems using machine learning techniques. These studies have investigated the application of machine learning to accurately predict crop according to the given inputs.

A method that recommended crops and fertilizers to farmers in order to help them boost crop productivity was put forth by Shinde et al. [2]. Additionally, it enabled the farmers to buy the recommended fertilizers directly through the app. Random Forest algorithm has been used to efficiently and precisely identify the crop for recommendation. Based on previously purchased fertilizer, the Apriori algorithm has been used to produce sets of regularly bought items that have assisted farmers in choosing the fertilizer for the suggested crop. Random Forest produced results that were more accurate and efficient. Rajak et al. [3] suggested an approach that used voting procedures while looking at the soil's characteristics, such as depth, water retention capacity, sewerage, erosion, etc., in order to build a model for the proposal. This approach advocated cultivating the crop. This method was developed using a dataset from a soil testing lab, a soil database, and a crop database. The recommendation system merged the majority vote method with Support Vector Machine (SVM) and Artificial Neural Networks (ANN) in order to accurately suggest crops.

Using a variety of Big Data Analytics and Machine Learning, Doshi et al. [4] established a intelligent Agro-Consultant software that helps farmers decide the crop to produce based on a couple of parameters, including season, location of the farm, soil specifications, and circumstances. For prediction, a variety of algorithms, including ANN and K-Nearest Neighbors (KNN), were applied. This model projected the best soil and meteorological conditions for an appropriate crop. Additionally, the Linear Regression method was used to forecast rainfall.

A recommendation system was put up by Kuanr et al. [5] to give farmers an option concerning a specific crop while considering factors like weather, moisture, and time. The algorithm determined whether the crop selection was acceptable for that season by using the farmer's location, the names of the crops, and a certain season as input. Additionally, fertilizers, chemicals, and equipment's were recommended. Effective information modification has been achieved using a fuzzy logic-based system.

In order to solve the issue of pests and illnesses harming crops during their growth season, Kumar et al. [6], created a model that offered a solution by recommending the best crops and pest management methods in accordance with the pests that can impact the suggested crops. When a user enters certain soil-specific information into the system, the model predicts which crop will be produced using similar soil characteristics. In this model, a decision tree and a linear SVM algorithm were used to estimate the crop, with the SVM method producing the results more effectively and precisely. This model's prediction accuracy of 89.66% is insufficient because it only took into account a small number of soil properties.

A framework proposed by Suresh et al. [7] maps the yield and soil data to estimate a range of potential harvests for the soil, as well as details on the supplements the soil is lacking for the particular crop. As a result, the user is free to select the yield that will be planted. In this way, the model aids in providing details to novice ranchers. This system employs a Supervised Machine Learning Algorithm to make more precise and productive harvest recommendations. The framework keeps track of the proper harvests based on the soil type and leaves it up to the ranchers to choose the yield to plant.

Banerjee et al. [8] proposed a system using fuzzy logic in which different fuzzy rule concepts were constructed for every crop to facilitate speedier concurrent processing. The corresponding cultivation index was produced using several soil features, and rainfall data as inputs. The system's performance has been evaluated for eight of West Bengal's most important crops. The system's average accuracy was tested at 92.14 percent, which is higher than equivalent existing systems. This method will enable farmers to select crops more successfully and accurately, leading to higher yield and a more robust economy.

Ujjainia et al. [9] presented a model-integrated type of concept known as the ensemble technique, which closed the gaps in the crop prediction process. By analyzing the various biosystem factors, it has become clear that the technology employed to create the tool for predicting crop yield is extremely diverse. Because a biosystem's properties vary depending on its location, a single algorithm technique is insufficient to fulfil the demand for crop forecasting. When considering the device to be utilized in a worldwide setting, the ensemble method was proven to be the optimum technique for crop prediction.

3 Methodology

3.1 Collection of Dataset

The dataset file includes the fundamental soil properties required to choose the crop to be planted [10]. We made use of several websites that offered comprehensive details about the crops. Rice, maize, chickpeas, kidney beans, pigeon beans, moth beans, mung beans, black gram, lentil, water melon, pomegranate, musk melon, banana, papaya, mango, coconut, grapes, apple, orange, coffee, cotton and jute are among the crops taken into account in this model. The dataset is publicly available in the Kaggle website. The dataset used for training shows the total number of occurrences of each crop that are accessible. Each crop has 100 data values in a dataset with 2200 total data values. The attributes taken into account include pH, average rainfall, temperature, and the amount of nitrogen (N), phosphorus (P), and potassium (K) in the soil. These soil characteristics are crucial for predicting crop yields. The nutrients that crop require in the highest amounts are N, P, and K. Consequently, they are often considered the most significant nutrients. Because it affects the availability of crucial nutrients, pH is significant. Rainfall is a crucial element in predicting any crop. Since each crop has a different need for water, we took this factor into account while projecting the crop. One of the main elements influencing how quickly plants develop is temperature. Therefore, for the aforementioned reasons, it is crucial to take into account soil characteristics when forecasting a palatable crop seed.

3.2 Classification Based Crop Prediction

There are varieties of ML algorithms which uses specific methods and techniques to improve accuracy in classification and prediction applications in different domains. Facts that are structured or unstructured can be classified [11]. The overall architecture of the proposed model is illustrated in Fig. 1. In this paper we have used 9 different classifiers, the details of all these classifiers are explained below:

- Random Forest: Being an ensemble learning method that is widely used to tackle both classification and regression issues, the model using this method must be trained to make predictions by passing the test features through each randomly generated tree's rules. As a result, for the identical test feature, each random forest will predict a different target. Then, votes are calculated based on each anticipated aim. The algorithm's final prediction is the target with the most anticipated votes. The ability of the random forest approach to handle missing values effectively plus the fact that the classifier can never overfit the model makes this algorithm a great choice [12].
- K-Nearest Neighbors (KNN): It's a non-parametric technique for forecasting. The expected value in this case is class membership. Finding the k nearest neighbors for each freshly incoming instance is the first step in the KNN technique. The event is classified based on the majority vote of these neighbors. The second stage is projecting a label for the new instance based on the label sets of its k neighbors [13].
- Naive Bayes: The Naive Bayes classifier is a statistical technique that is used to ascertain the encoding relationships and predictive relationships between different characteristics. It is predicated on the independent assumptions that every simplified element in a class will independently influence the probability in another class with

regard to its chance of occurrence and non-occurrence. Because it only needs a short training dataset to predict results, the most up-to-date classifier and exact supervised learner are used to instruct the classification model quickly and efficiently. Even if the nature is simpler, the proportion of more precise and effective results in difficult real-world situations has improved the expected outcomes.[14].

- Decision Tree: This method of supervised learning uses a tree to express class labels and attribute representations. At this stage, the record's attribute is compared to the root attribute, and a new node is reached depending on the outcome. This comparison is performed until a leaf node with a predicted class value is reached. As a result, modeling a decision tree is very helpful for forecasting [15].

- Logistic Regression: A basic linear model called logistic regression uses a logistic function to build the model. Using the given dataset's relationship trends, it divides the data into distinct classes. It can classify unknown data records relatively quickly and is simple to deploy and train. Its default assumption, nevertheless, is that dependent and independent variables have a linear connection which in some cases can limit the model's performance[16].

- Support Vector Classifier (SVC): A collection of algorithms known as Support Vector Machines, or SVM, evaluate data for regression and classification. It is memory-efficient and represents various classes on a single plane iteratively to reduce error. As a result, if the error in basic linear regression persists, it is one of the finest algorithms to apply. However, its performance suffers when dealing with noisy information and huge datasets because there is a probability that classes would overlap [17].

- XGBoost: The XGBoost Classifier is a high-performance and scalable library specifically developed for efficiently training machine learning models. It employs ensemble learning, combining predictions from weak models to generate a more robust prediction. Notably, XGBoost excels in effectively managing missing values, enabling it to handle real-world datasets without extensive pre-processing. Moreover, XGBoost incorporates built-in parallel processing capabilities, facilitating the training of models on large datasets within a reasonable time frame [18].

- CatBoost: The open-source categorical boosting toolkit known as Catboost was developed by Yandex. CatBoost can be applied to forecasting, ranking, recommendation systems, personal assistants, and classification in addition to regression and classification. It manages categorical features by employing sorted target statistics [19].

- LightGBM: It is a gradient boosting framework centered on decision trees, aimed at improving model performance while minimizing memory usage. It addresses the constraints of the widely-used histogram-based technique in Gradient Boosting Decision Tree (GBDT) frameworks, like EFB, through the implementation of two innovative strategies: gradient-based one-side sampling and exclusive feature bundling. These approaches collaboratively enhance the model's efficiency, giving it a competitive advantage over other GBDT frameworks [20] (Fig. 1).

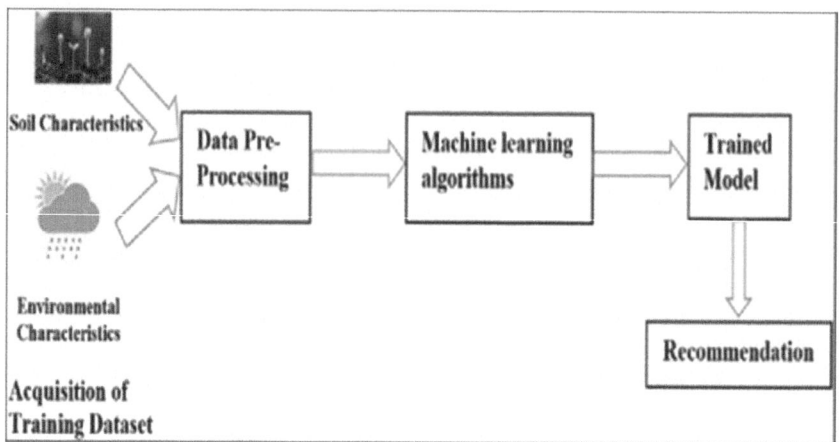

Fig. 1. Architecture of the Proposed Model

4   Result and Analysis

To analyze the overall performance of these techniques on the dataset, an extensive experimentation is conducted using Google COLAB platform. The performance of all these ML classifiers is evaluated using various evaluation matrices.

- Confusion Matrix: A confusion matrix act as a summary of a ML model's performance on a set of test data. In order to predict category labels for input examples, classification models frequently use this method to evaluate their accuracy. In respect to the test data, the matrix offers a thorough overview of the model's true positive (TP), true negative (TN), false positive (FP), and false negative (FN). In multi-class classification, the matrix's dimensions normally take the form of a n x n matrix, whereas in binary classification, the matrix takes the shape of a 2×2 table [21].
- Accuracy (AC): It is used to calculate the effectiveness of a model, which represents the proportion of rightly predicted occurrences out of total occurrences.
- Precision (P): It act as an indicator of the accuracy of a model's positive predictions by measuring the ratio of TP predictions to the total number of positive predictions generated by the model.
- Recall(R): It evaluates the capability of a classification model to identify all pertinent instances within a dataset. It is calculated by dividing the number of TP occurrences by the sum of TP and FN occurrences.
- F1-Measure: It is employed to assess the comprehensive performance of a model. It is calculated as the harmonic mean of precision and recall, joining these two metrics into a single measure.

All the metrics gathered from distinct models are displayed in the table below. Figure 2's bar graph, which compares all of the metrics from the tests and simulations, illustrates this comparison. Xgboost classifier provided the highest % of all the metrics, including precision, recall, f1 score, and accuracy, as shown by the bar graph. Figure 3 provides the confusion matrix for the xgboost framework (Table 1 and Fig. 2).

Table 1. Comparisons of Different Classifiers

CLASSIFIERS	PRECESSION	RECALL	FI SCORE	SUPPORT	ACCURACY
RANDOM FOREST	0.99	0.99	0.99	660	99.09
KNN	0.99	0.99	0.99	660	98.79
NAÏVE BAYES	0.96	0.96	0.96	660	99.55
DECESSION TREE	0.96	0.96	0.96	660	99.39
LOGISTIC REGRESSION	0.96	0.96	0.96	660	95.76
SVC	0.99	0.99	0.99	660	98.64
XGBOOST	1.00	1.00	1.00	440	99.77
LIGHTGBM	0.99	0.99	0.99	440	98.86
CATBOOST	0.99	0.99	0.99	440	99.31

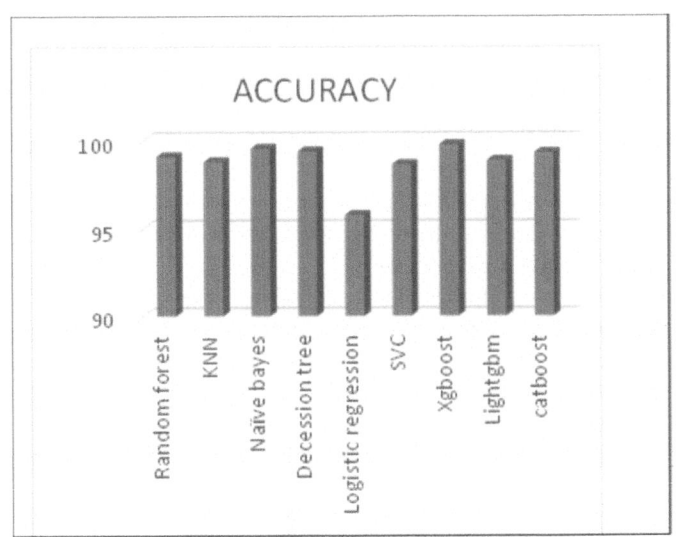

Fig. 2. Accuracy of Different ML Models

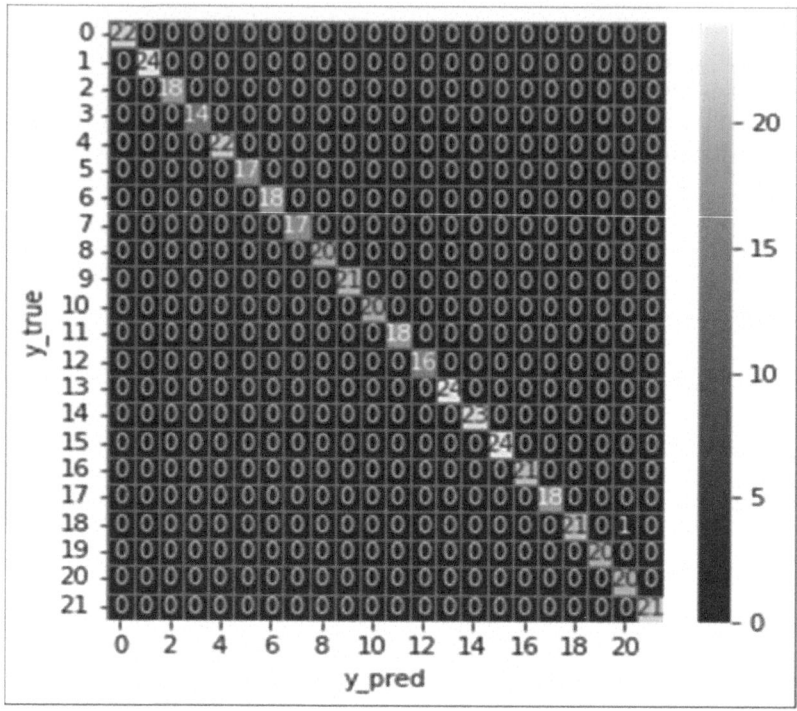

Fig. 3. Confusion Matrix of XGboost Classifier

5   Conclusion

By utilizing well-known classification algorithms such as Random Forest, KNN, Nave bayes, Decision tree, Logistic regression, SVC, Xgboost, Lightgbm, and catboost, we have provided the research prospects for choosing the right crop. Several simulations with hyperparameters set in accordance with the accuracy and sensitivity criteria were conducted in order to develop the model. We have found that the comparative analyses of these algorithms have different levels of accuracy, which makes it difficult to evaluate the effectiveness and efficiency of forecasts. However, with an accuracy of 96.77%, the XGboost algorithm offered the best results.

In the future, this work can be advanced by taking into account a wider diversity of crops. As per the availability of data, the current research is only applicable to 22 harvests. To appropriately estimate the soil fertility, rainfall, and water availability, future studies might take a closer look at specific geographic regions. Moving forward, we intend to create a crop recommendation system that agriculturalists and farmers can utilise to make better-informed choices that will boost output and boost the economy.

References

1. Pudumalar, S., Ramanujam, E.: Crop Recommendation System for Precision Agriculture. In: IEEE Eighth International Conference on Advanced Computing (ICoAC)) (2016)

2. Shinde, M., Ekbote, K., et al.: Crop recommendation and fertilizer purchase system. Int. J. Comput. Sci. Inf. Technol. **7**(2), 665–667 (2016)
3. Rajak, R.K., Pawar, A., et al.: Crop recommendation system to maximize crop yield using machine learning technique. Int. Res. J. Eng. Technol. **4**(12), 950–953 (2017)
4. Doshi, Z., Nadkarni, S., Agrawal, R., Shah, N.: Agroconsultant: Intelligent crop recommendation system using machine learning algorithms. In: 2018 Fourth International Conference on Computing Communication Control and Automation (ICCUBEA). IEEE, pp. 1–6 (2018)
5. Kuanr, M., Rath, B.K., Mohanty, S.N.: Crop recommender system for the farmers using mamdani fuzzy inference model. Int. J. Eng. Technol. **7**(4.15), 277–280 (2018)
6. Kumar, A., Sarkar, S., Pradhan, C.: Recommendation system for crop identification and pest control technique in agriculture. In: 2019 International Conference on Communication and Signal Processing (ICCSP), pp. 0185–0189. IEEE (2019)
7. Suresh, G., Senthil Kumar, A., Lekashri, S., Manikandan, R.: Efficient crop yield recommendation system using machine learning for digital farming. Int. J. Mod. Agric. **10**(1) (2021). ISSN: 2305–7246
8. Banerjee, G., Sarkar, U., Ghosh, I.: A fuzzy logic-based crop recommendation system. In: Proceedings of International Conference on Frontiers in Computing and Systems, Advances in Intelligent Systems and Computing, vol. 1255, Springer, Singapore (2021)
9. Ujjainia, S., Gautam, P., Veenadhari, S.: A crop recommendation system to improve crop productivity using ensemble technique. Int. J. Innovative Technol. Exploring Eng. (IJITEE), **10**(4) (2021). ISSN: 2278–3075
10. Kaggle: Crop Recommendation Dataset (2020). https://www.kaggle.com/datasets/atharvaingle/crop-recommendation-dataset. Accessed August 2023
11. Gandhi, N., Armstrong, L.J., Petkar, O., Tripathy, A.K.: Rice crop yield prediction in India using support vector machines. In: IEEE International Joint Conference on Computer Science and Software Engineering, pp. 1–5. Khon Kaen (2016)
12. How the random forest algorithm works in machine learning. http://dataaspirant.com/2017/05/22/random-forest-algorithm-machine
13. k-nearest neighbors' algorithm-Wikipedia. https://en.wikipedia.org/wiki/K-nearest_neighbors_algorithm. Accessed Feb 2018
14. Wahbeh, A.H., et al.: A comparison study between data mining tools over some classification methods. Int. J. Adv. Comput. Sci. Appl. **8**(2), 18–26 (2011)
15. How Decision Tree Algorithm works. http://dataaspirant.com/2017/01/30/how-decision-tree-algorithm-works/. Accessed Feb 2018
16. Attaluri, S., Batcha, N., Mafas, R.: Crop plantation recommendation using feature extraction and machine learning techniques. J. Appl. Technol. Innovation **4**(4) (2020). (e -ISSN: 2600–7304)
17. Parikh, D., Jain, J., Gupta, T., Dabhade, R.: Machine learning based crop recommendation system. Int. J. Adv. Res. Sci. Commun. Technol. (IJARSCT) **6**(1) (2021)
18. Liew, X., Hameed, N., Clos, J.: An investigation of XGBoost-based algorithm for breast cancer classification. 2666–8270/© 2021 The Author(s). Published by Elsevier Ltd. This is an open access article under the CC BY-NC-ND license (http://creativecommons.org/licenses/by-nc-nd/4.0/),2021
19. Wang, B., Wang, Y., Qin, K., Xiao, O.: Detecting transportation modes based on LightGBM classifier from GPS trajectory data. Detecting transportation modes based on LightGBM classifier from GPS trajectory data. In: IEEE 2018 26th International Conference on Geoinformatics (2018)
20. Ibrahim, A., Ridwan, R., Muhammed, M., et al.: Comparison of the catboost classifier with other machine learning methods Int. J. Adv. Comput. Sci. Appl. (IJACSA), **11**(11) (2020)
21. Heydarian, M., Doyle, T.E., Samavi, R.: MLCM: multi-label confusion matrix. IEEE Access **10**, 19083–19095 (2022)

Human Fall Detection Using Transfer Learning-Based 3D CNN

Ekram Alam[1,3]($\boxtimes$)(iD), Abu Sufian[2](iD), Paramartha Dutta[3](iD), and Marco Leo[4](iD)

[1] Department of Computer Science, Gour Mahavidyalaya, Old Malda, India
ealam@ieee.org
[2] Department of Computer Science, University of Gour Banga,
English Bazar, India
[3] Department of Computer and System Sciences, Visva-Bharati University,
Santiniketan, India
[4] National Research Council of Italy, Institute of Applied Sciences and Intelligent
Systems, Lecce, Italy

Abstract. Unintentional or accidental falls are one of the significant health issues in senior persons. The population of senior persons is increasing steadily. So, there is a need for an automated fall detection monitoring system. This paper introduces a vision-based fall detection system using a pre-trained 3D CNN. Unlike 2D CNN, 3D CNN extracts not only spatial but also temporal features. Proposed model leverages the original learned weights of a 3D CNN model pre-trained on the Sports1M dataset to extract the spatio-temporal features. Only the SVM classifier was trained, which saves the time required to train the 3D CNN. Stratified shuffle five split cross-validation has been used to split the dataset into training and testing data. Extracted features from the proposed 3D CNN model were fed to an SVM classifier to classify the activity as fall or ADL. Two datasets, GMDCSA and CAUCAFall, were utilized to conduct the experiment. The source code for this work can be accessed via the following link: https://github.com/ekramalam/HFD_3DCNN.

Keywords: Fall Detection · 3D CNN · Fall Datasets · Transfer Learning

1 Introduction

According to a report by the "World Health Organization (WHO)" [1], the global population of individuals aged 60 years or above is projected to increase from 1 billion in 2020 to 1.4 billion in 2030 and is expected to rise further to 2.1 billion by 2050. The steady rise in the global population of senior citizens poses a significant challenge in caring for seniors. Accidental or unintentional falls represent a significant health concern among senior individuals, often resulting in severe injuries, permanent damage, escalated healthcare expenses, and, in some cases, even death if timely medical assistance is delayed [2]. An automated

© The Author(s), under exclusive license to Springer Nature Switzerland AG 2025
M. Majumder et al. (Eds.): ICCTE 2023, CCIS 2376, pp. 102–113, 2025.
https://doi.org/10.1007/978-3-031-81935-3_9

monitoring system can be exceedingly beneficial for detecting human falls among senior citizens or individuals with illnesses who live alone.

Among various systems designed for detecting falls, vision-based and wearable sensor-based fall detection systems are two major approaches. Wearable sensors, while promising in many aspects, present several disadvantages for detecting human falls. Wearable sensors can be annoying and uncomfortable to wear continuously, leading to hesitance in embracing this. Prolonged use may result in side effects like skin irritation or discomfort, potentially causing ill effects on health, particularly for those with sensitive skin or pre-existing conditions. Additionally, frequent battery charging can be a hassle for seniors, making these devices less user-friendly and less practical for those with limited mobility or cognitive challenges.

In the vision-based approach, there is no need to attach the sensor to the body. Vision sensors usually do not require frequent battery charging, further enhancing convenience and long-term viability. Vision sensors are becoming increasingly prevalent in various settings, including homes, hospitals, nursing homes, industries, and public spaces, generating vast amounts of continuous data. This widespread availability and self-sustaining nature make vision sensor-based methods highly promising for human fall detection. Vision sensor-based approaches are also cost-effective [3].

Machine Learning (ML) based techniques, especially data-driven approaches like Deep Learning (DL), work very well to process visual data. Traditional 2D Convolutional Neural Networks (CNNs) [4,5] is a popular DL technique in computer vision. 2D CNN works frame by frame. It considers only one frame at a time. One frame provides only spatial information. 2D CNN works well in image data, but 2D CNN is not so good for video data. Though spatial information is important to detect falls, temporal information is also required for a better understanding of fall patterns. The intricate nature of human falls requires models that can effectively capture both temporal and spatial features. To extract the temporal and spatial features (spatio-temporal), 3D CNN [6] is a better option [7]. 3D CNNs have proven to be a superior approach [8–10] for human fall detection.

3D CNN-based models demonstrate remarkable results, but like other DL techniques, they also necessitate extensive datasets, extended training time, and substantial computing resources. These issues can be addressed by using transfer learning [11]. We used a 3D CNN model [8] pre-trained on the Sports-1M dataset [12]. A linear "Support Vector Machine (SVM)" [13,14] classifier was employed to classify an activity as a "fall" or an "activity of daily living (ADL)".

The key highlights of this research are outlined below.

- **3D-CNN:** 3D-CNN gives good results because it extracts spatio-temporal features. We have used 3D CNN in this work.
- **Transfer Learning:** Transfer learning has been used in this work. This work uses a modified pre-trained C3D [8] model.

- **Reduced Training Time:** We used the original weights of C3D in our modified C3D model. Only the SVM classifier was trained. So, training time reduced significantly.
- **Data Adaptability:** One of the essential features of deep learning models like 3D CNN is data generalization. A pre-trained model on Sports-1M was utilized for this work.
- **Tested on recent complex datasets:** We tested our work on two recent complex datasets [15,16].
- **Good Results:** Even though we used the original weight of the C3D model without any training on the fall datasets [15,16], the outcomes are promising as depicted in Table 2, and Table 3.

The remaining sections of this manuscript are outlined as follows. Section 2 furnishes a concise review of relevant literature. Section 3 outlines the datasets and the method employed in this work. The experiment's outcome is detailed and analyzed in Sect. 4. Finally, Sect. 5 encapsulates the findings of this paper and underscores potential avenues for future research.

2   Literature Review

The detection of human falls can be achieved through the utilization of wearable sensors or vision sensors. Many fall detection systems have been implemented using wearable sensors [17–20], but vision sensor based systems are more suitable for this task [8–10]. Brief descriptions of recent and notable fall detection systems utilizing vision sensors and deep learning are provided below.

Rezaee et al. [21], and Arun et al. [22] used CNN to detect human falls. Rezaee et al. [21] used thermal datasets on the modified version of ShuffleNet [23]. Inturi et al. [24] introduced a fall detection system that utilizes pose estimation [25]. A pre-trained AlphaPose [26] model was leveraged to estimate the pose (joint key points). CNN and LSTM were employed to process the key points and to find the results. Saurav et al. [27] introduced a human fall detection technique using many variants models consisting of CNN & ConvLSTM, CNN & LSTM, and 3D CNNs. Patel et al. [28] proposed a human fall detection using "Long-term Recurrent Convolutional Network (LRCN)", which is a hybrid version of LSTM and CNN.

Fei et al. [29] introduced a two-streams model for fall detection. One stream used optical flow, and the other one used pose estimation. Graph Convolutional Networks (GCN) [30] and CNN were used for pose and optical flow, respectively. Egawa et al. [31] introduced a fall detection system utilizing a spatial-temporal convolutional neural network with attention mechanisms based on graphs. Amsaprabhaa et al. [32] introduced a multi-modal skeletal gait feature fusion-based human fall detection method. Two types of features were extracted using 1D-CNN and spatio-temporal graph convolution network, which were finally combined to get the results.

Chen et al. [33] and Osigbesan et al. [34] introduced a human fall detection system using pose estimation. In Chen et al. [33], at first, the 2D pose was estimated, which was transformed into the 3D pose. This 3D pose data was provided as input to the fall detection network to get the final outputs. Their fall detection network is a fully convolutional architecture with residual blocks. Osigbesan et al. [34] introduced a fall detection method for aviation maintenance personnel. They used 3D-CNN and LSTM to process the pose data.

Li et al. [35] presented a human fall detection network using future frame prediction. Wu et al. [36] employed GAN (generative adversarial network) [37] to detect human falls. Zi et al. [38] introduced a fall detection system, especially for low-lighting environments. Dual illumination, YOLOv7, and Deep SORT techniques were used to process the data.

Leal et al. [39] presented a human fall detection technique by combining a 3D-CNN and an RNN. Ha et al. [40] introduced a human fall detection model using 3D-CNN and a mixture of experts. Alanzi et al. [41] presented a human fall detection method utilizing four branch-3D-CNN (4S-3D-CNN). They divided the input video into 32-frame groups. Segmentation was done to extract humans from these frames. Segmented frames were converted to four fused images using image fusion, which were finally passed to 4S-3DCNN to get the results.

Many fall detection systems have been implemented using 2D CNNs, with only a few utilizing 3D CNNs. 3D CNNs are known to perform better with visual data, but they often come with higher training times. To address this challenge, we adopted a strategy of leveraging a pre-trained 3D CNN. This approach allows our system to extract spatio-temporal features and save training time.

3 Materials and Method

This section is divided into three subsections. Subsection 3.1 discusses the two datasets "GMDCSA" and "CAUCAFall" briefly. Subsection 3.2 provides a brief description of 3D-CNN. Subsection 3.3 describes the methodology of this work.

3.1 Dataset

We utilized two datasets, "GMDCSA" [25] and "CAUCAFall" [16], to evaluate our experiment. The GMDCSA dataset comprises 16 fall videos and 16 non-fall (ADL) videos. The fall video's length exhibits minimum, maximum, mean, mode, and median values of 4 s, 6 s, 5.06 s, 6 s, and 5 s, respectively. Similarly, the ADL video's length showcases minimum, maximum, mean, mode, and median values of 3 s, 12 s, 6.5 s, 6 s, and 6 s, respectively. The frames per second (fps) value for GMDCSA is 30. The GMDCSA was generated by performing ADL and fall activities by a single individual. The primary objective of creating this dataset is to detect false positives. Many ADL activity in this dataset involves sleeping activity, which is very similar to fall activity and difficult to identify as ADL by a model.

The CAUCAFall dataset, introduced by Guerrero et al. [16], encompasses fall and ADL activities conducted by ten distinct subjects. This dataset incorporates variations in gender (five males and five females), weights, heights, ages, and

Table 1. GMDCSA and CAUCAFall datasets details

Dataset	Type	Length (in Seconds)					fps	NFl	NAd	NSb	NCm
		Min	Max	Mean	Mode	Median					
GMDCSA	Fall	4	6	5.06	6	5	30	16	16	1	1
GMDCSA	ADL	3	12	6.5	6	6	30	16	16	1	1
CAUCAFall	Fall	5	13	8.18	9 & 10	10	20	50	50	10	1
CAUCAFall	ADL	4	13	9.14	10	10	20	50	50	10	1

subject outfits. The dataset was created in varying lighting conditions. Some of the video sequences involve occlusions. Every subject performs five falls and five non-fall activities, which makes a total of 100 video sequences, 50 "fall" and 50 "non-fall" video sequences. The minimum, maximum, mean, mode, and median values of the length of the fall video sequences of CAUCAFall datasets are 5 s, 13 s, 8.18 s, 9 & 10 s, and 10 s, respectively. The minimum, maximum, mean, mode and median values of the length of the ADL video sequences of CAUCAFall datasets are 4 s, 13 s, 9.14 s, 10 s, and 10 s respectively.

Table 1 provides a brief overview of both datasets, where NFl, NAd, NSb, and NCm represent the number of fall video sequences, the number of ADL video sequences, the number of subjects, and the number of cameras employed, respectively.

3.2   3D CNN

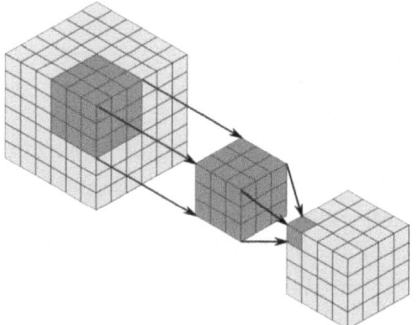

Fig. 1. 3D Connvolution

The filter size in 3D convolution and pooling consists of three components, namely "d × h × w," where 'h' and 'w' are analogous to the dimensions used in 2D CNNs, representing height and width. The additional component 'd' signifies the depth, indicating the number of frames or images. The output of the 3D convolution is not 2D data. Instead, it generates a 3D cuboid output as shown in Fig. 1. Similar to 3D convolution, 3D pooling works. 3D Pooling layer downsamples input 3D data. 3D CNNs use 3D convolution and 3D pooling to process the 3D data (video) as 3D volumes, allowing them to comprehend motion dynamics by extracting spatio-temporal features.

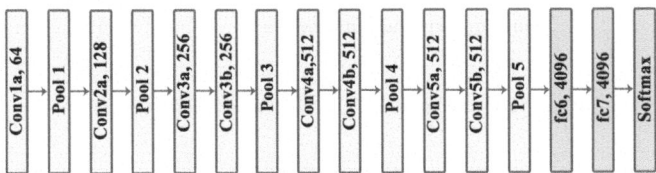

Fig. 2. Architecture of the C3D model

Tran et al. [8] presented a 3D-CNN with eight convolutions layers (Conv1a, Conv2a, Conv3a, Conv3b, Conv4a, Conv4b, Conv5a, Conv5b), five pooling layers (Pool1, Pool2, Pool3, Pool4, Pool5), two fully connected layers (fc6, fc7), and a Softmax classifier, illustrated in Fig. 2. They named their 3D-CNN model as C3D. The numbers of filters for the convolution layers Conv1a, Conv2a, Conv3a, Conv3b, Conv4a, Conv4b, Conv5a, and Conv5b were 64, 128, 256, 256, 512, 512, 512, 512 respectively. The size of each kernel for the "convolutional layer" was $3 \times 3 \times 3$. The kernel size for the first "pooling layer" (Pool1) was $1 \times 2 \times 2$, and for the rest of the "pooling layers" it was $2 \times 2 \times 2$. We used the modified version of this C3D model in this work.

3.3 Method

We used the modified version of the C3D model "pre-trained" on the Sport-1M dataset as a feature extractor. The Sport-1M dataset contains over one million video clips that cover a wide range of sports and physical activities, including various sports games, exercises, and outdoor activities. This dataset contains 487 classes related to sports and actions, making it suitable to use in action recognition problems like human fall detection. The architecture of the proposed model of this work is shown in Fig. 3. This 3D CNN model comprises eight "convolutional layers", five "pooling layers", one "fully connected" layer [4], and an SVM classifier.

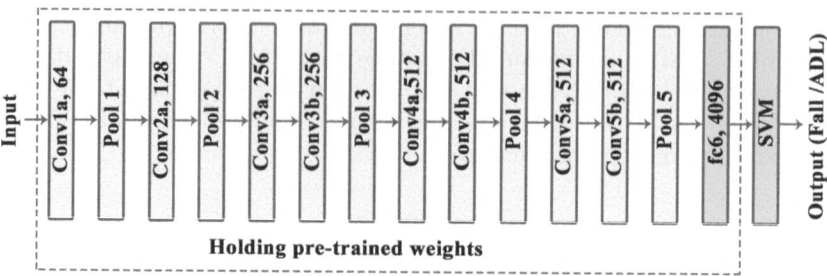

Fig. 3. Proposed 3D CNN model architecture

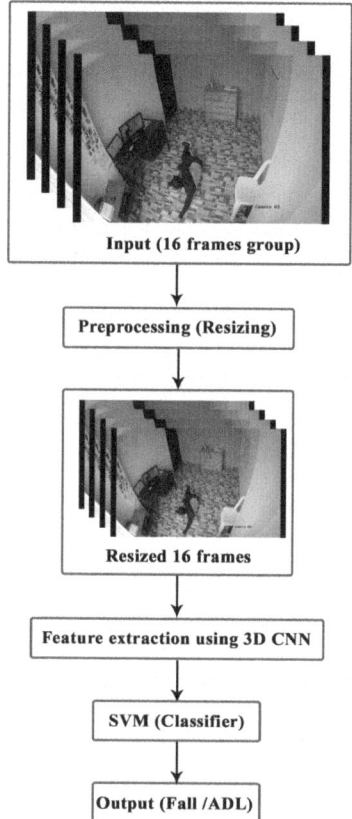

Fig. 4. Proposed work methodology

The proposed methodology to detect human falls is illustrated in the flow diagram depicted in Fig. 4. At first, the input video is divided into groups (chunks) of 16 frames. These chunks are resized to 112 × 112. The resized chunks of videos are fed as the input to the proposed 3D CNN model to extract the features. Extracted features by the proposed 3D CNN model were given as input to a linear SVM classifier to classify a frame group (chunk) as fall or ADL. The original weight of the C3D model was used to extract the features by our proposed 3D CNN model without doing any retraining. Only the SVM was trained. We used the "stratified shuffle split cross-validation technique" [42] to evaluate this work on two recent datasets (GMDCSA and CACUCAFall). We divided both datasets into five splits. For each split, 70% of the dataset was utilized for training, while 30% was designated for testing.

4 Result

We evaluated the proposed model to detect human falls on two datasets, "GMD-CSA" and "CAUCAFall". The "stratified shuffle split cross-validation technique" [42] was used to divide the dataset into five splits. Figure 5a and Fig. 5b show the confusion matrices of the experiment conducted on GMDCSA and CAUCAFall datasets, respectively. The GMDCSA dataset's confusion matrices show no false negative value for any split but some false positives for each split except the first split. The false positive values for split 1, split 2, split 3, split 4, and split 5 are 0, 2, 4, 3, and 3, respectively. The GMDCSA dataset contains many sleeping activities as ADL. That is why some ADL activity was detected as fall activity.

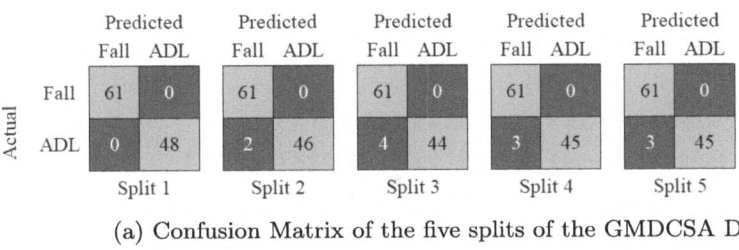

(a) Confusion Matrix of the five splits of the GMDCSA Dataset

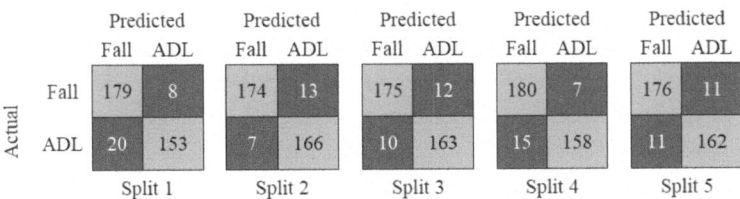

(b) Confusion Matrix of the five splits of the CAUCAFall Dataset

Fig. 5. Confusion Matrix of the five splits of the GMDCSA and the CAUCAFall Dataset

The confusion matrices of the proposed experiment on the dataset 'CAU-CAFall' contain some false positive as well as false negative values because this dataset is more complex than the GMDCSA dataset. CAUCAFall dataset is developed by performing the "fall" and "ADL" activities by ten different subjects.

We calculated sensitivity, specificity, precision, accuracy, F1 Score, False Positive Rate (FPR), and False Negative Rate (FNR) to measure the performance of this experiment. The values of these metrics [43] of each of the five splits are presented in Table 2 and Table 3 for the datasets GMDCSA and CAUCAFall, respectively. The average values of these metrics are also provided in the mentioned tables (Table 2 and Table 3).

Table 2. Evaluation metrics of the experiment on the dataset 'GMDCSA'

Metric	Split1	Split2	Split3	Split4	Split5	Average Values
Sensitivity	100	96.83	93.85	95.31	95.31	96.26
Specificity	100	100	100	100	100	100
Precision	100	100	100	100	100	100
Accuracy	100	98.17	96.33	97.25	97.25	97.8
F1 Score	100	98.39	96.83	97.60	97.60	98.08
FPR	0	4.17	8.33	6.25	6.25	5
FNR	0	0	0	0	0	0

Table 3. Evaluation metrics of the experiment on the dataset 'CAUCAFall'

Metric	Split1	Split2	Split3	Split4	Split5	Average Values
Sensitivity	95.72	93.05	93.58	96.26	94.12	94.55
Specificity	88.44	95.95	94.22	91.33	93.64	92.72
Precision	89.95	96.13	94.59	92.31	94.12	93.42
Accuracy	92.22	94.44	93.89	93.89	93.89	93.67
F1 Score	92.75	94.57	94.09	94.24	94.12	93.95
FPR	11.56	4.04	5.78	8.67	6.35	7.28
FNR	4.27	6.95	6.41	3.74	5.88	5.45

5   Conclusion and Future Scope

In this work, we utilized transfer learning to employ the pre-trained weights of the 'C3D model' up to the first "fully connected layer" (fc6). The extracted features were used as input for an SVM to classify the results as either a fall or an ADL. The GMDCSA and CAUCAFall datasets were employed to evaluate the proposed model. The average values of sensitivity, specificity, precision, accuracy, and F1 score for the five splits for the GMDCSA dataset were 96.26, 100, 100, 97.8, and 98.08, respectively. For the CAUCAFall dataset, the average values of sensitivity, specificity, precision, accuracy, and F1 score were 94.55, 92.72, 93.42,

93.67, and 93.95, respectively. While the results are promising, it is important to note that the datasets were generated by healthy individuals, not actual senior citizens. It would be more useful to create a new dataset containing actual fall data and test the model on this real dataset. Additionally, this work can be extended to implement an end-to-end 3D-CNN model for fall detection.

Disclosure of Interests. The authors have no competing interests to declare that are relevant to the content of this article.

References

1. Ageing and health—who.int. https://www.who.int/news-room/fact-sheets/detail/ageing-and-health. Accessed 24 Aug 2023
2. Martínez-Villaseñor, L., Ponce, H., Brieva, J., Moya-Albor, E., Núñez-Martínez, J., Peñafort-Asturiano, C.: Up-fall detection dataset: a multimodal approach. Sensors **19**(9), 1988 (2019)
3. Paulauskaite-Taraseviciene, A., et al.: Geriatric care management system powered by the IoT and computer vision techniques. In: Healthcare, vol. 11, p. 1152. MDPI (2023)
4. Alam, E., Sufian, A., Das, A.K., Bhattacharya, A., Ali, M.F., Rahman, M.H.: Leveraging deep learning for computer vision: a review. In: 2021 22nd International Arab Conference on Information Technology (ACIT), pp. 1–8. IEEE (2021)
5. Ghosh, A., Sufian, A., Sultana, F., Chakrabarti, A., De, D.: Fundamental concepts of convolutional neural network. In: Recent Trends and Advances in Artificial Intelligence and Internet of Things, pp. 519–567 (2020)
6. Maturana, D., Scherer, S.: Voxnet: a 3D convolutional neural network for real-time object recognition. In: 2015 IEEE/RSJ International Conference on Intelligent Robots and Systems (IROS), pp. 922–928. IEEE (2015)
7. Liu, Y., et al.: Dynamic gesture recognition algorithm based on 3D convolutional neural network. Comput. Intell. Neurosci. **2021** (2021)
8. Tran, D., Bourdev, L., Fergus, R., Torresani, L., Paluri, M.: Learning spatiotemporal features with 3D convolutional networks. In: Proceedings of the IEEE International Conference on Computer Vision, pp. 4489–4497 (2015)
9. Vrskova, R., Hudec, R., Kamencay, P., Sykora, P.: Human activity classification using the 3DCNN architecture. Appl. Sci. **12**(2), 931 (2022)
10. Alanazi, T., Muhammad, G.: Human fall detection using 3D multi-stream convolutional neural networks with fusion. Diagnostics **12**(12), 3060 (2022)
11. Sufian, A., Alam, E., Ghosh, A., Sultana, F., De, D., Dong, M.: Deep learning in computer vision through mobile edge computing for IoT. In: Mobile Edge Computing, pp. 443–471 (2021)
12. Karpathy, A., Toderici, G., Shetty, S., Leung, T., Sukthankar, R., Fei-Fei, L.: Large-scale video classification with convolutional neural networks. In: Proceedings of the IEEE Conference on Computer Vision and Pattern Recognition, pp. 1725–1732 (2014)
13. Noble, W.S.: What is a support vector machine? Nat. Biotechnol. **24**(12), 1565–1567 (2006)
14. Cervantes, J., Garcia-Lamont, F., Rodríguez-Mazahua, L., Lopez, A.: A comprehensive survey on support vector machine classification: applications, challenges and trends. Neurocomputing **408**, 189–215 (2020)

15. Alam, E., Sufian, A., Dutta, P., Leo, M., Hameed, I.A.: GMDCSA-24: a dataset for human fall detection in videos. Data Brief, 110892 (2024)
16. Guerrero, J.C.E., España, E.M., Añasco, M.M., Lopera, J.E.P.: Dataset for human fall recognition in an uncontrolled environment. Data Brief **45**, 108610 (2022)
17. Mekruksavanich, S., Jantawong, P., Jitpattanakul, A.: Deep learning approaches for HAR of daily living activities using IMU sensors in smart glasses. In: 2023 Joint International Conference on Digital Arts, Media and Technology with ECTI Northern Section Conference on Electrical, Electronics, Computer and Telecommunications Engineering (ECTI DAMT & NCON), pp. 474–478. IEEE (2023)
18. Seenath, S., Dharmaraj, M.: Conformer-based human activity recognition using inertial measurement units. Sensors **23**(17), 7357 (2023)
19. Andrade-Ambriz, Y.A., Ledesma, S., Ibarra-Manzano, M.-A., Oros-Flores, M.I., Almanza-Ojeda, D.-L.: Human activity recognition using temporal convolutional neural network architecture. Expert Syst. Appl. **191**, 116287 (2022)
20. Tahir, A., et al.: IoT based fall detection system for elderly healthcare. In: Internet of Things for Human-Centered Design: Application to Elderly Healthcare, pp. 209–232. Springer (2022)
21. Rezaee, K., Khosravi, M.R., Moghimi, M.K.: Intelligent elderly people fall detection based on modified deep learning deep transfer learning and IoT using thermal imaging-assisted pervasive surveillance. In: Intelligent Healthcare: Infrastructure, Algorithms and Management, pp. 113–132. Springer (2022)
22. Arun, D., Sumukh Subramanya, H., Goel, T., Tanush, N., Nayak, J.S.: Video-based elderly fall detection using convolutional neural networks. In: Proceedings of Third International Conference on Intelligent Computing, Information and Control Systems: ICICCS 2021, pp. 803–814. Springer (2022)
23. Zhang, X., Zhou, X., Lin, M., Sun, J.: ShuffleNet: an extremely efficient convolutional neural network for mobile devices. In: Proceedings of the IEEE Conference on Computer Vision and Pattern Recognition, pp. 6848–6856 (2018)
24. Inturi, A.R., Manikandan, V., Garrapally, V.: A novel vision-based fall detection scheme using keypoints of human skeleton with long short-term memory network. Arab. J. Sci. Eng. **48**(2), 1143–1155 (2023)
25. Alam, E., Sufian, A., Dutta, P., Leo, M.: Real-time human fall detection using a lightweight pose estimation technique. In: Proceedings of the Computational Intelligence in Communications and Business Analytics (CICBA-2023) Conference (2023). In press
26. Fang, H.-S., et al.: AlphaPose: whole-body regional multi-person pose estimation and tracking in real-time. IEEE Trans. Pattern Anal. Mach. Intell. (2022)
27. Saurav, S., Saini, R., Singh, S.: Vision-based techniques for fall detection in 360° videos using deep learning: dataset and baseline results. Multimedia Tools Appl. **81**(10), 14173–14216 (2022)
28. Patel, V., Kaple, S., Satpute, V.R.: Indoor human fall detection using deep learning. In: International Conference on Advancements in Interdisciplinary Research, pp. 235–242. Springer (2022)
29. Fei, K., Wang, C., Zhang, J., Liu, Y., Xie, X., Tu, Z.: Flow-pose net: an effective two-stream network for fall detection. Vis. Comput. **39**(6), 2305–2320 (2023)
30. Zhang, S., Tong, H., Xu, J., Maciejewski, R.: Graph convolutional networks: a comprehensive review. Comput. Soc. Netw. **6**(1), 1–23 (2019)
31. Egawa, R., Miah, A.S.M., Hirooka, K., Tomioka, Y., Shin, J.: Dynamic fall detection using graph-based spatial temporal convolution and attention network. Electronics **12**(15), 3234 (2023)

32. Amsaprabhaa, M., et al.: Multimodal spatiotemporal skeletal kinematic gait feature fusion for vision-based fall detection. Expert Syst. Appl. **212**, 118681 (2023)
33. Chen, Z., Wang, Y., Yang, W.: Video based fall detection using human poses. In: CCF Conference on Big Data, pp. 283–296. Springer (2022)
34. Osigbesan, A., et al.: Vision-based fall detection in aircraft maintenance environment with pose estimation. In: 2022 IEEE International Conference on Multisensor Fusion and Integration for Intelligent Systems (MFI), pp. 1–6. IEEE (2022)
35. Li, S., Song, X.: Future frame prediction network for human fall detection in surveillance videos. IEEE Sens. J. (2023)
36. Wu, L., et al.: Video-based fall detection using human pose and constrained generative adversarial network. IEEE Trans. Circ. Syst. Video Technol. (2023)
37. Creswell, A., White, T., Dumoulin, V., Arulkumaran, K., Sengupta, B., Bharath, A.A.: Generative adversarial networks: an overview. IEEE Signal Process. Mag. **35**(1), 53–65 (2018)
38. Zi, X., Chaturvedi, K., Braytee, A., Li, J., Prasad, M.: Detecting human falls in poor lighting: object detection and tracking approach for indoor safety. Electronics **12**(5), 1259 (2023)
39. Leal, J., Moayyed, H., Vale, Z.: Detection of human falls via computer vision for elderly care–an I3D/RNN approach. In: International Symposium on Distributed Computing and Artificial Intelligence, pp. 113–122. Springer (2023)
40. Ha, T.V., Nguyen, H.M., Thanh, S.H., Nguyen, B.T.: Fall detection using mixtures of convolutional neural networks. Multimedia Tools Appl. 1–28 (2023)
41. Alanazi, T., Babutain, K., Muhammad, G.: A robust and automated vision-based human fall detection system using 3D multi-stream CNNs with an image fusion technique. Appl. Sci. **13**(12), 6916 (2023)
42. Szeghalmy, S., Fazekas, A.: A comparative study of the use of stratified cross-validation and distribution-balanced stratified cross-validation in imbalanced learning. Sensors **23**(4), 2333 (2023)
43. Alam, E., Sufian, A., Dutta, P., Leo, M.: Vision-based human fall detection systems using deep learning: a review. Comput. Biol. Med. **146**, 105626 (2022)

Anomaly Detection in Respiratory Events Using Machine Learning

Arundhati Roy$^{(\boxtimes)}$ ⓘ and Sriparna Saha ⓘ

Department of Computer Science and Engineering, Maulana Abul Kalam Azad University of Technology, West Bengal, Haringhata, India
arundhati.cbz@gmail.com, sahasriparna@gmail.com

Abstract. The measurement of respiratory rate (RR) holds utmost importance as it is closely associated with major respiratory ailments. In this study, a public dataset with the objective of developing a robust predictive model is utilized. By calculating Mean Square Error, Degree of prediction is analyzed followed by the measurement of R^2 values. Numerous comparison study with various regression models molding into different statistical techniques, predict the superiority of Random Forest in prediction of the several breathing patterns. The predictive model evolves as it learns from newly detected anomalies and adapts to changing patterns in the respiratory data. This ongoing feedback loop enhances its predictive capabilities over time.

Keywords: Regression · Respiration Rate · Mean Square Error · R^2 Value · Statistical Techniques · Hyperparameter Tuning · Sleep Apnea · Machine Learning · Random Forest

1 Introduction

Breathing is a complex process influenced by the central nervous system, guided by various chemoreceptors and baroreceptors, maintaining equilibrium and countering respiratory disorders like hypoxemia and hypercarbia [1, 2]. Respiratory rate (RR), measured in breaths per minute (bpm), is a vital clinical indicator, revealing metabolic misunderstandings leading to hypoxemia or hypercarbia. Anomaly detection in respiration involves identifying abnormal breathing patterns, monitoring key parameters like RR, depth, and regularity [2–5]. Algorithms detecting anomalies like tachypnea, bradypnea, irregular patterns, apnea, or desaturation are crucial in intensive care, sleep medicine, and remote monitoring for conditions like COPD and asthma, enabling early intervention and improved patient outcomes [6, 7].

Scientists have dedicated a decade to diligently minimizing such errors, resulting in various approaches for obtaining comprehensive individual respiratory data, crucial for further investigations. Researchers, including Egizio *et al.*, utilize regression to correct respiratory sinus arrhythmia's (RSA) impact, revealing a robust positive correlation between respiratory belt and pneumotachograph derived regression results [8–10]. Traditional multiple linear regression for respiratory belt calibration yields inaccurate

M. Majumder et al. (Eds.): ICCTE 2023, CCIS 2376, pp. 114–126, 2025.
https://doi.org/10.1007/978-3-031-81935-3_10

airflow predictions during changing breathing styles [11, 12]. Seppänen *et al.* introduce an improved, belt signal-based calibration method, outperforming standard regression and Liu *et al.*'s approach in subject specific and subject-independent assessments [7, 10]. They enhance respiratory belt calibration with an optimized FIR filter bank, significantly increasing accuracy, elevating R^2 values (piezo: 9%, inductive: 10%), and reducing RMSE (piezo: 36%, inductive: 43%) compared to the standard method, thereby advancing respiratory monitoring through regression analysis [13]. Sharma *et al.* introduces a comfortable, over-clothing wearable radio-frequency sensor system for noninvasive health monitoring. With high correlation to reference devices, the sensors accurately track heart rate, respiratory rate, volume, and unique breathing patterns, demonstrating potential for convenient and effective diagnostics [14]. Scholten, A.W.J. *et al.* evaluates a wireless, non-adhesive belt's non-inferiority in monitoring infant heart rate (HR) compared to standard electrocardiography (ECG). Results show promising HR monitoring performance, while exploratory analysis suggests moderate agreement for respiratory rate (RR) but challenges in detecting apnea/tachypnea [2] Anomaly detection of respiration, machine learning techniques like classification, regression, clustering, dimensionality reduction, deep learning, and ensemble methods are utilized for various purposes [11, 15, 16]. Swift evaluation of respiratory patterns is crucial in various emergency medical scenarios. This technology holds potential for various crucial medical scenarios, including identifying sleep apneas in home settings and overseeing respiratory events in mechanically ventilated patients within intensive care units. Here we have used the regression analysis method on the dataset collected from the public domain to analyze the breathing signal [17]. Regression analysis facilitates continuous monitoring of respiration, making it suitable for real-time data streams, and provides insights into how specific features impact respiration [18]. Additionally, it can integrate with alarms or alerts, enhancing healthcare monitoring systems and enabling timely intervention in respiratory-related conditions [19, 20].

Early detection of respiratory anomalies is crucial for health monitoring, and regression models offer the advantage of setting customized thresholds for triggering alerts or investigations. They provide quantitative assessment through metrics like Mean Squared Error (MSE) and R-squared (R^2), allowing for precise anomaly quantification [19]. Such as in the concerned dataset, we have used total four regression models aid in modeling normal respiratory behavior by establishing relationships between respiratory features and air in liters per minute [17, 20]. The significance of this regression analysis lies in its ability to detect anomalies by identifying deviations from the expected behavior. [20].

The article is structured in the following manner: Sect. 2 provides a brief introduction to the respiration belt, while Sect. 3 delineates the methodologies employed in the experiments, comprising three subsections. Section 4 elucidates the experimental setup and results. Section 5 delves into the discussion and conclusions. Lastly, in Sect. 6 potential future scopes are drawn.

2 Sensor

The respiration belt is a medical device worn around the chest or abdomen containing sensors to monitor respiratory movements during sleep, physical activities, and clinical settings, aiding in diagnosing disorders and treatment evaluation [14, 17]. The chest belt which is used for the data acquisition for the Kaggle Dataset [17] is used as the respiration monitoring sensor is pictorially depicted in Fig. 1.

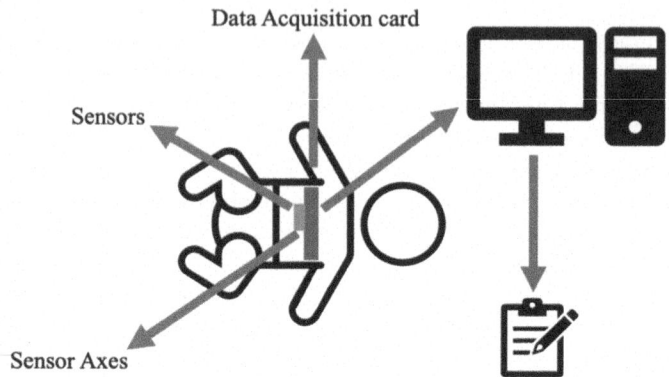

Fig. 1. Chest belt used for data acquisition

3   Methodology

In this research, the methodology consists of two core phases: regression analysis and statistical examination, aiming to uncover the intricate relationships within the acquired data, providing valuable insights into respiratory monitoring and analysis.

3.1   Data Acquisition

Researchers use a chest belt to monitor breathing signals and compare them with Douglas Bag (DB) data [17]. Key features extracted include average amplitude (a) between troughs and peaks and the fundamental period (p). Their main goal is to establish the function $f(x)$ that maps the feature pair (a, b) to DB-derived airflow within a time window, where a exhibits quadratic or cubic ties to ventilation, and p inversely correlates with airflow.

3.2   Regression Models

Linear Regression. Linear regression (LR) models relationships between variables using a straight line [11, 12]. In our dataset with two independent variables and one dependent variable, LR establishes a linear relationship:

$$Y = \beta_0 + \beta_1 x_1 + \beta_2 x_2 + \varepsilon \tag{1}$$

where, Y is the predicted value, x_1 and x_2 are the independent variables, and β_0, β_1, and β_2 are coefficients. LR aims to find the best-fitting line by minimizing the squared differences between predicted and actual Y values. During model training, the coefficients are determined, enabling predictions of Y based on x_1 and x_2 in our dataset [17, 21]

Support Vector Regression. Support vector regression (SVR) extends support vector machines to regression tasks. It's robust, capable of handling both linear and nonlinear relationships through kernels and focuses on minimizing error margins. This approach involves utilizing data points within this defined boundary for making predictions [22]. Here we have a dataset with two features (independent variables) denoted as X_1 and X_2, and the target variable (dependent variable) as Y. The SVR model is represented by the following equation

$$Y = f(X_1, X_2) = w.X + b \qquad (2)$$

where, Y is the predicted target variable, X_1, and X_2 are the feature variables. $(X_1 and X_2)$ are the feature variables. $X = [X_1, X_2]$ is the feature vector. $w = [w_1, w_2]$ represents the weight vector, which is learned by the SVR algorithm, b is the bias term or intercept, also learned by the SVR algorithm.

Polynomial Regression. Polynomial regression (PR) captures nonlinear relation-ships with polynomial equations. The polynomial regression's general expression takes the form,

$$f(x) = c_0 + c_1 x + c_2 x^2 + .. + c_n x^n \qquad (3)$$

where, n represents the degree of the polynomial, and c signifies a collection of coefficients. In polynomial regression, the goal is to discover an nth-degree polynomial function that provides the best possible approximation to our dataset [17, 22, 23].

Random Forest Regression. Random forest (RF) provides feature importance scores, and is effective for both linear and nonlinear relationships [16]. There are no such equations available for this model. Here is the general algorithm for the RF model [16, 19].

Training dataset $D = \{(x_1, y_1), (x_2, y_2), \ldots, (x_n, y_n)\}$ where number of trees T, maximum depth D and Output: RF model F. D contains n data points, and each data point is represented as a tuple (x_i, y_i), where x_i is the set of independent variables or features associated with the i-th data point, and y_i is the corresponding dependent variable or target value. So, x_i represents the feature vector for the i-th data point, and y_i is the corresponding target value [22, 23].

Initialize an empty list of decision trees, $F = []$. For $t \leftarrow 1$ to T we have to do the following. First, we must sample a bootstrap dataset [22, 23] D_t from D. Then train a decision tree f_t with a maximum depth of D on D_t. Then Add f_t to F. That will return finally to,

$$F(x) = \frac{1}{T} \sum_{t=1}^{T} f_t(x) \qquad (4)$$

In our RF model, the dataset, comprising 1432 samples, two independent variables ('peaks_diff' and 'period'), and one dependent variable ('oxigen_per_lit'), is dealt with [17, 20, 21]. The model is constructed following a formula that initializes an empty list of decision trees, $'F'$, , and then proceeds to have an ensemble of decision trees created. For each tree (from 1 to the specified number of trees, $'T'$ – set as 100 in our code), a bootstrap dataset ($'D'_t$) is sampled from the original training dataset ($'D'$) using random selection with replacement. On this bootstrap dataset, a decision tree ($'f'_t$) is trained with a specified maximum depth ($'D'$). Once trained, $'f'_t$ is added to the list of decision trees, $'F'$. . This process is repeated for all the decision trees in the RF. The final model ($'F'$) is comprised of this ensemble of decision trees, which collectively work to provide robust predictions for 'oxigen_per_lit' based on the 'peaks_diff' and 'period' features, reducing overfitting and improving predictive accuracy [22, 23].

The training decision trees which are involved in splitting dataset based on a specific criterion follows this equation,

$$Split(D, F, criteria) = \arg \min_{f \in F} \sum_{D_v} \frac{|D|}{|D_v|} criteria(D_v) \tag{5}$$

RF is an ensemble technique, consist of multiple individual decision trees created during training. To arrive at the final prediction, the results from all these trees are combined—classification predictions involve selecting the mode of the classes, while regression predictions entail computing the mean prediction (Fig. 2).

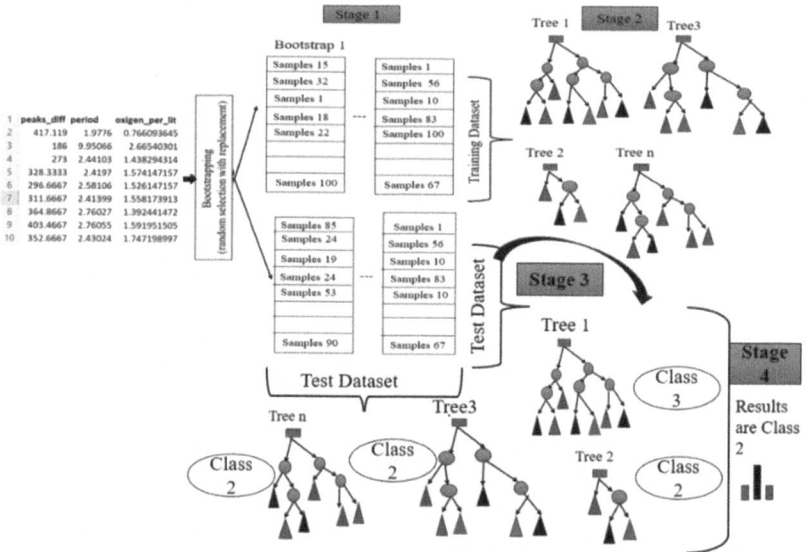

Fig. 2. Block diagram of the working function of RF regression model.

This ensemble approach reduces the risk of overfitting and improves the model's ability to generalize to unseen data. In the following equation,

$$Y_{pred} = \frac{1}{N} \sum_{i=1}^{N} Y_{tree_i} \tag{6}$$

where, Y_{pred} represents the final predicted target value. N is the number of decision trees in the forest. $Y_{tree\,i}$ is the prediction made by the ith decision tree. The equation for R^2 is:

$$R^2 = 1 - \frac{\sum_{i=1}^{n}(Y_i - \widehat{Y}_i)^2}{\sum_{i=1}^{n}(Y_i - \overline{Y})^2} \tag{7}$$

The equation for MSE is,

$$MSE = \frac{1}{n}\sum_{i=1}^{n}(Y_i - \widehat{Y}_i)^2 \tag{8}$$

where, n is the number of data points in the validation or test set. Y_i is the actual target value for the i-th data point. $\widehat{Y}$ the predicted target value for the i-th data point. $\overline{Y}$ is the mean of the actual target values.

3.3 Hyperparameter Tuning

This paper's methodology emphasizes hyperparameter tuning for enhancing machine learning models. It involves identifying critical hyperparameters, defining search spaces, and systematically exploring options through grid and randomized search methods [24, 25]. Data partitioning and performance metrics like MSE or R^2 guide rigorous model assessment. Iterative training and validation lead to the optimal hyperparameter set, crucial for avoiding underfitting and overfitting. Contemporary automated tools and cloud platforms simplify this vital process, ensuring peak model performance. Hyperparameter tuning, often denoted as (θ) can be mathematically expressed as the process of finding the optimal set of hyperparameters (θ^*), for a machine learning model (M) by minimizing or maximizing a performance metric (P).

$$\theta^* = \arg\max_{\theta \in \Theta} P(\theta) \tag{9}$$

3.4 Parameters Used for Statistical Analysis

Statistical analysis, in our context, is a robust methodology for exploring and interpreting data related to respiration and anomaly detection. It encompasses techniques like regression analysis, hyperparameter tuning, and feature importance assessment to build accurate predictive models. This analytical approach empowers data-driven decision-making and enhances our understanding of complex respiratory patterns.

Cross Validation. In the process of evaluating our machine learning model, we employ the robust technique of cross-validation in conjunction with Root Mean Squared Error (RMSE) [15, 16]. This approach serves as a rigorous stress test to thoroughly assess the model's capabilities. Cross-validation with RMSE is vital for gauging the model's performance on unseen data. Our dataset is partitioned into multiple subsets or 'folds', typically utilizing a 5-fold cross-validation strategy [22, 23]. During each iteration, the model is trained on a combination of these folds and subsequently validated on the reserved one. RMSE takes center stage as the key metric for evaluating the consistency between the model's predictions and actual target values. Unlike Mean Squared Error (MSE), RMSE offers a more interpretable measure. The RMSE equation,

$$RMSE = \sqrt{\frac{1}{n}\sum_{i=1}^{n}(actual_i - Predicted_i)^2} \qquad (10)$$

guides our assessment, where 'n' represents the data points, 'Σ' signifies summation over each data point, '$actual_i$' denotes the true target values, and '$predicted_i$' reflects the model's predictions for the i-th instance. Lower cross-validated RMSE values indicate that, on average, the model's predictions closely align with reality. This meticulous evaluation process is pivotal for selecting the best-performing model, a critical step in practical machine learning deployments.

Residual Analysis. In the residual analysis phase, we meticulously assess the RF regression model's performance by scrutinizing its residuals. These residuals, calculated as the differences between actual values (Y_{test}) and predicted values (Y_{test_pred}) for each data point,

$$\varepsilon_i = Y_i - Y_{test-Pred_i} \qquad (11)$$

offer critical insights. We have visualized their distribution using a histogram with a kernel density estimate (KDE) [16], allowing us to evaluate whether they conform to a normal distribution centered around zero, a pivotal assumption in regression models. Additionally, we examine homoscedasticity [10], a key assumption for regression model reliability, by plotting residuals against predicted values (Y_{test_pred}) for the i-th instance. This analysis not only ensures the model adherece to underlying assumptions but also provides crucial insights into its predictive performance and potential areas for improvement.

4   Experiments and Results

The results obtained from implementing multiple regression models on the dataset provide valuable insights into each model's performance in predicting the dependent variable, "air in liters per minute". Here is a concise summary of the results from each implemented regression model. Moving on we are going to see how we have achieved figuring out the 'best' model via regression analysis, hyperparameter tuning as well as several statistical analysis techniques. The following experiments are organized into three parts. Summarizing our findings and interpretations, we gain insights into the impact of specific features, the model's predictive power, and any statistically significant relationships we've discovered in this paper.

4.1   Model's Performance Using MSE and R^2

Superiority in model's performance is indicated by lowest MSE values (less prediction error) and the highest R^2 (a better fit to the data). Table 1 shows that RF regression

Table 1. Performance analysis of different regression models

Regression model	MSE		R^2	
	Validation Data	Test Data	Validation Data	Test Data
Linear Regression	2.26	2.14	0.37	0.39
Support Vector Regression	2.48	1.95	0.30	0.44
Polynomial Regression	1.75	1.87	0.51	0.46
Random Forest Regression	1.39	1.65	0.61	0.53

consistently achieves the highest R^2 values and the lowest MSE values for both the validation and test data.

LR model yields the lowest R^2 values and the highest MSE values for the test data, whereas SVR exhibits the lowest R^2 value and the highest MSE value for the validation dataset. These results indicate that RF regression excels in capturing the dataset's underlying patterns, making it a promising model for predicting oxygen levels (Fig. 3).

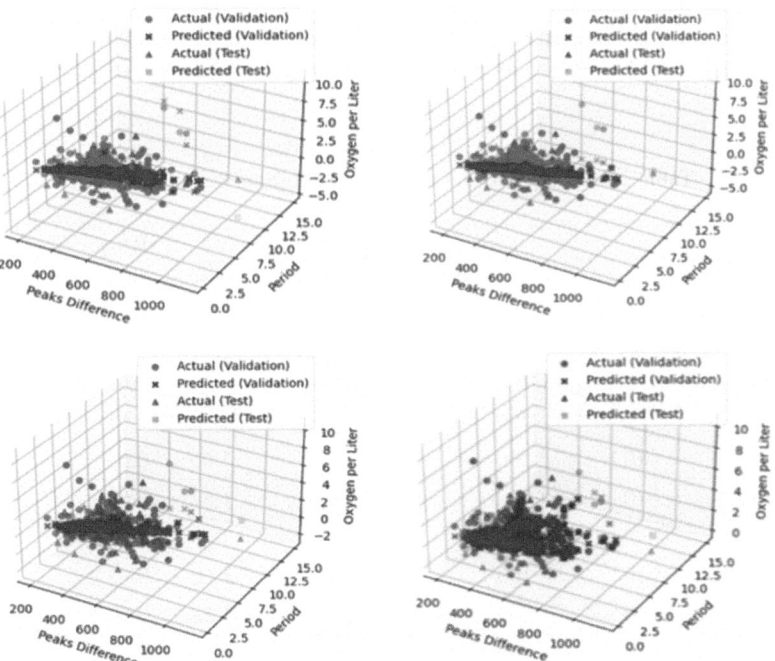

Fig. 3. 3D scatter plot for data visualization: LR (Top Left), SVR (Top Right), PR (Bottom Left), RF (Bottom Right).

4.2 Hyperparameter Tuning

Table 2 shows 10 random instances of the hyperparameter tuning for a RF Regression model on our dataset that explore different combinations of hyperparameters, including the number of estimators, maximum depth of the trees, minimum samples required to split an internal node, and minimum samples required to be a leaf node [22, 23].

The Table highlights that the hyperparameter configuration from Instance 63, consisting of {'max_depth': 20, 'min_samples_leaf': 1, 'min_samples_split': 10, 'n_estimators': 100}, provides the best overall performance by minimizing Mean Squared Error (MSE) on both the validation and test datasets. This configuration achieved a mean MSE of 1.16 on the validation data and 1.58 on the test data. Its remarkable balance between the two datasets showcases its robustness and ability to generalize effectively to unseen data, making it a strong choice for further applications and analyses [22, 23].

4.3 Statistical Techniques

Statistical techniques can help to interpret the results of our regression model and understand the relationships between variables. Here's a workflow that includes statistical techniques to analyse RF regression model.

Cross-Validation for Robustness. We have performed a cross-validation study on our hyper tuned RF model to ensure its robustness and its performance across different folds of data.

Table 2. Hyperparameter tuning of parameters for RF model

Hyperparameters				MSE		R^2	
Max depth	Min. Sample leaf	Min. Sample split	n estimators	Validation data	Test data	Validation data	Test data
10	1	2	100	1.18	1.61	0.62	0.54
10	1	2	200	1.18	1.59	0.61	0.54
20	1	10	200	1.15	1.60	0.62	0.54
20	1	10	300	1.16	1.57	0.63	0.55
10	1	5	300	1.67	1.58	0.63	0.55
10	1	10	100	1.15	1.60	0.63	0.54
30	4	10	200	1.16	1.58	0.63	0.54
30	4	10	300	1.17	1.56	0. 63	0.55
30	2	5	100	1.18	1.62	0.62	0.54
30	2	10	300	1.16	1.59	0.63	0.55

Table 3. Cross-validation study

Regression Model	Root Mean Square Error
Random Forest	1.07
Polynomial	1.80
SVR	2.00
Linear	2.05

From the following Table 3, we can see that our predicted regression gives RMSE value of approximately 1.068 which indicates a better fit of model to the data. The comparison study also indicates the RMSE of hyper tuned RF regression is significantly lower than alternative models or methods, so our model is performing reasonably well in making predictions.

Residual Analysis. Calculating the residuals, the results give visualization of their Residuals and check for the homoscedasticity (constant variance) for the RF model's predictions on the test data. In such case, we consider two different plots: one of them is Residual Distribution plot as depicted in Fig. 4, which shows the relationship between the predicted values and residuals exhibiting any pattern as the predicted values change. Here it shows the residuals approximately normally distributed with a mean of zero.

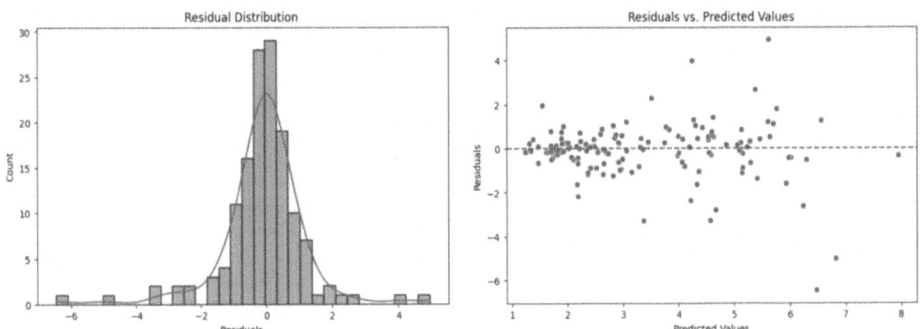

Fig. 4. Residual analysis for proposed model using RF regression

The plot of residual vs. independent variables it creates scatter plot and there is no systematic pattern observed (see Fig. 4). The presence of a red dashed line at $y = 0$ signifies the ideal scenario. Showing a clear pattern or trend indicates heteroscedasticity (non-constant variance) which is not present in our system. Such results confirm the reliability of the proposed regression model.

Statistical Tests. The provided tabular result presents the outcomes of several statistical tests and performance metrics for comparing the RF model against three other regression models: LR, SVR, and PR.

Table 4. Performance Metrics for Different Regression Models.

Model	Mean MSE	Mean MAE	Mean R-squared	Mean F-statistics
Random Forest	1.26	0.73	0.65	757.67
Linear Regression	2.30	1.14	0.36	207.31
Support Vector Regression	2.60	1.09	0.28	143.50
Polynomial Regression	1.71	0.90	0.53	425.06

In comparing the RF model with LR, SVR, and PR. T-tests (in Table 4) highlight the RF's superior performance. It exhibits significantly lower Mean Squared Error (MSE) and Mean Absolute Error (MAE) with p-values under 0.05, indicating its stronger predictive accuracy. Moreover, RF achieves significantly higher R-squared values, signifying its enhanced capability to explain variance. The F-Statistic tests reaffirm the RF's superiority in explaining variance, showing statistically significant differences when compared to the other models.

Table 5. Statistical Comparison of RF with Other Regression Models

Competing algorithm	Control Algorithm = Random Forest							
	T-statistics				p-Value			
	MSE	MAE	R-squared	F-statistics	MSE	MAE	R-squared	F-statistics
Linear Regression	−5.65	−8.55	7.72	3.76	0.0048	0.0010	0.0015	0.0198
Support Vector Regression	−6.51	−6.55	8.87	4.12	0.0029	0.0028	0.0009	0.0145
Polynomial Regression	−2.70	−3.95	3.60	2.81	0.0536	0.0168	0.0226	0.0483

The Table 5 displays mean performance metrics for RF, LR, SVR, and PR models. RF shows the lowest Mean Squared Error (MSE) and Mean Absolute Error (MAE), indicating superior predictive accuracy. It also has the highest R-squared and F-statistic scores, showcasing its excellence in explaining dataset variance.

5   Discussion and Conclusion

This study encompassed a thorough exploration of regression techniques and statistical methodologies applied to a breathing dataset, with the goal of constructing a resilient predictive model. The goal is to find the best regression model for predicting oxygen per Liter levels. Our findings highlight the methodology's value in enhancing oxygen level predictions, even in real-world scenarios with potential data quality issues.

The intricate interplay between anomaly detection and the development of predictive machine learning models in respiratory events is a powerful partnership. In healthcare and environmental monitoring, these two elements not only complement but also reinforce each other. Anomaly detection, facilitated by a respiration belt, enhances the dataset's reliability by identifying irregular breathing patterns, leading to timely healthcare notifications. This feedback loop refines the predictive model and ensures its resilience. In return, the model benefits from clean data and adapts to changing patterns, offering promising solutions for respiratory monitoring, healthcare, and environmental analysis.

6 Future Scope

Our forthcoming efforts will be dedicated to unravelling the intricacies of respiratory data, working tirelessly to enhance our capacity for anomaly detection, and ultimately striving for a more comprehensive and accurate approach to managing respiratory events. This path represents a promising avenue for future research endeavours.

References

1. Liu, H., Allen, J., Zheng, D., Chen, F.: Recent development of respiratory rate measurement technologies (2019)
2. Scholten, A.W.J., et al.: Cardiorespiratory monitoring with a wireless and nonadhesive belt measuring diaphragm activity in preterm and term infants: a multicenter non-inferiority study. Pediatr. Pulmonol. (2023)
3. Scholten, A.W.J., et al.: Multicentre paired non-inferiority study of the cardiorespiratory monitoring performance of the wireless and non-adhesive Bambi® belt measuring diaphragm activity in neonates: study protocol. BMJ Paediatr. Open 6 (2022)
4. Deshpande, G., Schuller, B.W.: Application of speech-derived breathing patterns in detecting respiratory disorders (2023)
5. McClure, K., Erdreich, B., Bates, J.H.T., McGinnis, R.S., Masquelin, A., Wshah, S.: Classification and detection of breathing patterns with wearable sensors and deep learning. Sensors 20, 1–13 (2020)
6. Kondo, T., Minocchieri, S., Baldwin, D.N., Nelle, M., Frey, U.: Noninvasive monitoring of chest wall movement in infants using laser. Pediatr. Pulmonol. 41, 985–992 (2006)
7. Liu, S., Gao, R., He, Q., Staudenmayer, J., Freedson, P.: Improved regression models for ventilation estimation based on chest and abdomen movements. Physiol. Meas. 33, 79–93 (2012)
8. Piuzzi, E., Pisa, S., Pittella, E., Podestà, L., Sangiovanni, S.: Wearable belt with built-in textile electrodes for cardio—respiratory monitoring. Sensors 20, 1–15 (2020)
9. Tobin, M.J.: Breathing pattern analysis. Intensive Care Med. 18, 193–201 (1992)
10. Loveridge, B., West, P., Anthonisen, N.R., Kryger, M.H.: Single-position calibration of the respiratory inductance plethysmograph. J. Appl. Physiol. Respir. Environ. Exerc. Physiol. 55, 1031–1034 (1983)
11. Zhang, S., Yu, X., Luiz, S., Junior, N., Chen, J.J.: Generating dynamic carbon-dioxide traces from respiration-belt recordings: feasibility using neural networks and application in functional magnetic resonance imaging (2023)
12. Clarke, M.E.: Garment-based respiration and pulse oximetry sensing using a stitched sensor and chest mounted pulse oximetry sensor. A Thesis Submitted to The Faculty of The Univeristy of Minnesota (2023)

13. Poole, K.A., Thompson, J.R., Hallinan, H.M., Beardsmore, C.S.: Respiratory inductance plethysmography in healthy infants: a comparison of three calibration methods. Eur. Respir. J. **16**, 1084–1090 (2000)
14. Sharma, P., Hui, X., Zhou, J., Conroy, T.B., Kan, E.C.: Wearable radio-frequency sensing of respiratory rate, respiratory volume, and heart rate. NPJ Digit. Med. **3**, 1–10 (2020)
15. Pegoraro, J.A., Lavault, S., Wattiez, N., Similowski, T., Gonzalez-Bermejo, J., Birmelé, E.: Machine-learning based feature selection for a non-invasive breathing change detection. BioData Min. **14** (2021)
16. Islam, M.Z., Martin, B., Gotcher, C., Martinez, T., O'Hara, J.F., Ekin, S.: Non-contact respiratory anomaly detection using infrared light-wave sensing (2023)
17. Breathing Data from a Chest Belt. https://www.kaggle.com/datasets/sagarsen/breathing-data-from-a-chest-belt. Accessed 16 Nov 2023
18. Merritt, C.R., Nagle, H.T., Grant, E.: Textile-based capacitive sensors for respiration monitoring. IEEE Sens. J. **9**, 71–78 (2009)
19. Alita, D., Putra, A.D., Darwis, D.: Analysis of classic assumption test and multiple linear regression coefficient test for employee structural office recommendation. IJCCS (Indonesian J. Comput. Cybern. Syst.) **15**, 295 (2021)
20. Vanegas, E., Igual, R., Plaza, I.: The effect of measurement trends in belt breathing sensors. Eng. Proc. **6**, 4–8 (2021)
21. Leicht, L., Vetter, P., Leonhardt, S., Teichmann, D.: The PhysioBelt: a safety belt integrated sensor system for heart activity and respiration. In: 2017 IEEE International Conference on Vehicular Electronics and Safety, ICVES 2017, pp. 191–195 (2017)
22. IEEE Engineering in Medicine and Biology Society, Cerutti, S., Patton, J., Annual international conference of the IEEE Engineering in Medicine and Biology Society 37 2015.08.25–29 Milano, EMBC 37 2015.08.25–29 Milano: 2015 37th Annual International Conference of the IEEE Engineering in Medicine and Biology Society (EMBC) 25–29 August 2015, Milano/Conference chairs Sergio Cerutti, Jim Patton, ed. in chief for conference editorial board
23. Rehman, M., et al.: Improving machine learning classification accuracy for breathing abnormalities by enhancing dataset. Sensors **21** (2021)
24. Egizio, V.B., Eddy, M., Robinson, M., Jennings, J.R.: Efficient and cost-effective estimation of the influence of respiratory variables on respiratory sinus arrhythmia. Psychophysiology **48**, 488–494 (2011)
25. Bar-Yishay, E., Putilov, A., Einav, S.: Automated, real-time calibration of the respiratory inductance plethysmograph and its application in newborn infants. Physiol. Meas. **24**, 149–163 (2003)

Federated Learning to Speed Up Pre-processing of Large Data Sets

Lhamu Sherpa and Nandan Banerji[✉]

Department of Computer Science and Engineering, Sikkim Manipal Institute of Technology, Sikkim Manipal University, Gangtok, Sikkim, India
nandannitdgp@gmail.com

Abstract. Machine learning algorithms are data-generated algorithms, if a modest amount of historical data is available, the predictions or recommendations can be inferred easily. The learning ecosystem involves pre-processing with the standard encoding of categorical data, feature scaling, null value treatment, and a few more. The task of pre-processing is very crucial as no data is clean, but most are useful. For small data sets, the time taken for pre-processing is within some considerable limit; but for large data sets, it may significantly grow and sometimes take longer than the expected limit. With respect to the single server model, if the federation is formed among a number of available systems and the available data is distributed, the pre-processing task is likely to be accelerated. It can be achieved through parallel execution by sharing the total data among the federation peers in a judicious manner. In this work, the proposed method adopts federated learning to accelerate the pre-processing time on large data sets. The results are compared with the classical machine learning approach; which reveals that the federated learning approach outperforms the classical counterpart.

Keywords: Supervised Machine Learning · Large Data sets · Federated Learning · Preprocessing

1 Introduction

With the accelerated growth of Artificial Intelligence (AI) and its necessary libraries/modules of learning algorithms, the application diversity increases [1,2]. Learning algorithms are rapidly being used in various applications, including emergencies and daily life. Real-time data from various devices helps predict a better tomorrow. Systems train this data to produce better results, with accuracy ranging from precision to accuracy. Large data, such as big data, is generated using classical machine learning approaches and a central learning server [3,4]. Experts gather and train this data to improve accuracy and precision. However, essentially, before learning, the need to be pre-processed by several internal stages as shown in Fig. 1 below.

L. Sherpa and N. Banerji—All authors contributed equally to this work.

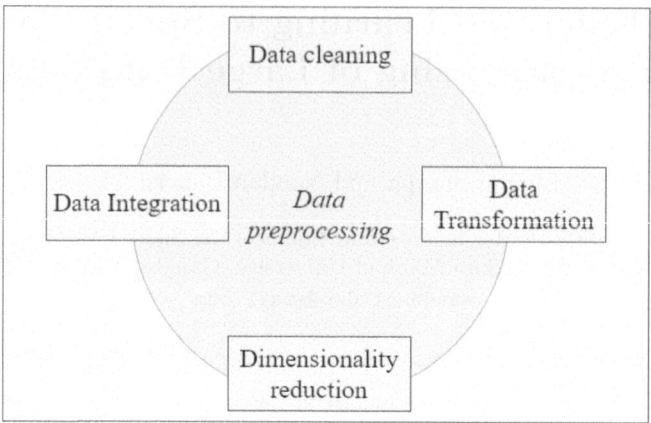

Fig. 1. Data Pre-processing stages.

The stages involve data cleaning (to correct inaccurate, poorly formatted, or otherwise messy data), data integration (it is useful when the data is gathered from multiple distributed sources), data transformation (treatment of categorical data or scale of the data in relative measure) and finally data reduction (dimension reduction if necessary).

The classical single-server learning model faces challenges in pre-processing large data sets from distributed sources. While small data sets can be managed by a single learning server, large datasets from distributed sources can hinder further learning stages. Data security and sharing through public channels also pose challenges. Major issues during pre-processing in this model include computation and memory demands. The authors of paper [5] have shown a convergence comparison between federated and classical machine learning.

1. Unavailability of better capacity servers.
2. Predefined deadline-based real-time applications.
3. Privacy-preserving for data on local sites.

With the recent advent of the Federated Learning (FL) approach, the above problems can be addressed and resolved. FLs are machine learning settings in which multiple client devices (e.g. smartphones, handheld devices) collaboratively form federations/teams under a central server's orchestration. In [6], the researchers suggested some approaches to build/form a federation. The total available data is shared among the federation members in a judicious manner. The pre-processing and other necessary tasks are done on each peer node of the federation with parallel execution; which is likely to save some time for considerably large data sets. Different approaches in Federated Learning are Data-Parallel and Model Parallel.

2   Proposed Methodology

The proposed method identifies a federation of moderately capable devices with minimal memory, processing, and resources for communication via wired or wireless means as shown in Fig. 2. It assumes the federation is built over fixed infrastructure or an ad-hoc scenario like the Internet of Things (IoT) with scalable or temporally available devices. The proposed system analyzes pre-processing stages and their required times in various variations. Data is shared among federation peers using various strategies. Convergence time is calculated for a single server-based learning model and the proposed federated model. Performance is compared to scaling and encoding data for training and testing stages [4,7].

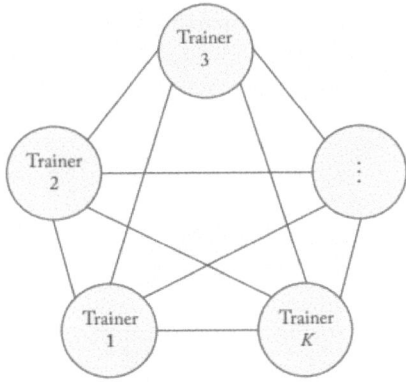

Fig. 2. A P2P architecture for federated learning infrastructure.

The proposed system analyzes pre-processing stages and their required times, sharing data among federation peers. It calculates convergence time for server-based and federated models, comparing performance in scaling and encoding data for training and testing stages [8,9].

- Equi-partition: It is the most Straight-Forward and democratic approach where the data is divided into equal-sized blocks, and each block is assigned to a certain federation member as shown in Fig.3. By ensuring that each device receives an equal amount of data. This strategy can help the machine learning model become less biased and more federal.
- Random approach: This approach randomly distributes data across the federation as shown in Fig. 4. Since it guarantees that each device receives a different collection of data, this method might be helpful for data sets with variable degrees of complexity or heterogeneity. However, it might result in partitions that are quite imbalanced; which might cause under-burden or over-burden the members. The node having less capability might get excess data than its capacity; also vice versa.

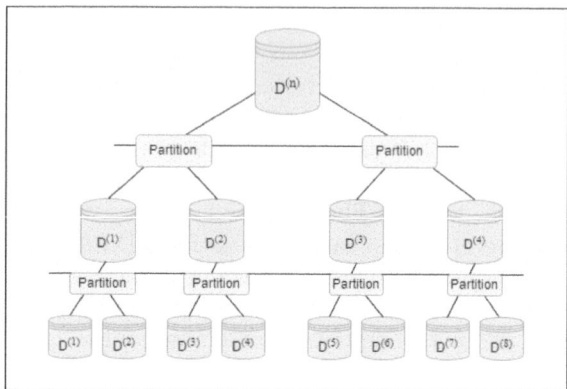

Fig. 3. Equi-partition Model

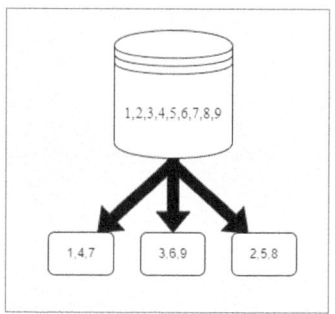

Fig. 4. Random Partition approach

- Ordered/Round-Robin approach: In ordered/round-robin data is divided sequentially among federation members. This method can be effective for data sets with a uniform distribution and guarantees that each device receives an equal quantity of data in a homogeneous environment.
- In the utility-based approach: The utility of a node or federation member is determined by its remaining memory, processing capability, and available bandwidth as shown in Fig. 5. Sometimes for wireless remote devices, the remaining energy is also considered as another parameter. The utility is calculated as a weighted value with the following equation.

$$\bar{R}_i = w_1.\bar{R}_{c_i} + w_2.\bar{R}_{m_i} + w_3.\bar{R}_{b_i}, \tag{1}$$

where, $\bar{R}_{c_i}$, $\bar{R}_{m_i}$ and $\bar{R}_{b_i}$ are respectively the normalized values of remaining processing power, memory and battery life of the node n_i and w_1, w_2 and w_3 are the weights that correspond to the resource components. The weight assignment must satisfy the condition

$$w_1 + w_2 + w_3 = 1 \tag{2}$$

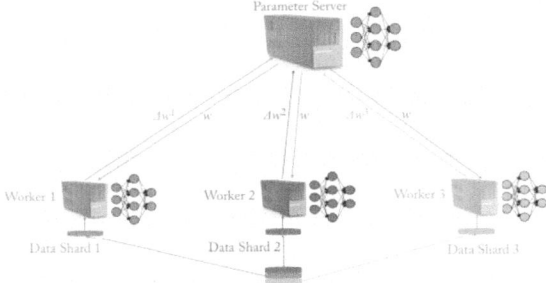

Fig. 5. Utility-based approach.

2.1 Solution Strategy

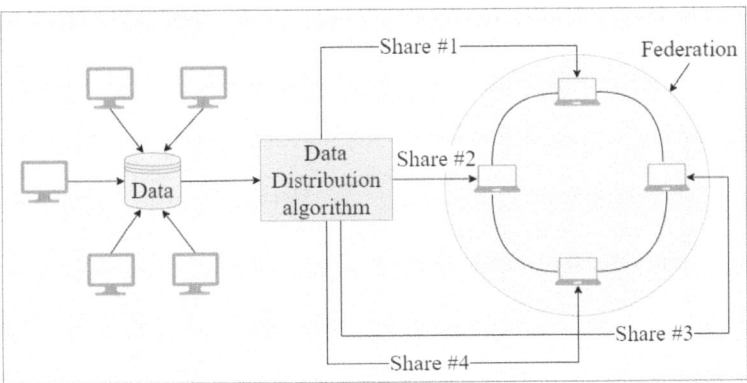

Fig. 6. Proposed System Model

The proposed work concentrated on data pre-processing stages by leveraging the Federated Learning approach in comparison to the classical Machine Learning approach. In Fig. 6, we can see that the data is collected from heterogeneous sources and then it is passed through a data distribution/sharing algorithm, to distribute the data among the federation members. Here in the design for simplicity, we approached an equipartition distribution, where each federation member gets equal shares of data. But it can also be set up with any data-sharing approaches mentioned above. Distributed approaches prioritize data distribution and privacy concerns, ensuring nodes don't reveal their data without compromising training or pre-processing loss. To measure that, we chose two major performance metrics as discussed below.

2.2   Performance Metrics

- Data pre-processing cost is a variable cost affecting the size of information in data sets [4, 10]. Treatments are used to refine and prepare data for supervised learning algorithms, affecting the variable data size. The proposed approach is likely to be more effective in determining data pre-processing costs.
- Applying Federated Learning, the proposed work tests data pre-processing with decentralization modes and compares it with a centralized approach using Federated Learning [2].

3   Results and Discussion

The results were obtained from a test environment with 4 standard personal computers with 8 GB primary memory connected via Ethernet. Each system had 4 concurrent threads and was implemented using Python libraries like numpy, pandas, sklearn, and dask. Table 1 shows the data set description. Pre-processing stages for analysis involved classical single server-based learning and the proposed federated learning approach, with feature scaling and encoding using standard averaging methods for null value treatment.

Table 1. Dataset description

Dataset Source	dataset Size	row	cols
https://www.kaggle.com/datasets/ wenruliu/adult-income-dataset	3754 MB	3256K	15

The observations were made from pre-processing stages being executed with classical single server-based learning and the proposed federated learning approach. The major pre-processing steps involved in analysis are feature scaling and encoding with null value treatment by standard averaging method (though there are few more methods available, it can be adopted with any such).

Figure 7 shows the preprocessing time (especially encoding and scaling) of the dataset within a single server approach as usually done in classical ML scenarios. The test is conducted with 40 epochs and the results are shown. It is found that each epoch/trial takes an average time for preprocessing.

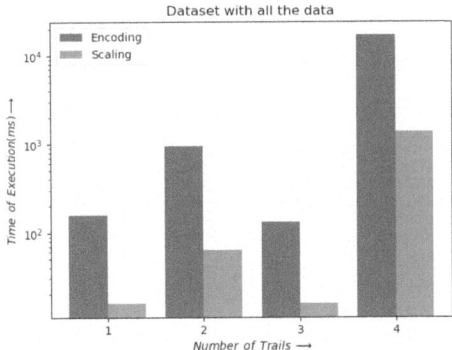

Fig. 7. All data in one place

In Fig. 8, the results are shown in two consecutive graphs for a federation of size 2, where the total data is segregated into two equal shares, and the same preprocessing task is conducted for both shares concurrently in different systems acting as the members of the federation. The obtained result shows that the time required for each share is comparatively minimal.

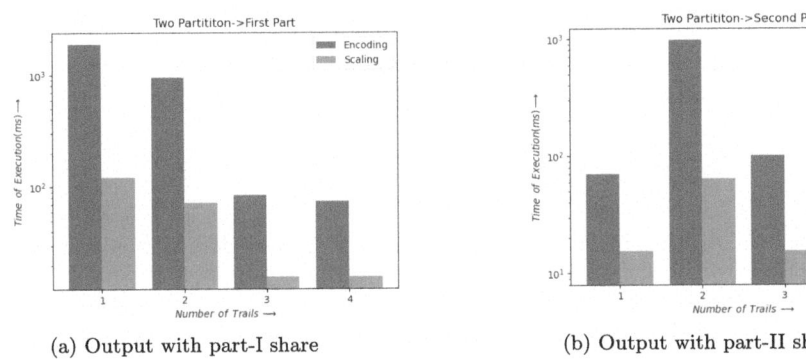

(a) Output with part-I share   (b) Output with part-II share

Fig. 8. Total data is shared with two members of the federation, each having equal half shares.

Figure 9 shows that increased federation size effectively reduces the preprocessing time for a single server model in ML learning, potentially extending to overall learning procedures.

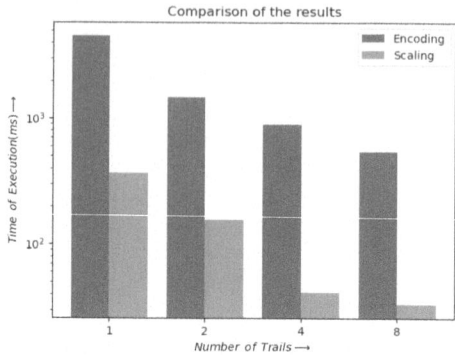

Fig. 9. Comparison with one, two, four, and eight member federations outputs.

4 Conclusion

The work suggests federated learning for pre-processing large datasets, distributing the workload across multiple systems or devices, reducing the burden, and improving performance. It explores methods like utility-based, round-robin, random, and linear methods, and highlights advantages like incorporating moderately capable devices and efficient resource utilization.The work proposes a federated learning approach for pre-processing large datasets using the Python Dask framework. This strategy enhances performance and shortens the machine-learning process by allocating tasks to locally accessible computers.

References

1. Verbraeken, J., Wolting, M., Katzy, J., Kloppenburg, J., Verbelen, T., Rellermeyer, J.S.: A survey on distributed machine learning. CoRR abs/1912.09789 (2019). https://arxiv.org/abs/1912.09789
2. Zhang, D., Tsai, J.: Machine learning and software engineering. Softw. Qual. Control **11**, 87–119 (2003). https://doi.org/10.1023/A:1023760326768
3. Hu, K., et al.: Federated learning: a distributed shared machine learning method. Complexity **2021**, 8261663 (2021). https://doi.org/10.1155/2021/8261663
4. Asad, M., Moustafa, A., Ito, T.: Federated learning versus classical machine learning: a convergence comparison. CoRR abs/2107.10976 (2021). https://arxiv.org/abs/2107.10976
5. Yang, Q., Liu, Y., Chen, T., Tong, Y.: Federated machine learning: concept and applications. CoRR abs/1902.04885 (2019). https://arxiv.org/abs/1902.04885
6. Kairouz, P., et al.: Advances and open problems in federated learning. CoRR abs/1912.04977 (2019). https://arxiv.org/abs/1912.04977
7. Banerji, N.: Distributed machine learning–an intuitive approach. In: Bhattacharyya, D., Saha, S.K., Fournier-Viger, P. (eds.) Machine Intelligence and Soft Computing. AISC, pp. 9–15. Springer, Singapore (2022). https://doi.org/10.1007/978-981-16-8364-0_2

8. Li, T., Sahu, A.K., Talwalkar, A., Smith, V.: Federated learning: challenges, methods, and future directions. CoRR abs/1908.07873 (2019). https://arxiv.org/abs/1908.07873

9. Rahman, M.S., Rivera, E., Khomh, F., Guéhéneuc, Y., Lehnert, B.: Machine learning software engineering in practice: an industrial case study. CoRR abs/1906.07154 (2019). https://arxiv.org/abs/1906.07154

10. Lorenzoni, G., Alencar, P., Nascimento, N.M., Cowan, D.D.: Machine learning model development from a software engineering perspective: a systematic literature review. CoRR abs/2102.07574 (2021). https://arxiv.org/abs/2102.07574

Flood Susceptibility Zonation Using Geospatial Frequency Ratio and Artificial Neural Network Techniques within Himalayan Terai Region: A Comparative Exploration

Deepanjan Sen[1](✉) ⓘ, Swarup Das[2], Sumon Dey[3] ⓘ, and Arindam Sarkar[4] ⓘ

[1] Department of Computer Applications, Dr. B. C. Roy Academy of Professional Courses, Durgapur, India
sen.deepanjan92@gmail.com
[2] Department of Computer Science and Technology, University of North Bengal, Darjeeling, India
[3] Department of Computer Science and Engineering, Akal College of Engineering and Technology, Eternal University, Rajgarh, Himachal Pradesh, India
[4] Department of Computer Science and Electronics, Ramakrishna Mission Vidyamandira, Howrah, India

Abstract. Flooding is a widespread natural disaster affecting environment, economy, infrastructure. This research involves implementing the Frequency Ratio (FR) and GIS-based Multilayer Feed Forward Artificial Neural Network (GMF-FANN) models on the lower bank of the Teesta River in the Himalayan foothills Terai Region of West Bengal. Data from historical flood reports, databases, satellite imagery, and field surveys have been adopted to develop training and testing datasets for flood susceptibility based on 10 flood affecting factors. The GMF-FANN model, trained using Conjugate Gradient Decent (CGD) algorithm, outperforms the FR model in Flood Susceptibility Zonation (FSZ) scenario with 91.5% and 79% accuracy assessment by Receiver Operating Characteristics (ROC) curve in both training and testing sets. The crucial findings of this study will undoubtedly aid local officials in developing appropriate long-term management plans to reduce future losses.

Keywords: Flood Inventory Map · Flood Susceptibility Zonation · Frequency Ratio · Artificial Neural Network · Akaike Information Criterion · Receiver Operating Characteristics

1 Introduction

Natural catastrophes such as landslides, wildfires, tsunamis, and floods inflict substantial financial and human losses globally. Floods have increased in frequency over the past few decades as a result of environmental deterioration, global warming, rising populations, and improper land utilization [1]. The Indian subcontinent, with 85% vulnerability, is one

M. Majumder et al. (Eds.): ICCTE 2023, CCIS 2376, pp. 136–148, 2025.
https://doi.org/10.1007/978-3-031-81935-3_12

of the most disaster-prone regions globally [2]. CRED's International disaster reports show that 38.5 million human lives have been lost in natural disasters globally between 1900–2023 within a span of 123 years, with 7 million in a single event type i.e., flood [3]. Flood has affected the Indian plain in every aspect - destruction of crops, houses, public utilities, human and cattle losses. A summarized statistical report based on flood damages of the Central Water Commission, Ministry of Jal Shakti, Govt. of India from 1953–2020 depicted in Table 1, provides the better imagery of the nation as well as the state where our proposed study area belongs to [4]. In the entire nation, river flooding is responsible for around 60% of the disasters, whereas severe rain and cyclones for the remaining 40%.

Table 1. Flood damages of India & West Bengal (1953–2020).

Sl. No	Location	Total Area Affected (m.ha.)	Population Affected (Millions)	Total Damages (Crores)	Total Human Lives Lost (Nos)
1	India	492.557	2198.788	437149.710	113943
2	West Bengal	50.903	254.568	63006.572	11035

Flooding has been a serious issue in rivers in northern West Bengal due to a variety of geo-environmental factors. A flood report on North Bengal of the colonial phase by P.C. Mahalanobish in the year 1927 clearly shows how strongly the Himalayan foreland terai region has been affected for the last 153 years (1870–2023) [5]. Research focuses on accurate flood-prone area identification and Flood Susceptibility Mapping (FSM) for flood prevention and management.

Advances in Remote Sensing (RS) data and GIS techniques have resulted in five key methodologies for creating maps that are susceptible to floods: statistical, physical, heuristics, machine learning as well as deep learning. These approaches are widely accepted for accurate flood analysis [6–8]. This study comprises the comparative exploration of integrated RS, GIS based Frequency Ratio (FR) and multilayer feed forward ANN models in the study area, focusing on accurate flood analysis.

2 Study Area

Most significant rivers in India, including the Ganga, Brahmaputra, Jhelum, Sutlej, and Teesta, have their originate from Himalayan glaciers, supporting over 1.3 billion people in Asia [9]. The trans-boundary river Teesta, a prominent right-bank channel of the river Brahmaputra, emerges from TsoLhamo Lake in the northern Sikkim Himalayas at an elevation of 6200 m. Cho-Lhamo Lake, Pahunri Glacier, and Khangse Glacier are thought to be the sources of the Teesta River [10].

The Teesta River, formed by the Lachen and Lachung river unites at Chungthang, has undergone several stream alterations since the early sixteenth century [11]. It initially flowed into the Ganga River until the 18th century, after the devastating flood of 1787, the canal migrated eastward and joined the Brahmaputra River [12]. The river conflu-ences with the river Rangit near Sikkim's Teesta Bazar before entering the West Bengal plain near Sevoke, where the Coronation Bridge unites India's the northeast territories [13]. The river passes through Jalpaiguri district blocks such as Mal, Jalpaiguri, and Maynaguri, as well as Cooch Behar district blocks such as Mekhliganj and Haldibari, where it crosses West Bengal's lengtheist road bridge, the 3.8-km-long Joyee Setu. The river ends its voyage in India and flows into Bangladesh through Dahagram, Rangpur Division. Being a trans-boundary river, the Teesta River basin's length and area distri-bution total roughly 414 km and 12,370 km^2, and it has been split into three primary area distributions (Fig. 1).

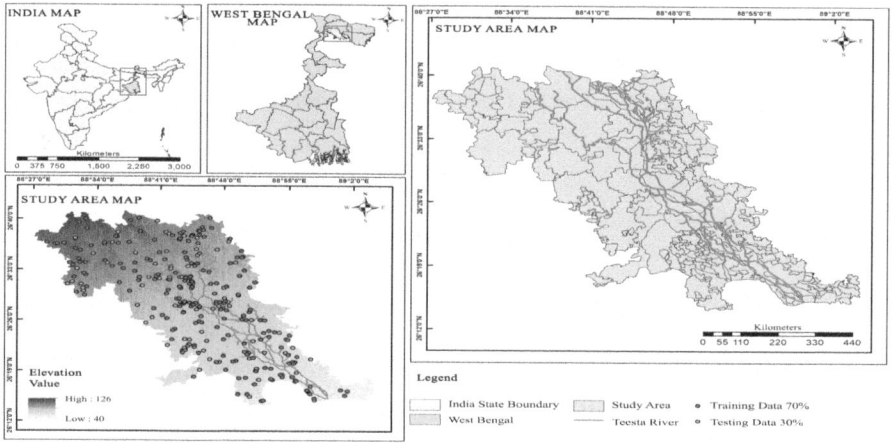

Fig. 1. Location: (a) India Map, (b) West Bengal Map, and (c) Study Area Map.

The Teesta River in West Bengal spans 123 km in length and 3294 km^2 in basin area [14]. The study examines the lower bank of the Teesta River in West Bengal, consisting of four blocks: Jalpaiguri, Maynaguri, Mekhliganj and Haldibari. Our study area is confined between 26°40'56.881"N to 26°13'52.511"N latitude and between 88°26'59.764"E to 89°3'46.947"E longitude with a span of 1260.460842 km^2.

3 Resources and Strategies

3.1 Methodologies Adopted

For assessing flood susceptibility in the proposed area, our present study was carried out in the subsequent five phases. The research methodology for the proposed study is shown in Fig. 2.

Step 1: Development of flood directory for Flood Inventory Map (FIM) preparation.
Step 2: Formation of spatial database of flood determinant factors.

Step 3: Assessing multicollinearity to determine the interdependencies among the floods' conditioning factors and remove interdependencies (if any).
Step 4: Application of FR and GMFFANN models for FSM preparation.
Step 5: Evaluate the efficacy of the created maps for flood susceptibility by employing the ROC curve both with training and testing dataset.

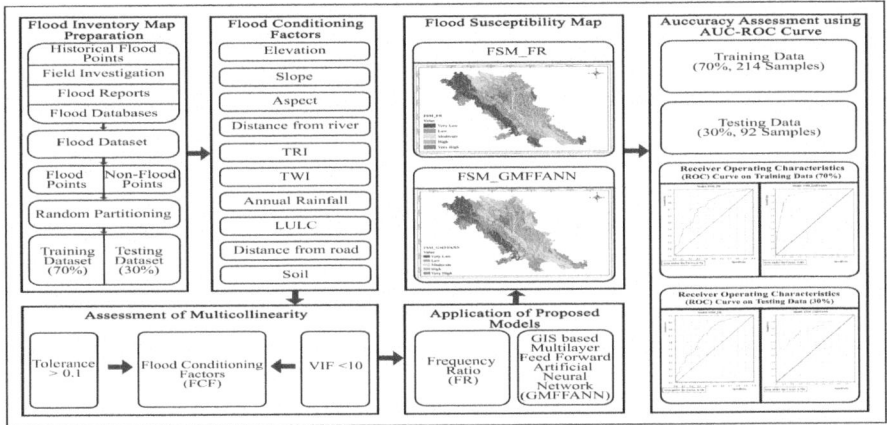

Fig. 2. Methodologies adopted for the proposed study.

3.2   Development of Flood Inventory Map

Flood susceptibility modelling involves analysing historical flood reports, databases, satellite images, and field surveys to create precise flood susceptibility maps [15]. The flood inventory map (FIM) is crucial in determining flood likelihood. Based on earlier studies, reports, field surveys and different flood databases, a cross examination using Google Earth Imagery was performed and 152 flood points have been identified with pinpoint location accuracy and have assigned 1 for this study area. Similarly, 152 non-flood data, assigned with 0, has been chosen on the Arc-GIS platform as required for binary analysis. Following random partitioning, 30% of the dataset has utilised for testing, while 70% has used for training.

3.3   Formation of Spatial Dataset

Flood susceptibility analysis relies on various parameters, each of which has a unique influence on mapping. Finding the appropriate parameters necessitates extensive literature study [6, 16], consultation with field specialists, and field observations. To construct thematic layers, this research has developed a geospatial database of 10 flood conditioning factors, as shown in Table 2.

Table 2. Flood Determinant Factors.

Sl. No	Flood Conditioning Factors	Spatial Details	Spatial Dataset Types	Classification Methods	Data Source
1	Elevation	30 × 30 m	SRTM DEM	Natural Breaks	USGS Earth Explorer
2	Slope	30 × 30 m	SRTM DEM	Natural Breaks	USGS Earth Explorer
3	Aspect	30 × 30 m	SRTM DEM	Natural Breaks	USGS Earth Explorer
4	Distance from river	30 × 30 m	SRTM DEM	Natural Breaks	USGS Earth Explorer
5	Topographic Roughness Index	30 × 30 m	SRTM DEM	Natural Breaks	USGS Earth Explorer
6	Topographic Wetness Index	30 × 30 m	SRTM DEM	Natural Breaks	USGS Earth Explorer
7	Annual Rainfall	0.25 × 0.25 degree	High Resolution Gridded Data	Natural Breaks	India Meteorological Department (IMD)
8	LULC	10 × 10 m	ESA Sentinel-2 imagery	Supervised Classification	ESRI
9	Distance from roads	1:50000	Reference Topographic Map	Natural Breaks	OpenStreetMap
10	Soil	1: 500000	Polygon data (vector data)	Soil Classification	FAO World Soil Data

3.4 Flood Determinant Factors

The comprehensive details of the 10 flood determinant factors for the study area are described in the following. Delineated maps created using ArcGIS 10.8 IDE are represented by Fig. 3 and 4.

Elevation is an important component in defining flood-prone places, with flat, low-land areas having a higher flood risk. Slope is used in hydrological computations to calculate water intensity, amplitude, and percolation. Aspects assess slope direction, which influences features such as water flow direction and precipitation. Flood spread and intensity are influenced by distance from the river, with limited river storage capacity increasing flooding chances [17].

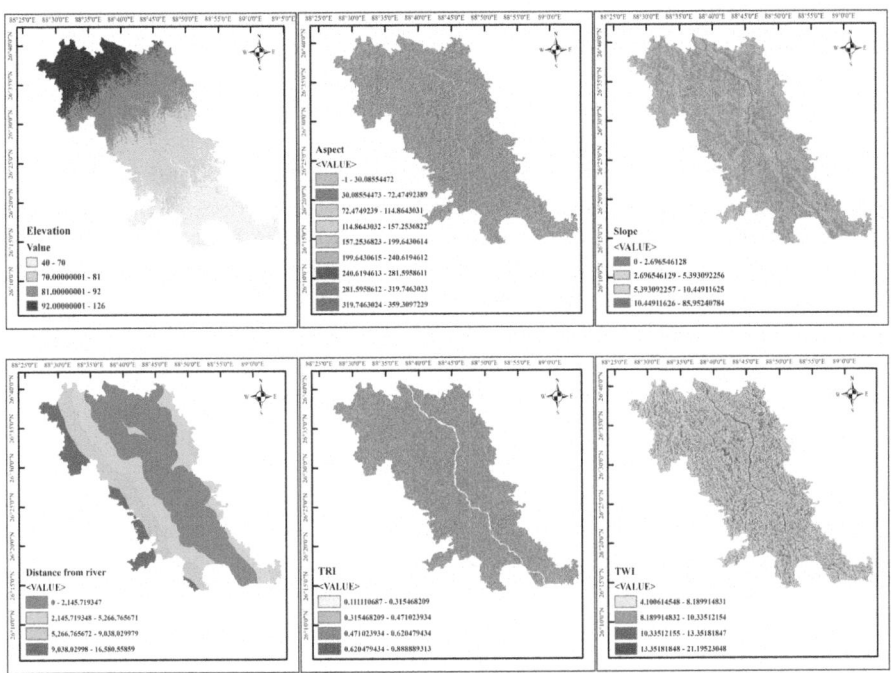

Fig. 3. Thematic Maps of the Determinant Factors for Flood (Set-1)

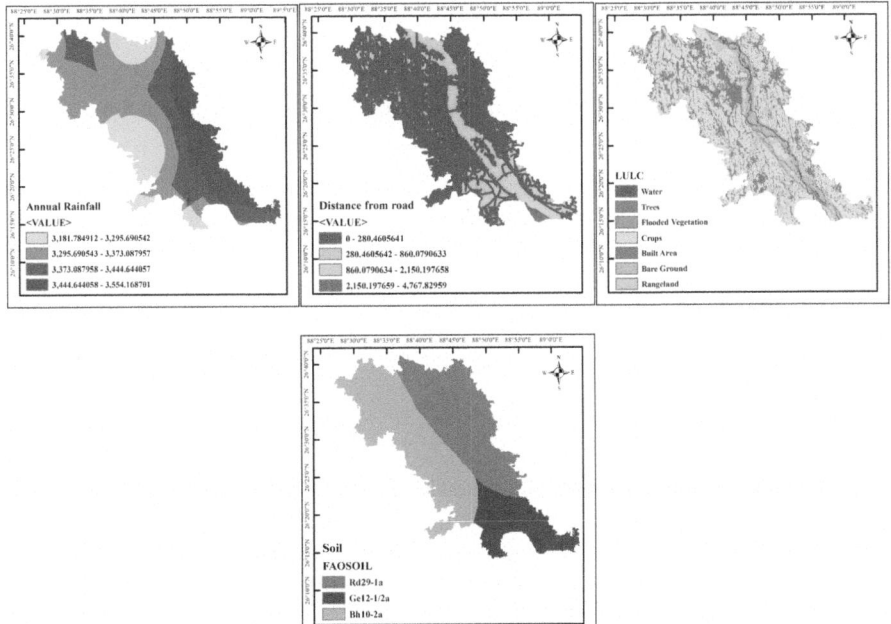

Fig. 4. Thematic Maps of the Determinant Factors for Flood (Set-2)

The Topographic Roughness Index (TRI) assesses land slope and undulation, with higher roughness in mountainous places and lower roughness in plains. The topographic wetness index (TWI) measures how topography affects flow accumulation in basins, with higher values indicating more flood potential. Rainfall is connected to floods, with the chance of flooding increasing as rainfall increases.

Land cover types influence hydrological processes such as runoff production, evaporation, transpiration, and infiltration. The research region comprises seven types of land cover: water, trees, flooded vegetation, crops, developed areas, bare ground, and rangeland. Distance from roads is critical when analysing flash flood-prone regions since roads diminish penetration and make areas nearest to highways more vulnerable. Soil types have a substantial influence on water inundation, regulating infiltration, percolation, and runoff [18].

3.5 Multicollinearity Checks

Multicollinearity is a statistical problem that develops when intercorrelated variables reduce the precision of models [19]. The elimination of multicollinearity among flood causative components can be achieved by focusing on the Variance Inflation Factor (VIF), or alternatively, new characteristics should be developed. Through the use of VIF values and tolerance, this study examined 10 flood-controlling parameters. When VIF values exceed 10 and tolerance is below 0.1, this indicates that the data set has a multi-collinearity problem (Table 3).

Table 3. Assessment of multi-collinearity on flood determinant factors

SP	E	SL	A	DFR	TRI	TWI	AR	LULC	DR	S
T	0.641	0.928	0.986	0.754	0.976	0.926	0.895	0.923	0.798	0.605
VIF	1.560	1.077	1.014	1.326	1.025	1.080	1.118	1.083	1.253	1.653

SP, Susceptibility parameters; T, Tolerance; VIF, Variance Inflation Factor; E, Elevation; SL, Slope; A, Aspect; DFR, Distance from river; TRI, Topographic roughness index; TWI, Topographic Wetness Index; AR, annual rainfall; LULC, Land use and land cover; DR, Distance from road; S, Soil.

3.6 Exploited Models

This study employs Frequency Ratio (FR), a bivariate statistical approach. Another investigation focused on the use of GIS-based Multilayer Feed Forward Artificial Neural Network (GMFFANN), which is trained using the Conjugate Gradient Decent (CGD) Algorithm, to enhance trend analysis and feature extraction from complex or vague data [20]. This research also produces a Flood Susceptibility Map (FSM) for the studied region.

Frequency Ratio (FR)

The core idea of bivariate statistical techniques is that "the path to the future lies in the past and present." A bivariate statistical method known as FR analysis, investigates the link between flood zones and the factors that cause floods. The value of Frequency Ratio is derived as the proportion of floods in a factor class to its area. Higher FR levels above 1 connects more significantly with flood incidence, while lower values below 1 interacts less strongly [21].

The Frequency Ratio (FR) and Flood Susceptibility Index (FSI) is represented mathematically as follows:

$$FR = \frac{\dfrac{Number\ of\ flood\ pixels\ for\ each\ factor}{Total\ number\ of\ floods\ occurance\ in\ study\ area}}{\dfrac{Number\ of\ class\ area\ pixels\ of\ the\ factor}{Total\ number\ of\ pixel\ in\ study\ area}} = \frac{\%FS}{\%A} \tag{1}$$

where %A denotes the factor class's area as a percentage of overall area and %FS denotes the percentage of flood in the factor class.

$$FSI = \sum_{f=1}^{n} FR_f \tag{2}$$

GIS Based Multilayer Feed Forward Artificial Neural Network (GMFFANN)
Multi-Layer Perceptron (MLP) and Training Algorithm

MLP is an effective approach for handling non-linear classification issues owing to its simple, flexible structure and high processing capacity. Employing various training techniques, it can convert input data into predictable results. The classical back propagation approach, which proved successful in feed forward multilayer networks but has limitations including convergence to local minima and slow learning speed, is one alternative learning strategy for neural networks. To increase effectiveness and speed, second order methods and second derivatives have been developed. For advanced training of multilayer perceptron network models, the batch-based technique Conjugate Gradient Descent is preferable. It is simpler to use yet produces more accurate predictions since it employs linear search and weight modification once every iteration [22].

Data Pre-processing and Development of Novel Network Architecture

Prior to building the network, pre-processing the flood dataset is crucial. For data pre-processing, 11 columns are provided, including 10 input columns and 1 output column. After pre-processing, the encoding parameters of 8 input columns are classed as "Numerical," while the other 2 input columns and 1 output column are classified as "Categorized". 17 columns are designated as final input and 1 column as final output for network designing.

The number of neurons in hidden layers significantly impacts neural network design. Two layers are suitable for the suggested network design, as underfitting occurs when there are insufficient neurons to identify complex signals, and overfitting occurs when the network's information processing capability is too high to train every neuron in the hidden layers. In order to find the best neural network architecture that is the best fit for the data, out of 60 identified networks, each trained 501 iterations on the input data. The best architecture, 17-8-4-1, with the highest Akaike Information Criterion (AIC) value

144   D. Sen et al.

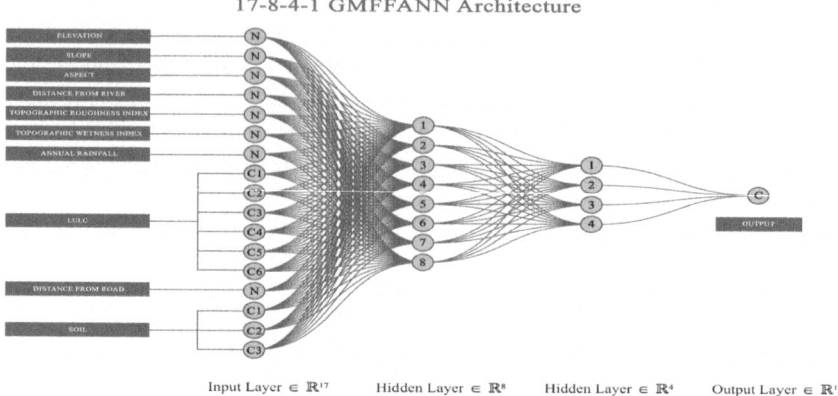

Fig. 5. 17-8-4-1 GMFFANN architecture for the proposed study area.

of -1926.637734, as shown in Fig. 5, has been opted for the research. AIC is a statistical measure used to evaluate a model's fitness to data, calculating the relative information value based on the maximum probability estimate and the number of distinct variables [23]. The formula for AIC is:

$$AIC = 2K - 2\ln L \qquad (3)$$

where, K is the number of distinct variables used and L is the log-probability estimate

4   Results and Discussion

4.1   Feature Importance

One advantage of ANN-based models is that they prioritise the input influencing features depending on their contribution to the modelling process and incorporate interaction effects among the independent elements. Table 4 shows the priority that the CGD algorithm assigned to the input parameters during training.

Table 4. Feature Importance during training of GMFFANN architecture

FI	E	SL	A	DFR	TRI	TWI	AR	LULC	DR	S
%	12.9996	0.2994	9.2752	17.1766	6.4234	1.8283	11.1223	20.4814	0.2677	20.1261

FI, Feature Importance; T, Tolerance; VIF, Variance Inflation Factor; E, Elevation; SL, Slope; A, Aspect; DFR, Distance from river; TRI, Topographic roughness index; TWI, Topographic Wetness Index; AR, annual rainfall; LULC, Land use and land cover; DR, Distance from road; S, Soil.

4.2   Flood Susceptibility Map and Its Spatial Distribution

Applying the FR model, the exhibits investigated and categorized the study area into five classes from extreme low to extreme high zones susceptible to flooding: 298.18 km^2 (19.7303%), 398.592 km (26.3745%), 242.98 km^2 (16.0778%), 449.50 44 km^2 (29.7433%), and 122.023333 km^2 (8.07417%), respectively.

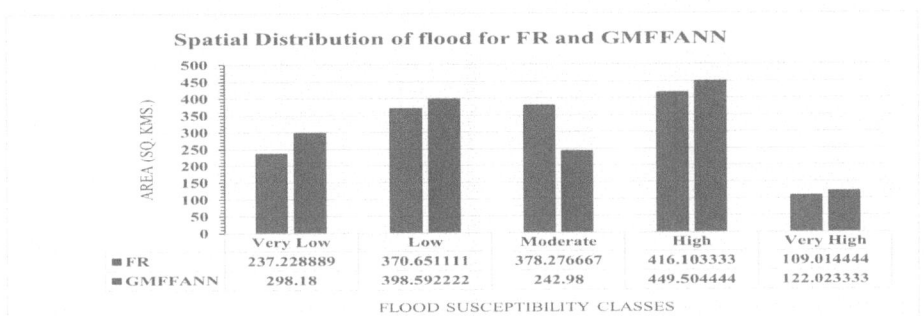

Fig. 6. Spatial Distribution of flood for FR and GMFFANN

The study region's 237.228889 km^2 (15.6973%), 370.651111 km^2 (24.5257%), 378.276667 km^2 (25.0303%), 416.103333 km^2 (27.5333%), and 109.014444 km^2 (7.21341%) km^2 were found to be, respectively, extremely low to extreme high zones, sensitive to flooding according to the GMFFANN model. As illustrated in Fig. 6. The graphical depiction of the flood dispensation of the research region is separated into five susceptibility classes, along with their corresponding area in square kilometers. Figure 7 shows the FSM of two suggested models.

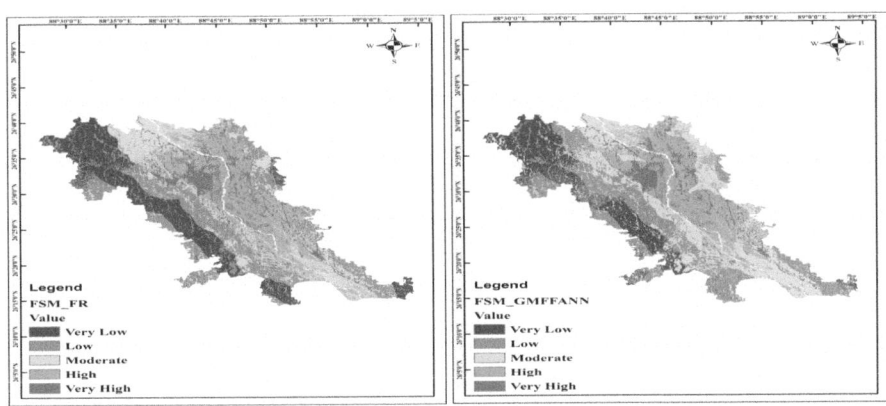

Fig. 7. Flood Susceptibility Map with FR and GMFFANN

146   D. Sen et al.

4.3 Accuracy Assessment of the Utilized Models

The flood susceptibility maps generated by the proposed models must be verified in order to prove scientifically relevant. Model validity was assessed using the receiver operating characteristics (ROC) curve and the area under the curve (AUC) value. The value of the AUC has been employed to quantify the model's prediction capability and is used to validate and compare models [24].

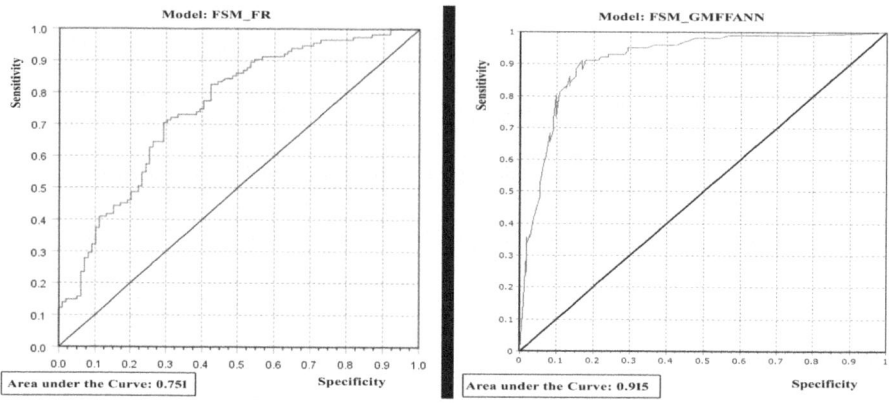

Fig. 8. Comparative exploration of ROC curve on Training Dataset

The findings showed that the GMFFANN model outperforms the traditional FR models in both the training and testing sets, by 91.5% and 79%, respectively. Whereas the FR model resulted 75.1% and 73.1% both for training and testing data respectively. Figure 8 and 9 elaborate the obtained AUCROC curves of training and testing dataset for the proposed models.

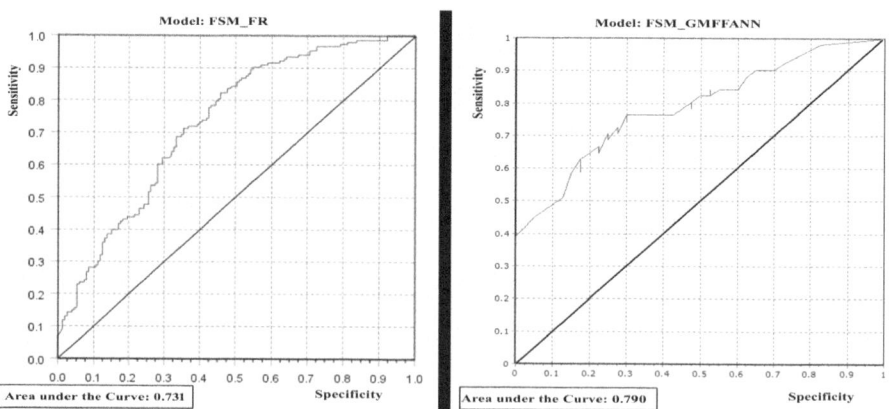

Fig. 9. Comparative exploration of ROC curve on Testing Dataset

5   Conclusion and Future Directives

Flood susceptibility techniques aim to accurately predict and minimize flood scenarios by adopting essential safety measures. This study delineates the flood susceptibility zone (FSZ) in the Eastern Indian districts of Jalpaiguri and Cooch Behar using an integrated RS and GIS approach-based FR and GMFFANN method. Machine learning models outperform independent statistical and conventional techniques for flood susceptibility research. The lower Teesta River basin in the Himalayan foothill zone has experienced significant flood episodes due to topographic variation and human interference [25].

The study demonstrates that the rapidly changing dynamic Himalayan foreland basin, as well as the repercussions of channelization and large-scale flood-plain alteration, are major contributors to recurring floods. The FR and GMFFANN technique sensitivity studies are innovative and can be used to different places throughout the world. Studies of flood simulation can examine the flood's extent, momentum, and time frame as well as the dynamics of the groundwater and surface waters. These frameworks may be used for complete flood susceptibility and risk management in the future if socio-demographic and financial aspects are included.

Land management authorities and government organizations will require a detailed flood susceptibility framework in order to put long-term mitigation strategies into effect [26]. This research might help policymakers and implementation authorities acquire basic knowledge on floods, such as their magnitude and hazard zones. A regional assessment of the likelihood of flooding can provide precise flood potentiality at the local level.

Data Availability Statement.   The corresponding author will, if required, provide necessary data available for research purposes upon request.

References

1. Ghosh, S., Saha, S., Bera, B.: Flood susceptibility zonation using advanced ensemble machine learning models within Himalayan foreland basin. Nat. Hazards Res. **2**, 363–374 (2022). https://doi.org/10.1016/j.nhres.2022.06.003
2. Parthasarathy, K.S.S., Saravanan, S., Deka, P.C., Devanantham, A.: Assessment of potentially vulnerable zones using geospatial approach along the coast of Cuddalore district, East coast of India. ISH J. Hydraulic Eng. **28**, 422–432 (2020). https://doi.org/10.1080/09715010.2020.1753250
3. CRED: EM-DAT. http://www.emdat.be. Accessed 30 July 2023
4. Central Water Commission, Ministry of jal shakti, Department of Water Resources, River Development and Ganga Rejuvenation, GoI. https://cwc.gov.in/. Accessed 30 July 2023
5. Mahalanobnis, P.C.: Report on rainfall and floods in North Bengal 1870–1922. Government of Bengal, Irrigation Department Publication. Calcutta (1927)
6. Arabameri, A., Rezaei, K., Cerdà, A., Conoscenti, C., Kalantari, Z.: A comparison of statistical methods and multi-criteria decision making to map flood hazard susceptibility in Northern Iran. Sci. Total. Environ. **660**, 443–458 (2019). https://doi.org/10.1016/j.scitotenv.2019.01.021
7. Mosavi, A., Ozturk, P., Chau, K.: Flood prediction using machine learning models: literature review. Water **10**, 1536 (2018). https://doi.org/10.3390/w10111536

8. Bentivoglio, R., Isufi, E., Jonkman, S.N., Taormina, R.: Deep Learning Methods for Flood Mapping: A Review of Existing Applications and Future Research Directions. Copernicus GmbH (2022)

9. Singh, D.S., Tangri, A.K., Kumar, D., Dubey, C.A., Bali, R.: Pattern of retreat and related morphological zones of Gangotri Glacier, Garhwal Himalaya India. Quat. Int. **444**, 172–181 (2017). https://doi.org/10.1016/j.quaint.2016.07.025

10. Singh, V., Goyal, M.K.: Analysis and trends of precipitation lapse rate and extreme indices over north Sikkim eastern Himalayas under CMIP5ESM-2M RCPs experiments. Atmos. Res. **167**, 34–60 (2016). https://doi.org/10.1016/j.atmosres.2015.07.005

11. Mukhopadhyay, S.C.: The Tista Basin: A Study in Fluvial Geomorphology (1984)

12. Majumdar, R.C.: History of Ancient Bengal. G. Bharadwaj, Calcutta (1971)

13. Rudra, K.: Rivers of the Ganga-Brahmaputra-Meghna Delta. Springer, Cham (2018)

14. Rahaman, M.M.: Abdullah - Al - Mamun: hydropower development along Teesta river basin: opportunities for cooperation. Water Policy **22**, 641–657 (2020). https://doi.org/10.2166/wp.2020.136

15. Antzoulatos, G., et al.: Flood hazard and risk mapping by applying an explainable machine learning framework using satellite imagery and GIS data. Sustainability. **14**, 3251 (2022). https://doi.org/10.3390/su14063251

16. Das, S., Gupta, A.: Multi-criteria decision based geospatial mapping of flood susceptibility and temporal hydro-geomorphic changes in the Subarnarekha basin India. Geoscience Front. **12**, 101206 (2021). https://doi.org/10.1016/j.gsf.2021.101206

17. Janizadeh, S., et al.: Prediction success of machine learning methods for flash flood susceptibility mapping in the tafresh watershed. Iran. Sustain. **11**, 5426 (2019). https://doi.org/10.3390/su11195426

18. Rahmati, O., Pourghasemi, H.R., Zeinivand, H.: Flood susceptibility mapping using frequency ratio and weights-of-evidence models in the Golastan Province Iran. Geocarto Int. **31**, 42–70 (2015). https://doi.org/10.1080/10106049.2015.1041559

19. Dey, S., Das, S.: Assessment of slope instability in Darjeeling Himalayan region: comparative evaluation of bi-variate statistical methodologies. In: 2022 International Conference on Artificial Intelligence and Data Engineering (AIDE). IEEE (2022) https://doi.org/10.1109/aide57180.2022.10060013

20. Chowdhury, D.R., Chatterjee, M., Samanta, R.K.: An artificial neural network model for neonatal disease diagnosis. Int. J. Artif. Intell. Expert Syst. (IJAE) **2**, 96–106 (2011)

21. Dey, S., Das, S.: Comparative exploration of statistical techniques for landslide identification and zonation mapping and assessment: a critical review. In: Communications in Computer and Information Science, pp. 44–55. Springer, Cham (2022). https://doi.org/10.1007/978-3-031-22485-0_5

22. Chowdhury, D.R., Sen, D.: Artificial neural network based trend analysis and forecasting model for course selection. Int. J. Comput. Sci. Eng. **5**, 20–26 (2017). https://doi.org/10.5281/zenodo.5226838

23. Akaike, H.: A new look at the statistical model identification. IEEE Trans. Autom. Control **19**, 716–723 (1974). https://doi.org/10.1109/tac.1974.1100705

24. Costache, R., et al.: Novel hybrid models between bivariate statistics, artificial neural networks and boosting algorithms for flood susceptibility assessment. J. Environ. Manag. **265**, 110485 (2020). https://doi.org/10.1016/j.jenvman.2020.110485

25. Rudra, R.R., Sarkar, S.K.: Artificial neural network for flood susceptibility mapping in Bangladesh. Heliyon **9**, e16459 (2023). https://doi.org/10.1016/j.heliyon.2023.e16459

26. Khoirunisa, N., Ku, C.-Y., Liu, C.-Y.: A GIS-based artificial neural network model for flood susceptibility assessment. Int. J. Environ. Res. Public Health **18**, 1072 (2021). https://doi.org/10.3390/ijerph18031072

A Fertilizer Recommendation System Using an Assembly of Regressors Coupled with Nature-Inspired Optimization Algorithms

Uditendu Sarkar[1], Gouravmoy Banerjee[2], and Indrajit Ghosh[2(✉)]

[1] National Informatics Centre, Ministry of Electronics and Information Technology, Government of India, Jalpaiguri 735101, West Bengal, India
[2] Department of Computer Science, Ananda Chandra College, Jalpaiguri 735101, West Bengal, India
ighosh2002@gmail.com

Abstract. The application of fertilizers in precise doses is a challenging task for the sustained production of crops worldwide. Due to the scarcity of soil scientists, rural farmers apply fertilizers in a blanket dose that hinders crop growth and yield production. Though several fertilizer recommendation systems have been proposed in the literature, they suffer from major limitations. To overcome the limitations, this paper aims to design a robust and efficient fertilizer recommendation system to suggest precise doses of fertilizer for a targeted crop. The system was designed using a regressor assembly of eight regressors for three major fertilizers: urea, single super phosphate, and muriate of potash for two major crops, potato and paddy, cultivated in three districts, Nadia, Hooghly, and Burdwan, in the state of West Bengal, India. In the first step, the performance of each regressor in the assembly was evaluated in terms of three well-known statistical metrics: R^2, *RMSE*, and *MAE*, and the best three were selected using Friedman's rank test. In the second step, to get better accuracy, each of these three regressors was hooked up with one of two nature-inspired hyperparameter optimization algorithms, called Particle Swarm Optimization and Bat. Finally, the outperforming regressor-optimizer couple recommends the precise doses of the fertilizers. Authentic soil health card data, collected from the Dept. of Agriculture, Government of India, was used for training, validation, and testing of the regressors. The experimental results showed that the random forest regressor coupled with the Bat algorithm outperformed the others irrespective of the locations, crops, and fertilizers.

Keywords: Fertilizer recommendation system · Machine learning regressors · Nature-inspired optimization · Hyperparameter optimization

1 Introduction

Estimation of precise doses of fertilizers for a crop considering existing soil nutrients is a very difficult task, especially for a rural farmer in India. In most cases, they often tend to rely on their own experience and judgment without having any scientific knowledge

M. Majumder et al. (Eds.): ICCTE 2023, CCIS 2376, pp. 149–159, 2025.
https://doi.org/10.1007/978-3-031-81935-3_13

about the existing nutrients and other relevant parameters of their soil. They apply a blanket dose of fertilizer for a crop based on its availability in the local market or as suggested in ready reckoners. However, the suggestions given in the ready reckoners are just a rough estimate of the quantity of fertilizers to be applied for a specific variety of crop. Such imprecise application of fertilizers hinders the normal growth of the crop, preventing a higher yield from its full potential. This yield loss directly impacts the rural agricultural economy. Specifically, in India, where the population is growing at a steady pace and the land under cultivation continues to diminish, this yield loss threatens future food security [1, 2]. Moreover, an indiscriminate application of fertilizers has an adverse effect on the soil property. On the other hand, the application of precise doses of essential fertilizers maintains good soil health and returns a better crop yield [3]. Therefore, scientific approaches must be adopted to recommend precise doses of fertilizers.

Regression is a powerful method for understanding the relationship between the independent and dependent variable(s). By exploring the relationship between independent and dependent variable(s) using a function, regressors predict the values of the output variable(s) based on the available inputs. With the advent of artificial intelligence, quite a few robust and intelligent machine learning regressors have been devised for diverse domains, such as the ridge, lasso, elastic-net, stochastic gradient descent, support vector, decision tree, random forest, gradient boost, Adaboost, k-nearest neighbor, and neural network [4, 5]. Several regressors were suggested to solve problems in different subdomains of agriculture, such as soil moisture prediction [6], yield prediction [7], agricultural energy consumption [8], irrigation scheduling, crop quality management, estimation of evapotranspiration, soil moisture content, and groundwater content [9].

The prevailing fertilizer recommendation systems designed using machine learning techniques can be classified into three categories: (1) systems that recommend the type of fertilizer to be applied; (2) systems that only recommend the blanket dosage of a fertilizer; and (3) systems that recommend site-specific fertilizers. The systems belonging to the first category only recommend the fertilizers be applied without suggesting the precise doses of the fertilizers needed for a particular crop, including Priya and Ramesh [10], Raviraja et al. [11], Supriya and Nagarathna [12].

As an improvement over the first category, the second group of systems was developed to recommend blanket doses of fertilizers. This category includes the models proposed by Suchitra and Pai [13], Reshma and Arvindhar [14], Garg et al. [15], etc. However, these systems suffer from major limitations, as they suggest only a blanket dose (lumpsum quantity) using ready reckoners without considering the nutrient content of soil and targeted crops.

The third group of systems was made as an improvement over the previous ones. They can suggest exact amounts of fertilizer that are needed based on the site. The models developed by Broner and Comstock [16], Tkatek et al. [17], Moreno et al. [18], and Haban et al. [19] are examples of such systems. Although each of these models provides precise doses of fertilizers, they suffer from significant limitations, such as low accuracy, a small dataset, recommendations in terms of nonstandard units, no performance evaluation, etc.

This paper aims to design a robust and efficient fertilizer recommendation system to suggest the precise doses of three major fertilizers: urea, single super phosphate

(SSP), and muriate of potash (MOP). Instead of arbitrarily choosing a single regressor, the system was designed with a regressor assembly of eight popular regressors. The performance of each regressor in the assembly was evaluated in terms of three popular metrics: coefficient of determination (R^2), root mean square error ($RMSE$), and mean absolute error (MAE). Based on the empirical values of these metrics, the best three regressors were selected using Friedman's rank test.

Optimization of model hyperparameters is a very fruitful technique to achieve better accuracy. For decades, many natural phenomena have inspired the development of numerous optimization algorithms for solving complex problems [20]. For better performance, the hyperparameters of the best three regressors were optimized with two popular nature-inspired optimization algorithms, Particle Swarm optimization (PSO) and Bat (BAT). Finally, the outperforming regressor-optimizer couple was used for region-specific recommendation of precise doses of the fertilizers for potato and paddy.

For empirical validation, the proposed system was applied to two varieties of potato (Jyoti Kufri High and Jyoti Kufri Low) and three varieties of paddy (IET 4094, Boro 4789, and IET 4097) cultivated in three districts, Nadia, Hooghly, and Burdwan, in the state of West Bengal, India. Training, validation, and testing of the regressors were done using authentic soil health card (SHC) data, collected from the Dept. of Agriculture, Government of India. The experimental results depicted that, for the study region, the random forest regressor coupled with the Bat algorithm outperformed (with average values of $R^2 = 0.99$, $RMSE = 5.26$, and $MAE = 2.23$) the others, irrespective of the targeted crops and fertilizers used.

2 Materials and Methods

2.1 Study Region and Dataset Collection

For the case studies, three districts, Nadia, Hooghly, and Burdwan, located in the Gangetic Plain in the state of West Bengal, India, were selected as the study region due to the availability of authentic data. The study region is one of the most fertile areas in the state of West Bengal, belonging to the Gangetic alluvium agro-climatic zone [21], as presented in Fig. 1.

The major soil classes in the study region are laterite, red, vindya, gangetic alluvium, and gravelly [22]. The gross cropped area of the region is about 692.79 thousand hectares, with a cropping intensity of 227.48% [23]. Potato and paddy are the two major cash crops cultivated in this region [24].

The system was designed to consider six region-specific independent variables to suggest the precise dose of a fertilizer. Three macronutrients, nitrogen (N), phosphorous (P), and potassium (K), and the other three soil parameters, pH, electrical conductivity, and organic carbon, were considered inputs. The datasets were collected from the soil health card (SHC) repository of the Department of Agriculture and Farmers Welfare, Government of India [25]. Three macronutrients, N, P, and K, are measured in kg/ha. The pH of the soil signifies the logarithmic value of the hydrogen concentration. The electrical conductivity is measured in deci-siemens per meter (dS/m), and organic carbon is measured in percentage of mass. The number of samples in the datasets collected for three districts, Nadia, Hooghly, and Burdwan, is 1599, 9042, and 922, respectively.

Fig. 1. The geographical location of the study region.

2.2   Reference Doses of the Fertilizers

Ramamurthy and Velayutham [26] suggested a model for the recommendation of fertilizers based on soil nutrient content, especially for N, P, and K. The Indian Council for Agricultural Research (ICAR) modified this model through several field experiments and finally suggested the Soil Test Crop Response (STCR) model [27] to establish a correlation between soil nutrients, crop yield, and fertilizer dose. In the present study, the STCR model was used to provide the standard reference (targeted) doses of three fertilizers, N, P, and K, for training, validation, and testing of the regressors.

2.3   System Description

The recommendation of appropriate doses of fertilizer for a targeted crop varies with the region-specific soil status. To design a regressor-based fertilizer recommendation system, no single regressor can be anticipated to produce the best result. Multiple regressors are to be attempted, and the best one should be selected to give the output with the highest possible accuracy. Manual selection of the best one from a group of regressors by trial-and-error strategy is a very cumbersome, time-consuming, and unfeasible task. These limitations motivated us to design the proposed fertilizer recommendation system with an assembly of eight promising regressors. All eight regressors were trained and tested with the same region-specific soil dataset for a targeted crop. The performance of each regressor in the assembly was evaluated in terms of three popular metrics: root mean square error ($RMSE$), coefficient of determination (R^2), and mean absolute error (MAE), and the best three were automatically selected using Friedman's rank test [28], thus avoiding the trial-and-error strategy. A four-fold cross-validation was carried out with 75% data for training and the remaining 25% data for testing. All the regressors were implemented in Python 3.7 using the Scikit learn library package. The eight regressors, their implementation strategy, and the hyperparameters are presented in Table 1.

Based on Friedman's rank test, RF, DT, and AD regressors (with mean rank ranging from 1.1 to 3.0) were selected as the best three regressors. In the second step, the

Table 1. Regressors, implementation strategy, and the hyperparameters.

Regressors	Implementation and hyperparameter values
1. *Nearest neighbour regressor (NN)*	Implemented using the ball tree algorithm [29], and the number of neighbours was set to 5
2. *Multilayer perceptron regressor (MP)*	A neural network regressor, implemented with one hidden layer having 100 hidden neurons, trained using a stochastic gradient descent algorithm [30] and logistic activation units
3. *Support vector regressor (SV)*	Built using the LIBSVM library [31], with radial basis function kernel and gamma set to the inverse of the number of features multiplied by the variance of input
4. *Decision tree regressor (DT)*	Executed by using the CART (classification & regression tree) algorithm [32] with the mean squared error criterion and minimum pruning cost $= 0.1$
5. *Gradient boosting regressor (BO)*	An ensemble of decision trees [33] with the number of estimators $= 10$ and the minimum pruning cost $= 0.1$
6. *Extremely randomized tree regressor (ET)*	Extremely randomized tree-based ensemble regressor [34] with the number of estimators set to 100 and the minimum pruning cost $= 0.1$
7. *Random forest regressor (RF)*	A random forest ensemble [35] with the number of estimators set to 100 and the minimum pruning cost $= 0.1$
8. *AdaBoost regressor (AD)*	Proposed by Freund and Schapire in 1997 [36], with multiple decision trees and the number of estimators set to 100

hyperparameters of these best three regressors, RF, DT, and AD, were considered to be optimized using PSO [37] and BAT [38]. These two algorithms have the advantages of ease of implementation and quick convergence. The root mean square was considered the cost function, as described by Yang [39]. For PSO, the values of the two constants, C1 and C2 [37], were set to 2, and the initial weight was set to 0.9 with a constant decrement factor of 0.01 per iteration. In the case of BAT, the maximum frequency [38] was set to 2, and the loudness and pulse rates were initialized with random numbers. For both PSO and BAT, the total number of particles or microbats was set to 30 for a maximum of 10 generations. The SHC dataset was divided into three parts: 75% for training, 12.5% for validation, and 12.5% for testing of the regressor-optimizer couples. Four-fold cross-validation with five iterations was carried out. Finally, the performances of the optimized regressors were evaluated using R^2, *RMSE*, and *MAE,* and the best-optimized model was selected to recommend the precise doses of the fertilizers. The workflow diagram of the proposed system is presented in Fig. 2.

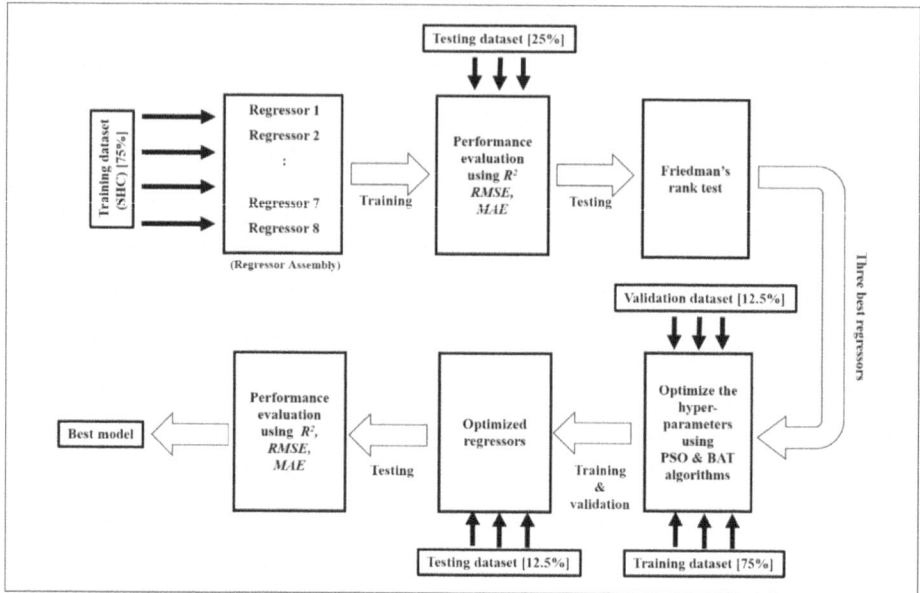

Fig. 2. The workflow diagram of the proposed system.

2.4 Performance Metrics

Three well-accepted performance metrics, coefficient of determination (R^2), root mean square error ($RMSE$) [40], and mean absolute error (MAE) were used to evaluate the performance at each step. R^2 and MAE are recognized as competent metrics for the performance evaluation of the regression models [40, 41]. These three metrics are defined as:

$$R^2 = 1 - \frac{\sum_{i=1}^{n}(y_i - \hat{y}_i)^2}{\sum_{i=1}^{n}(y_i - \overline{y})^2} \tag{1}$$

$$RMSE = \sqrt{\frac{1}{n}\sum_{i=1}^{n}(y_i - \hat{y}_i)^2} \tag{2}$$

$$MAE = \frac{1}{n}\sum_{i=1}^{n}(|y_i - \hat{y}_i|) \tag{3}$$

3 Results and Discussion

The empirical values of three performance metrics, R^2, *RMSE*, and *MAE*, using Eqs. 1–3 against eight regressors are presented in Table 2. The experimental results reveal that the RF regressor with its default parameters achieved the highest average value of R^2 (0.99) and the lowest average values of *RMSE* (7.02) and *MAE* (2.89), indicating the best recommendation of fertilizers. The SV regressor, on the other hand, produced the highest average value of *RMSE* (165.27), whereas the highest average value of *MAE* (132.21) was obtained for the MP regressor. Another observation is that the tree-based regressors, DT, and other ensemble trees such as an extremely randomized tree, random forest, and AdaBoost (except the gradient boost regressor), performed much better than the other regressors in terms of all three metrics.

Table 3 reveals that based on the evaluated performances, Friedman's rank test selected RF, DT, and AD regressors (with mean rank ranging from 1.1 to 3.0) as the best three regressors. These three regressors were then subjected to hyperparameter optimization using PSO and BAT algorithms. For the DT regressor, the minimum pruning cost of the hyperparameters was optimized, and for the AdaBoost and RF regressors, the number of estimators was optimized to minimize the *RMSE* as the objective function. After optimization, the values of R^2, *RMSE*, and *MAE* obtained against three regressors for all the varieties of paddy and potato are presented in Table 4. Table 4 reveals that for the study region, the RF regressor coupled with the Bat algorithm outperformed (with average values of $R^2 = 0.99$, $RMSE = 5.26$, and $MAE = 2.23$) the others irrespective of the targeted crops and fertilizers used. In addition, another observation is that the other regressors, when optimized with PSO and BAT, performed better than their base versions.

Table 2. The empirical values of three performance metrics, R^2, *RMSE*, and *MAE*.

Models	Potato (Jyoti Kufri High)			Potato (Jyoti Kufri Low)			Paddy (IET 4094)			Paddy (Boro 4789)			Paddy (IET 4097)			Average values		
	R^2	RMSE	MAE	R^2	RMSE	MAE	R^2	RMSE	MAE	R^2	RMSE	MAE	R^2	RMSE	MAE	R^2	RMSE	MAE
NN	0.96	43.91	30.26	0.96	45.97	29.40	0.95	24.89	15.35	0.88	79.53	39.08	0.79	52.77	24.05	0.90	49.41	27.62
MP	0.18	191.31	169.79	0.06	195.89	170.72	0.24	88.65	76.76	0.28	166.65	126.80	0.27	148.67	117.01	0.20	158.23	132.21
SV	0.21	188.20	167.24	0.17	195.80	157.76	0.18	95.84	65.58	0.22	190.52	97.18	0.04	156.02	108.55	0.17	165.27	119.26
DT	0.99	7.85	4.71	0.99	9.50	4.98	0.99	4.71	2.68	0.99	11.72	4.92	0.99	9.00	3.69	**0.99**	**8.57**	**4.19**
BO	0.86	73.22	64.10	0.86	72.52	61.87	0.87	35.51	30.45	0.86	67.33	50.97	0.87	58.49	46.03	0.86	61.41	50.68
ET	0.99	20.11	10.16	0.99	17.44	7.97	0.99	9.08	4.61	0.97	33.85	15.24	0.97	19.50	8.18	0.98	19.99	9.23
RF	0.99	6.13	3.05	0.99	7.49	3.37	0.99	2.65	1.50	0.99	9.82	3.44	0.99	9.02	3.10	**0.99**	**7.02**	**2.89**
AD	0.99	13.80	10.02	0.99	15.19	10.20	0.99	6.92	5.01	0.98	15.91	7.77	0.99	12.98	7.87	**0.99**	**12.96**	**8.17**

Table 3. Results of Friedman's rank test.

Models	R^2 Rank	*RMSE* Rank	*MAE* Rank	Mean Rank
Random forest (RF)	1.4	1.0	1.0	1.1
Decision tree (DT)	1.6	1.5	2.0	1.7
AdaBoost (AD)	3.2	2.5	3.4	3.0
Extremely ran. trees (ET)	3.8	3.3	3.6	3.6
Nearest neighbour (NN)	5.0	4.3	5.0	4.8
Boosting (BO)	6.0	4.8	6.0	5.6
Multi-layer perceptron (MP)	7.4	6.2	8.0	7.2
Support vector (SV)	7.6	6.3	7.0	7.0

Table 4. Values of the performance metrics for regressor-optimizer couples obtained against each crop variety.

Crops	Metrics	DT-BAT	DT-PSO	RF-BAT	RF-PSO	AD-BAT	AD-PSO
Potato (Jyoti Kufri High)	R^2	0.99	0.99	**0.99**	0.99	0.99	0.99
	RMSE	9.24	7.69	**3.85**	4.08	9.57	9.51
	MAE	5.33	4.60	**2.23**	2.37	7.57	7.48
Potato (Jyoti Kufri Low)	R^2	0.99	0.99	**0.99**	0.99	0.99	0.99
	RMSE	7.24	7.06	**4.43**	4.61	9.64	10.76
	MAE	4.63	4.27	**2.37**	2.48	7.52	8.64
Paddy (IET 4094)	R^2	0.99	0.99	**0.99**	0.99	0.99	0.99
	RMSE	4.40	3.89	**2.42**	2.52	4.83	5.10
	MAE	2.72	2.27	**1.22**	1.26	3.85	4.08
Paddy (Boro 4789)	R^2	0.99	0.99	**0.99**	0.99	0.99	0.99
	RMSE	11.87	11.12	**7.68**	8.52	10.99	11.57
	MAE	5.49	4.79	**2.95**	3.15	7.19	7.16
Paddy (IET 4097)	R^2	0.99	0.99	**0.99**	0.99	0.99	0.99
	RMSE	8.65	8.31	**7.95**	7.71	11.41	11.14
	MAE	3.76	3.52	**2.41**	2.49	7.24	7.14
Average values	R^2	0.99	0.99	**0.99**	0.99	0.99	0.99
	RMSE	8.28	7.61	**5.26**	5.49	9.29	9.62
	MAE	4.39	3.89	**2.23**	2.35	6.67	6.90

4   Conclusion

Recommending a precise dose of fertilizer for a targeted crop based on region-specific soil parameters is a major challenge worldwide. This work provides a robust and feasible solution to this problem. Designing a fertilizer recommendation system using a novel architecture with regressor assembly and nature-inspired optimization techniques selects the highest-performing regressor and always ensures to achieve the best recommendation, thus avoiding the hectic task of arbitrarily searching for the appropriate regressor. Our proposed system outperforms the other existing systems suggested in the literature.

The system suggests the precise doses of three major fertilizers that, in turn, help to reduce the indiscriminate use of fertilizers, maintain good soil health, and minimize the cost of production and environmental hazards. All these issues are very important for sustainable agriculture.

This fertilizer recommendation system can be implemented using a low-resource computing system that is affordable to rural farmers for estimating precise doses of three major fertilizers, N, P, and K, for two varieties of potato and three varieties of paddy cultivated as a cash crop in the Gangetic Plains of West Bengal.

The training, validation, and testing of the system are performed using region-specific data that makes it capable of being implemented in other similar agro-climatic regions without further structural modification. The only limitation of this system is the availability of soil parameters. However, designing an improved version of the system by using other hybrid models is our future target.

Acknowledgment. The authors would like to acknowledge the valuable suggestions and expertise of Mr. Sujit Pal, Joint Director, and Mr. Prabir Hazra, Deputy Director, Dept. of Agriculture, Government of West Bengal.

References

1. U. N. FAO, FAO Stat (Annual population). http://www.fao.org/faostat/en/#data/OA. Accessed 06 Aug 2020
2. World Bank Agricultural land (% of land area). https://data.worldbank.org/indicator/AG.LND.AGRI.ZS. Accessed 06 Aug 2020
3. Manna, M.C., et al.: Long-term effect of fertilizer and manure application on soil organic carbon storage, soil quality and yield sustainability under sub-humid and semi-arid tropical India. Field Crop Res **93**(2–3), 264–280 (2005)
4. Bonaccorso, G.: Machine Learning Algorithms. Packt Publishing Ltd., Birmingham (2017)
5. Gervasi, O., et al. (eds.): Proceedings of Computational Science and Its Applications–ICCSA 2020: 20th International Conference, Cagliari, Italy, 1–4 July 2020, Proceedings, Part VII, vol. 12255. Springer (2020)
6. Goap, A., Sharma, D., Shukla, A.K., Krishna, C.R.: Comparative study of regression models towards performance estimation in soil moisture prediction. In: Advances in Computing and Data Sciences: Second International Conference, ICACDS 2018, Dehradun, India, pp. 309–316 (2018)
7. Ruß, G., Kruse, R.: Regression models for spatial data: an example from precision agriculture. In: Proceedings of Advances in Data Mining. Applications and Theoretical Aspects: 10th Industrial Conference ICDM 2010, Berlin, Germany, pp. 450–463 (2010)

8. Karkacier, O., Goktolga, Z.G., Cicek, A.: A regression analysis of the effect of energy use in agriculture. Energy Policy **34**(18), 3796–3800 (2006)

9. Benos, L., Tagarakis, A.C., Dolias, G., Berruto, R., Kateris, D., Bochtis, D.: Machine learning in agriculture: a comprehensive updated review. Sensors **21**(11), 3758 (2021)

10. Priya, R., Ramesh, D.: Adaboost.RT based soil NPK prediction model for soil and crop specific data: a predictive modelling approach. In: Big Data Analytics: 6th International Conference BDA 2018, Warangal, India, pp. 322–331 (2018)

11. Raviraja, S., Raghavender, K.V., Sunagar, P., Ragavapriya, R.K., Kumar, M.J., Bharath, V.G.: Machine learning based mobile applications for autonomous fertilizer suggestion. In: 2022 4th International Conference on Inventive Research in Computing Applications (ICIRCA), Coimbatore, India, pp. 868–874 (2022)

12. Supriya, M.S., Nagarathna: A machine learning based crop and fertilizer recommendation system. **9**(7), 64–68 (2021)

13. Suchithra, M.S., Pai, M.L.: Improving the performance of sigmoid kernels in multiclass SVM using optimization techniques for agricultural fertilizer recommendation system. In: Soft Computing Systems: Second International Conference, ICSCS 2018, Kollam, India, pp. 857–868 (2018)

14. Reshma, S.J., Aravindhar, D.J.: A systematic approach of classifying soil & crop nutrient using machine learning algorithms. Int. J. Intell. Syst. Appl. Eng. **10**(2s), 174–179 (2022)

15. Garg, R., Aggarwal, H., Centobelli, P., Cerchione, R.: Extracting knowledge from big data for sustainability: a comparison of machine learning techniques. Sustainability **11**(23), 6669 (2019)

16. Broner, I., Comstock, C.R.: Combining expert systems and neural networks for learning site-specific conditions. Comput. Electron. Agric. **19**(1), 37–53 (1997)

17. Tkatek, S., Amassmir, S., Belmzoukia, A., Abouchabaka, J.: Predictive fertilization models for potato crops using machine learning techniques in Moroccan Gharb region. Int. J. Electr. Comput. Eng. (IJECE) **13**(5), 5942–5950 (2023)

18. Moreno, H.R., Garcia, O., Arias, L.A.: Model of neural networks for fertilizer recommendation and amendments in pasture crops. In: 2018 ICAI Workshops (ICAIW) Proceedings, Bogota, Colombia, pp. 1–5. IEEE (2018)

19. Haban, J.J.I., Puno, J.C.V., Bandala, A.A., Billones, R.K., Dadios, E.P., Sybingco, E.: Soil fertilizer recommendation system using fuzzy logic. In: 2020 IEEE Region 10 Conference (TENCON) Proceedings, Japan, pp. 1171–1175. IEEE (2020)

20. Binitha, S., Sathya, S.S.: A survey of bio inspired optimization algorithms. Int. J. Soft Comput. Eng. **2**(2), 137–151 (2012)

21. Ali, Sk.J.: Fertilizer recommendation for principal crops and cropping sequences of West Bengal. Department of Agriculture, Government of West Bengal, Kolkata (2005)

22. Das Gupta, H.K.: Production of edible oil in West Bengal study of possibilities in attaining self sufficiency, Ph.D. thesis, University of North Bengal (2017)

23. Agriculture Contingency Plan. http://agricoop.nic.in/agriculturecontingency/west-bengal?page=1. Accessed 06 Aug 2022

24. Prashnani, M., Singh, D.K., Joshi, R., Ray, S.S.: Understanding crop growing pattern in Bardhaman district of West Bengal using multi-date RISAT 1 MRS data. Int. Arch. Photogramm. Remote. Sens. Spat. Inf. Sci. **40**, 861–864 (2014)

25. Soil Health Card. https://soilhealth.dac.gov.in/. Accessed 07 Aug 2022

26. Ramamoorthy, B., Velayutham, M.: Soil Test Crop Response Correlation Work in India. World Soil Resources Report No. 41. Food and Agricultural Organization, Rome, pp. 96–100 (1971)

27. AICRP on Soil Test Crop Response Correlation. https://www.bckv.edu.in/index.php/en/aicrp-on-soil-test-crop-response-correlation-en. Accessed 07 Aug 2023

28. Derrac, J., García, S., Molina, D., Herrera, F.: A practical tutorial on the use of nonparametric statistical tests as a methodology for comparing evolutionary and swarm intelligence algorithms. Swarm Evol. Comput. 1(1), 3–18 (2011)
29. Omohundro, S.M.: Five Balltree construction algorithms, Technical Report, pp. 1–22. International Computer Science Institute, Berkeley (1989)
30. Bottou, L.: Stochastic gradient learning in neural networks. In: Proceedings of Neuro-Nimes, vol. 91, no. 8, p. 12 (1991)
31. Chang, C.C., Lin, C.J.: LIBSVM: a library for support vector machines. ACM Trans. Intell. Syst. Technol. (TIST) 2(3), 1–27 (2011)
32. Li, B., Friedman, J., Olshen, R., Stone, C.: Classification and regression trees (CART). Biometrics 40(3), 358–361 (1984)
33. Friedman, J.H.: Greedy function approximation: a gradient boosting machine. Ann. Stat. 2001, 1189–1232 (2001)
34. Geurts, P., Ernst, D., Wehenkel, L.: Extremely randomized trees. Mach. Learn. 63, 3–42 (2006)
35. Breiman, L.: Random forests. Mach. Learn. 45(1), 5–32 (2001)
36. Freund, Y., Schapire, R.E.: A decision-theoretic generalization of on-line learning and an application to boosting. J. Comput. Syst. Sci. 55(1), 119–139 (1997)
37. Kennedy, J., Eberhart, R.: Particle swarm optimization. In: Proceedings of ICNN 1995 International Conference on Neural Networks, vol. 4, pp. 1942–1948 IEEE (1995)
38. Yang, X.S.: A new metaheuristic bat-inspired algorithm. In: Nature Inspired Cooperative Strategies for Optimization (NICSO 2010), pp. 65–74. Springer, Heidelberg (2010)
39. Yang, X.S.: Nature-Inspired Optimization Algorithms. Elsevier (2014)
40. Aptula, A.O., Jeliazkova, N.G., Schultz, T.W., Cronin, M.T.: The better predictive model: high q2 for the training set or low root mean square error of prediction for the test set? QSAR Comb. Sci. 24(3), 385–396 (2005)
41. Cornejo-Bueno, L., Casanova-Mateo, C., Sanz-Justo, J., Salcedo-Sanz, S.: Machine learning regressors for solar radiation estimation from satellite data. Sol. Energy 183, 768–775 (2019)

State-of-the-Art in Feature Selection: Applications of the Slime Mould Algorithm

Taniya Chatterjee[1], Puja Bhakta[1], Mili Ghosh[1], and Debaditya Barman[2]([⊠]) [iD]

[1] Department of Computer Science and Technology, University of North Bengal, Darjeeling 734013, India
[2] Department of Computer and System Sciences, Visva-Bharati, Santiniketan 731235, India
debadityabarman@gmail.com

Abstract. Feature selection makes machine learning models less complex, more accurate, and more interpretable. The Slime Mould Algorithm (SMA) is a bio-inspired technique that leverages the demeanour of slime moulds to solve optimisation problems efficiently. This survey paper aims to deliver an extensive review of the recent research works involving the application of SMA as well as SMA fused with other optimisation methods to develop a solution for the feature selection problems. This involves summarizing and categorizing relevant research papers, methodologies, and findings. The survey paper consolidates and structures the existing knowledge about how the SMA is applied in the context of feature selection. Our work will help researchers and practitioners better comprehend the current advancement in this field.

Keywords: Feature Selection · Machine Learning · Slime Mould Algorithm · Nature Inspired Algorithms

1 Introduction

In the context of machine learning (ML), *features* refer to the input variables or attributes. Essentially, features are the properties of an object which are analyzed by ML algorithms in order to understand various patterns or relationships. Features can be numeric, categorical, or even text-based, depending on the nature of the problem.

Feature selection (FS) is a process in any ML pipeline that involves choosing a subset of relevant features from a given set of features in a dataset. The objective is to retain the most important and informative features while discarding those that may be redundant or irrelevant. The primary goal of any FS technique is to enhance the efficiency of ML algorithms by reducing dimensionality, increasing model interpretability, and mitigating issues related to overfitting and computational complexity. Selecting an appropriate feature selection technique is crucial, and often, the selection relies on the traits of the data and the specific goal(s) of the project. Slime Mould Algorithm (SMA) [22] is a bio-inspired optimisation algorithm which mimics the behaviour of a fascinating organism known as slime mould, specifically Physarum Polycephalum. This single-celled, amoeba-like organism exhibits remarkable capabilities in terms of path-finding, network optimisation, and adaptation. This algorithm is particularly unique due to its biomimicry of slime mould's behaviour and explore its potential in solving complex optimisation problems.

In this study, we provide a detailed assessment of the practical applications of the SMA algorithm within the context of feature selection. This entails examining the various ways in which the SMA has been utilized in FS-related tasks, summarizing the key findings and outcomes of these applications, and offering insights into the strengths, weaknesses, and potential areas of improvement or further research related to the ways SMA can be used for FS.

Across numerous real-world applications dealing with high-dimensional datasets, feature selection methods have found extensive application. In [18] SMA is coupled with fractional calculus and proposed fractional order based SMA. This method can be used to extract the best feature by avoiding local solutions and effectively identifying the search landscape by taking into account a historical record of the positions of the agents. Chemometric datasets are also evaluated using a metaheuristic algorithm in the feature selection classification process, which is recognized as a very promising method. This evaluation is done by the Marine Predators Algorithm (MPA) operators-based SMA method, which contributes to an accelerated convergence rate of the proposed SMAMPA [13], thereby mitigating the susceptibility to getting trapped in local optima.

We organize the rest of the paper as follows. Section 2 shows the background details of FS with three methods and slime mould algorithm with their working principle and describes some nature inspired optimisation algorithms for FS and also provides the role of the SMA in FS. Section 3 presents a discussion on different SMA-based FS algorithms. The challenges and future direction can be found in the Sect. 4. Finally, the Sect. 5 presents the conclusion.

2 Background

In any application related to machine learning (ML), image processing (IP), and data analysis (DA), feature selection (FS) is a crucial step aimed at identifying the most informative variables within a dataset. One unconventional yet intriguing approach to solve this problem is the SMA, a bio-inspired optimisation technique that leverages the natural behaviors of slime mould to identify and select relevant features efficiently.

2.1 Feature Selection

We can broadly categorized the existing techniques for FS into three main groups. These groups are 1) filter methods 2) wrapper methods 3) embedded methods. Each category has its own approach and characteristics.

Filter methods [27] evaluate the importance of a feature without using any ML algorithms. These methods select the most relevant ones by filtering irrelevant features to rank the features depending on any statistical measures or heuristics. Multivariate filter methods use different search strategies e.g., forward selection, backward elimination and bidirectional selection, etc., to generate a subset of features. In feature selection, there are various standard filter methods present, like the correlation among features, information gain, chi-square test, mutual information etc.

Wrapper methods [27] assess the features by training and evaluating an ML model multiple times with different feature combinations. This approach is broadly categorised into two types. The first is the sequential selection method, which starts with either an empty or complete set and then incrementally adds or removes features until the optimum value for a previously selected objective function(s) is achieved. The second one, the heuristic search method evaluates multiple subsets to optimize a previously selected objective function(s). So, these subsets of features are produced either by scouring the search space systematically or by devising solutions to a single or multi-objective optimisation problem.

Embedded methods [29] incorporate feature selection as a key part of the process of model training. These methods choose the relevant features while training the model depending on certain criteria or regularization techniques. Embedded methods offer a balanced approach that leverages the power of filter as well as wrapper methods while maintaining computational efficiency.

Effective feature selection is a critical step in any ML pipeline, but it has several criteria that need to be addressed to achieve an optimal result.

Challenges of Feature Selection: The key challenges of feature selection are: 1) Most of the real-life data are high-dimensional, which is often the cause of computational inefficiency and difficulty in finding the best feature subset. 2) Redundant features can confuse algorithms and hinder the optimal feature selection process. 3) Correlation among features can introduce various issues, such as multicollinearity, which makes the relevant feature selection difficult. 4) In a real-life dataset, a subset of features may interact. These interactions may only become prominent when the features are considered together. Since partial selection is probable, these interactions are difficult to preserve. 5) Feature selection's effectiveness relies on the quality of the problem dataset. Noisy or missing data often leads to suboptimal feature selection results. 6) In embedded methods and wrapper methods, choosing the right evaluation metric is crucial. It should align with the problem's goals, such as accuracy, precision, or F_1-score.

Criteria for Effective Feature Selection: The criteria for effective feature selection are: 1) The selected features should be relevant to the problem. They should contain valuable information that contributes to the model's ability to make accurate predictions. 2) Features should be as independent as possible. Highly correlated or redundant features can lead to multicollinearity issues and challenge model interpretation. 3) The model's generalization capability should improve by selecting a relevant subset of features while training. It should reduce overfitting and improve the model's performance on the validation or test set. 4) Features should be interpretable. This is important in some domains where the model's interpretability is a priority. 5) Feature selection should not introduce excessive computational overhead. 6) Feature selection methods should be robust to noise in the data and small changes in the dataset. They should not produce drastically different results with minor variations. 7) The effectiveness of feature selection should be validated using appropriate evaluation metrics and cross-validation techniques. It should not be based solely on training performance but should consider how the model performs on unseen data. 8) Feature selection methods should perform consistently with small and large datasets without significantly increasing computational burden.

2.2   Working Principle of SMA

Li et al. [22] proposed the Slime Mould Algorithm (SMA) in 2020. SMA is a meta-heuristic algorithm based on the diffusion and foraging behaviour of an eukaryotic organisms named slime mould. The mathematical model of SMA has four phases described as follows:

Phase 1 (Population Initialization): A population of slime mould contains N number of slime members. At first, each slime member's position $(X_i, i = 1, 2, \ldots, N)$ is generated randomly. An X_i is a vector in D dimension.

Phase 2 (Approaching Towards Food): Slime moulds can sense the odor of food in the atmosphere, guiding their search for food sources. Equation (1) is used to mimic its approaching behavior:

$$X(t+1) = \begin{cases} X_b(t) + v_g \times (W \times X_P(t) - X_Q(t)), & r < p \\ v_h \times X(t), & r \geq p \end{cases} \qquad (1)$$

where $X(t)$ represents slime mould's position at t^{th} iteration, $X_b(t)$ is the best position found at t^{th} iteration, $X_P(t)$ and $X_Q(t)$ are the positions of two randomly selected slime moulds from the population at the tth iteration, v_g and v_h are two coefficient vectors, W is a weighting factor, r represents a randomly generated number between -1 to 1, and p is a control parameter. When $r < p$ the population favors global exploration; whereas when $r \geq p$ it favours local exploitation. The control parameter p is computed using Eq. (2).

$$p = tanh|f(X_i) - f_{best}|, \; i = 1, 2, 3, \ldots, N \qquad (2)$$

where $f(X_i)$ represents the fitness of the ith slime mould and f_{best} is the best fitness obtained so far.

The weighting factor W can be computed using Eq. (3).

$$W = 1 \pm r \times \log(\frac{f_b(t) - f(X_i)}{f_b(t) - f_w(t)} + 1) \qquad (3)$$

where $f_b(t)$ and $f_w(t)$ denote the best and worst fitness found at t^{th} iteration, respectively. For the first part of the entire population, the *addition* is performed, and for the rest *subtraction* is performed [34] in Eq. (3).

The parameters v_g and v_h lie within the intervals of $[-a, a]$ and $[-b, b]$, respectively. Equation (4) and (5) are used to generate the values for a and b, respectively.

$$a = arctanh\left(\left(\frac{-t}{max\_iter}\right) + 1\right) \tag{4}$$

where max_iter represents the maximum number of iterations.

$$b = \frac{t}{max\_iter} \tag{5}$$

Phase 3 (Wrapping Food): In this phase, the *contraction* behavior of slime mould is modelled. As the slime mould consumes more food, the bio-oscillator produces a larger wave, and the vein grows thicker due to the rapid flow of cytoplasm. Equation (3) simulates the event of the level of food concentration vs. thickness of the vein, where the parameter r introduces uncertainty in the process. Equation (6) updates a slime mould's position based on the above phenomenons.

$$X(t+1) = \begin{cases} s \times (UB - LB) + LB, & s < z \\ X_b(t) + v_g \times (W \times X_P(t) - X_Q(t)), & r < p \\ v_h \times X(t), & r \geq p \end{cases} \tag{6}$$

where LB and UB represent the lower bound and upper bound of the feasible search region, respectively, s and r is a randomly generated value between $[0, 1]$, and the threshold z is set between $[0, 0.1]$. In [17,34], z has been set to 0.03. The parameter p determines the global exploration and local exploitation behavior.

Phase 4 (Oscillation): Slime mould optimizes its positions around the positions with higher food concentrations by modulating the flow of cytoplasm within their veins. When slime mould locates a high-quality food, W is used to mathematically predict the oscillation frequency when it is close to multiple food sources with different levels of odour concentrations. This value helps slime mould to move quickly towards food. In total, three parameters W, v_g, and v_h, capture the variations in the width of slime mould's vein-like structures. This strengthens the slime mold's capacity to identify the most favorable food source.

The Fig. 1 presents the flowchart of SMA.

Strength of SMA: SMA has demonstrated successful applications in diverse problem domains across various fields like routing [30], network design [10], resource allocation [33] etc. It's adaptability and simplicity make it a valuable tool for addressing different optimization challenges. Here, we'll discuss some of

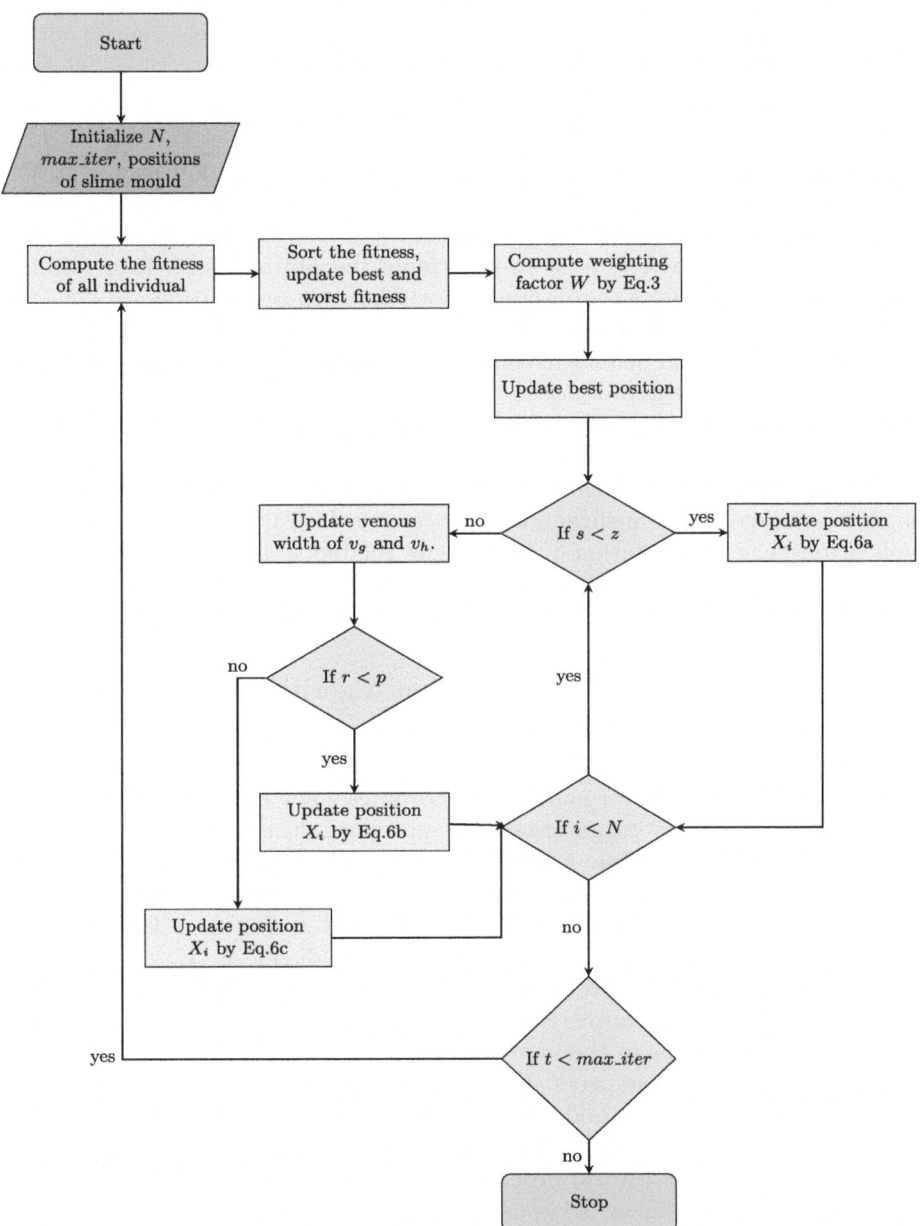

Fig. 1. Flow chart of SMA

the key strengths of the SMA: 1) It can be easily customized and adapted to specific problem domains and constraints. 2) We can easily modify SMA to tackle multi-objective or multi-criteria optimization problems by considering trade-off between multiple conflicting objectives. 3) Due to the inherent parallelism nature of SMA, it can be implemented using parallel computing architectures, which is beneficial for solving large optimization problems.

Real World Applications of SMA: In this section, we report some applications of SMA in various real-world problems. In [16], SMA technique is applied to optimize traffic signal timing, aiming to alleviate congestion and enhance traffic flow. In [6], SMA is used to determine the optimal path in a communication network, such as the internet or wireless sensor networks. In [28], the SMA method is utilized to discover an optimal solution by finding the trade-off between time, cost, and quality in construction projects. In managing petroleum reserves, forecasting oil production is a critical challenge. The study by [5] proposes an adaptive neuro-fuzzy inference system for the forecasting model, improved using SMA. Researchers continue to explore SMA's potential in new application areas and problem domains. One of the most important auxiliary diagnostic technology is medical image processing. In [23], the SMA technique is applied to image segmentation on the images related to breast cancer to obtain good results, and the paper also develops a multilevel image segmentation model that helps the research process.

2.3   Few Nature-Inspired Optimisation Algorithms Commonly Used for Feature Selection

- **Genetic Algorithm (GA):** GAs evolve a population of potential solutions (i.e., feature subsets) over multiple generations. They use techniques like parent selection, crossover, mutation, and survivor selection to create new generations of solutions while favouring fitter solutions. GAs can explore various feature combinations and converge towards a subset that maximizes the model's performance. GAs are successfully used as feature selectors in the handwritten digit recognition [25], breast cancer diagnosis [1].
- **Particle Swarm Optimisation (PSO):** PSO tries to mimic the social behaviour of a flock of birds referred as particles. In PSO, particles iteratively adjust their positions in the search space based on their own experience and the experience of their peers. Each particle of the population represents a potential feature subset, and the algorithm finds the subset of features which optimizes the chosen fitness metric. PSO has been used in the feature selection in numerous real-world problems, including document clustering [3,21], text analysis and mining [7], image processing [20] etc.
- **Ant Colony Optimisation (ACO):** ACO algorithm mimics the behavior of ants to solve complex optimisation problems. Ants leave a pheromone trail while searching for food. So, pheromone concentration is significantly high on the most promising path. In the ACO algorithm, agents leave pheromone trails as they explore different feature subsets. So, the pheromone levels on different subset of features represent the desirability of including those features in the subset [4]. ACO has also been used to solve numerous real-world problems like routing [11], scheduling [8], and DNA sequencing [9].

2.4   Role of the SMA in Feature Selection

SMA belongs to the category of bio-inspired optimisation algorithms. Like other nature-inspired algorithms (e.g., GA, PSO, ACO), SMA can be applied to solve feature selection (FS) problems [32], which searches for an optimal feature subset. It can help in exploring various feature combinations and selecting the subset that optimizes a specified criterion (e.g., classification accuracy, information gain, etc.). In the context of FS, SMA is employed to determine the most relevant features by modelling the given problem dataset as a network or graph. Features are prioritized to find the most promising path, where features can be considered nodes, and the SMA can explore paths that connect them.

3   Performance Evaluation and Comparison in Feature Selection Using Slime Mould Algorithm

This section provides an overview of various SMA-based approaches to solve feature selection (FS) problems. These approaches can be grouped into three: Hybrid SMA, Improved SMA, and Enhanced SMA. A hybrid SMA is an optimisation approach that combines the principles of SMA with other optimisation techniques or algorithms to leverage their strengths and improve overall performance. The goal of the hybridization is to make SMA more effective in solving specific problems. To create an improved version of the SMA, we can consider implementing various enhancements based on the problem domain. In enhanced SMA, additional features, capabilities, or techniques have been introduced to make the SMA more powerful or versatile. The methodology for hybrid, improved, enhanced based slime mould technique is described below:

For FS problems, Abdel et al. [2] suggested three binary versions of the hybrid SMA algorithm: binary SMA (BSMA), attacking-feeding strategy BSMA (AFBSMA), and two-phase mutation BSMA (TMBSMA). Additionally, AFB-SMA and TMBSMA are combined to develop FMBSMA. These algorithms have five steps: initialization, evaluation, transfer function, two-phase mutation, and attacking-feeding strategy. FS typically utilizes discrete variables, but here, the transfer function converts continuous values to discrete values in BSMA. Ewees et al. [14] proposed a hybrid strategy combining gradient-based optimizer (GBO) and SMA for global optimisation. This strategy balances exploration and exploitation. Zhou et al. [34] introduced a local dimensional mutation strategy known as LASMA. It significantly enhances the functionality of the original SMA optimizer for the FS technique. This hybrid approach can effectively balance exploration and exploitation capabilities. Nisha et al. [24] developed a hybrid slime mould-grey wolf optimisation technique for effective FS by incorporating a set of rules that could address the limitations of traditional FS techniques. Javidan et al. [19] applied SMA-based FS followed by the SVM classifier to diagnose illnesses of apple tree leaves which show similar symptoms.

Ewees et al. [12] developed a novel stochastic FS method by combining SMA with firefly algorithm to enrich exploration since it has a high capacity to identify

Table 1. A brief description of SMA-based feature selection problems.

Algorithm	Year	Advantages	Limitations	Datasets used
BSMA [2]: Hybrid	2021	Remove redundant and unresponsive information, increase classification accuracy	It works on specific datasets where binary algorithm is applicable	Twenty eight datasets from UCI repository Frank (2010)[a]
GBOSMA [14]: Hybrid	2023	Quick convergence strategy that balances both exploitation and exploration and escape from local optima	This algorithm takes a longer time to complete whole process	Twenty datasets from UCI repository
LASMA [34]: Hybrid	2023	Higher convergence accuracy	The LASMA takes a longer time for classifying big datasets	Eighteen disease related dataset
SMA-WOA [24]: Hybrid	2023	It provides high accuracy	This technique demonstrates slow convergence as well as low solving accuracy	Seven standard UCI benchmark datasets
SMA-SVM [19]: Hybrid	2023	This algorithm plays an important role to classify the group of disease	SVM can not support large size of data and large training time needed	Plantvillage dataset
SMAFA [12]: Improved	2021	Provide greater flexibility in exploring the search domain and increase diversity	Execution of the algorithm depends on a probability value	H1N1, Hepatitis and twenty datasets from UCI repository[b]
ISMA-kNN [32]: Improved	2021	Avoids getting stuck in local minima	kNN include slow learning and sensitivity to noisy data	Nine UCI benchmark disease datasets
GLSMA [26]: Improved	2022	Global optimisation, faster exploration and convergence speed, greatest classification precision	Greater time complexity than ISMA-kNN approach	Forteen UCI high dimensional gene dataset
ISMA [31]: Improved	2022	Fast convergence, enhance classifier's efficiency	Classification of medical dataset is highly challenging	Ten biomedical datasets[c]
DFSMA [17]: Improved	2022	Capable for global optimisation	The execution time is too long	Twelve biomedical datasets
FOSMA [18]: Enhanced	2021	Deal with uncertain dataset and high ability to improve classification precision	The fitness value depend on the selected features and target label	Twenty datasets from frank2010uci
SMAMPA [13]: Enhanced	2023	High accuracy and good stability	Execution depends on probability value	Biology, games, physics, and biomedical datasets
ABSMA [15]: Enhanced	2023	Increase diversity and escape from local optima	Separately implementation of mutation and combination causes inefficiency of ABSMA	The timber bridge, the three-story frame structure

[a] https://www.openml.org/search
[b] https://archive.ics.uci.edu/ml/index.php
[c] Dua D, Graff C (2017) UCI machine learning repository

viable places and have an optimal solution. Wazery et al. [32] used opposition-based learning (OBL) strategy to develop an effective SMA for solving the feature selection problem. OBL method helps to reach global optimum solution. There is no assurance that the subset of features chosen in a single run will be discovered. ISMA requires less time compared to conventional SMA. An improved SMA named GLSMA was proposed using Gaussian mutation and levy flight in [26] to resolve high-dimensional challenges in gene feature selection and continuous optimisation problems. In ISMA [31], solutions converge rapidly when the random guess is close to the optimal solution. However, the search process takes longer when the random guess is distant from the optimal solution. Then it looks for the solution in every direction using an OBL-based search strategy. To increase classification accuracy, the best solutions from the starting and opposite positions were chosen. Hu et al. [17] developed a dispersed foraging SMA (DFSMA) having a decentralized foraging strategy to enhance the original SMA for resolving optimisation issues. Then, the DFSMA was transferred to a binary space in order to develop the Binary DFSMA to identify features from high-dimensional biological data.

Ibrahim et al. [18] suggested a modification of SMA using fractional calculus to efficiently avoid local optima and find the search landscape by considering a historical positions of the agents. This fractional-order SMA has been used to extract features from a problem dataset to boost performance. Ewees et al. [13] proposed SMAMPA which is a modification of SMA using the marine predators algorithm (MPA). The MPA operators improve the efficiency of the local search. SMAMPA increases the convergence rate of the proposed algorithm while escaping local optimality. Ghiasi et al. [15] proposed advanced binary SMA (ABSMA) for FS in structural health monitoring problems. In the proposed ABSMA, the mutation and crossover operators improve diversity, avoid rapid convergence during optimisation, and conquer the deadlock.

In Table 1, we compare different SMA-based FS algorithms. The time complexities of LASMA [34], GLSMA [26], DFSMA [17] are $O(N(D(1+T(2+D))+\log N))$, $O(D+TN(1+3D+\log D+\log N))$, $O(ND(2T+1)+(2T+1)(N\log N)+2TN)$ where, N denotes the population size, D represents the number of decision variables and the maximum iteration is represented by T.

4   Future Direction

SMA's performance is susceptible to the choice of various parameters, such as movement rules, pheromone update rates, and population size. The value of these parameters is often selected depending on the properties of the problem at hand. Therefore, SMA may be prone to premature convergence. Thus, selecting an informative subset of features without getting stuck in local optima is a significant challenge. SMA may face scalability challenges when applied to feature selection tasks, especially for high-dimensional datasets with many features. The volume of the search space increases rapidly as the number of features increases, which can lead to high computational complexity. The above challenges raise a few questions: how can SMA's parameters be adaptively adjusted during the optimisation process to improve convergence and search efficiency? Can SMA effectively combine with other FS methods or ML algorithms to improve its performance and scalability?

There are several emerging trends and areas where further research is needed in the context of FS using SMA. First, research is required to adapt SMA for FS by handling large-scale datasets with a high degree of efficiency. Then, the robustness of SMA variants needs to be established for noisy data and outliers to ensure reliable FS methods in real-world situations. In addition to these considerations, it is essential to explore strategies for enhancing the convergence characteristics of SMA and effectively addressing local optima in feature selection scenarios.

5   Conclusion

This paper provides an extensive review of the existing research works on the various feature selection techniques based on Slime Mould Algorithm (SMA). Through our analysis, it is clear that SMA is frequently used with other strategies, demonstrating the versatility and benefits of this strategy while also emphasizing several drawbacks and potential difficulties. There have been attempts to investigate its limits and speed up convergence to address the local and global optimization problems related to SMA. Additionally, we observe that many modified SMA variants have been used, employing various parameters to address certain feature selection issues.

In various high-impact fields like healthcare, finance, environmental modeling, and others, where the choice of features can significantly affect important choices and results, the importance of SMA in feature selection becomes more apparent.

Despite the importance of SMA in feature selection, it's important to recognize that this is still a developing field of study. To maximize its potential, more research, testing, and algorithm improvements are required. To improve SMA's performance in feature selection algorithms, additional study is necessary for both academics and professionals, particularly concerning the search procedure and parameter control.

References

1. Aalaei, S., Shahraki, H., Rowhanimanesh, A., Eslami, S.: Feature selection using genetic algorithm for breast cancer diagnosis: experiment on three different datasets. Iran. J. Basic Med. Sci. **19**(5), 476 (2016)
2. Abdel-Basset, M., Mohamed, R., Chakrabortty, R.K., Ryan, M.J., Mirjalili, S.: An efficient binary slime mould algorithm integrated with a novel attacking-feeding strategy for feature selection. Comput. Industr. Eng. **153**, 107078 (2021)
3. Abualigah, L.M., Khader, A.T., Hanandeh, E.S.: A new feature selection method to improve the document clustering using particle swarm optimization algorithm. J. Comput. Sci. **25**, 456–466 (2018)

4. Al-Ani, A.: Feature subset selection using ant colony optimization. Int. J. Comput. Intell. (2005)

5. AlRassas, A.M., et al.: Advance artificial time series forecasting model for oil production using neuro fuzzy-based slime mould algorithm. J. Pet. Explor. Product. Technol. 1–13 (2022)

6. Arivunambi, A., Paramarthalingam, A.: Intelligent slime mold algorithm for proficient jamming attack detection in wireless sensor network. Glob. Trans. Proc. **3**(2), 386–391 (2022)

7. Bai, X., Gao, X., Xue, B.: Particle swarm optimization based two-stage feature selection in text mining. In: 2018 IEEE Congress on Evolutionary Computation (CEC), pp. 1–8. IEEE (2018)

8. Balaprakash, P., Birattari, M., Stützle, T., Yuan, Z., Dorigo, M.: Estimation-based ant colony optimization and local search for the probabilistic traveling salesman problem. Swarm Intell. **3**, 223–242 (2009)

9. Blum, C.: Beam-ACO-hybridizing ant colony optimization with beam search: an application to open shop scheduling. Comput. Oper. Res. **32**(6), 1565–1591 (2005)

10. Cai, Z., Xiong, Z., Wan, K., Xu, Y., Xu, F.: A node selecting approach for traffic network based on artificial slime mold. IEEE Access **8**, 8436–8448 (2020)

11. Dorigo, M., Stützle, T.: Ant Colony Optimization: Overview and Recent Advances. Springer (2019)

12. Ewees, A.A., et al.: Improved slime mould algorithm based on firefly algorithm for feature selection: a case study on QSAR model. Eng. Comput. 1–15 (2021)

13. Ewees, A.A., et al.: Enhanced feature selection technique using slime mould algorithm: a case study on chemical data. Neural Comput. Appl. **35**(4), 3307–3324 (2023)

14. Ewees, A.A., Ismail, F.H., Sahlol, A.T.: Gradient-based optimizer improved by slime mould algorithm for global optimization and feature selection for diverse computation problems. Expert Syst. Appl. **213**, 118872 (2023)

15. Ghiasi, R., Malekjafarian, A.: Feature subset selection in structural health monitoring data using an advanced binary slime mould algorithm. J. Struct. Integr. Maint. 1–17 (2023)

16. Hamza, M.A., et al.: Intelligent slime mould optimization with deep learning enabled traffic prediction in smart cities. Comput. Mater. Continua **73**(3) (2022)

17. Hu, J., et al.: Dispersed foraging slime mould algorithm: continuous and binary variants for global optimization and wrapper-based feature selection. Knowl.-Based Syst. **237**, 107761 (2022)

18. Ibrahim, R.A., Yousri, D., Abd Elaziz, M., Alshathri, S., Attiya, I.: Fractional calculus-based slime mould algorithm for feature selection using rough set. IEEE Access **9**, 131625–131636 (2021)

19. Javidan, S.M., Banakar, A., Vakilian, K.A., Ampatzidis, Y.: A feature selection method using slime mould optimization algorithm in order to diagnose plant leaf diseases. In: 2022 8th Iranian Conference on Signal Processing and Intelligent Systems (ICSPIS), pp. 1–5. IEEE (2022)

20. Khadhraoui, T., Ktata, S., Benzarti, F., Amiri, H.: Features selection based on modified PSO algorithm for 2D face recognition. In: 2016 13th international conference on computer graphics, imaging and visualization (CGiV), pp. 99–104. IEEE (2016)

21. Kushwaha, N., Pant, M.: Link based BPSO for feature selection in big data text clustering. Futur. Gener. Comput. Syst. **82**, 190–199 (2018)

22. Li, S., Chen, H., Wang, M., Heidari, A.A., Mirjalili, S.: Slime mould algorithm: a new method for stochastic optimization. Futur. Gener. Comput. Syst. **111**, 300–323 (2020)
23. Liu, L., et al.: Performance optimization of differential evolution with slime mould algorithm for multilevel breast cancer image segmentation. Comput. Biol. Med. **138**, 104910 (2021)
24. Nisha, S.S., Khan, A.A.R., Sathik, M.M.: Efficient feature selection using hybrid slime mould-grey wolf optimization. Math. Stat. Eng. Appl. **71**(4), 10492–10499 (2022)
25. Oliveira, L.S., Sabourin, R., Bortolozzi, F., Suen, C.Y.: Feature selection using multi-objective genetic algorithms for handwritten digit recognition. In: 2002 International Conference on Pattern Recognition, vol. 1, pp. 568–571. IEEE (2002)
26. Qiu, F., Guo, R., Chen, H., Liang, G., et al.: Boosting slime mould algorithm for high-dimensional gene data mining: diversity analysis and feature selection. Comput. Math. Methods Med. **2022** (2022)
27. Rodriguez-Galiano, V.F., Luque-Espinar, J.A., Chica-Olmo, M., Mendes, M.P.: Feature selection approaches for predictive modelling of groundwater nitrate pollution: an evaluation of filters, embedded and wrapper methods. Sci. Total Environ. **624**, 661–672 (2018)
28. Son, P.V.H., Khoi, L.N.Q.: Application of slime mold algorithm to optimize time, cost and quality in construction projects. Int. J. Constr. Manage. 1–12 (2023)
29. Stańczyk, U.: Feature evaluation by filter, wrapper, and embedded approaches. In: Feature Selection for Data and Pattern Recognition, pp. 29–44 (2015)
30. Takaoka, T., Sato, M., Otake, T., Asaka, T.: Novel routing method using slime mold algorithm corresponding to movement of content source in content-oriented networks. J. Signal Process. **23**(4), 173–176 (2019)
31. Vommi, A.M., et al.: Feature selection using improved slime mould algorithm for classification of medical datasets. In: 2022 Second International Conference on Next Generation Intelligent Systems (ICNGIS), pp. 1–6. IEEE (2022)
32. Wazery, Y.M., Saber, E., Houssein, E.H., Ali, A.A., Amer, E.: An efficient slime mould algorithm combined with k-nearest neighbor for medical classification tasks. IEEE Access **9**, 113666–113682 (2021)
33. Wu, X., Wang, Z.: Multi-objective optimal allocation of regional water resources based on slime mould algorithm. J. Supercomput. **78**(16), 18288–18317 (2022)
34. Zhou, X., et al.: Boosted local dimensional mutation and all-dimensional neighborhood slime mould algorithm for feature selection. Neurocomputing 126467 (2023)

MOODBYTBLB: Impact of Covid-19 Among Indians: A Sentiment Analysis Using Textblob

Sanchita Neogi and Rahul Karmakar[(✉)] [iD]

Department of Computer Science, The University of Burdwan, Bardhaman, India
rkarmakar@cs.buruniv.ac.in

Abstract. The whole world faced an extremely grave situation during the Covid-19 pandemic. In these difficult times, social media platforms like Twitter were vigorously used by people to share their emotions and experiences. For being able to identify these emotions we use sentiment analysis. Sentiment analysis is the process of identifying the sentiment of a text based on words used in it. Sentiment analysis mainly categories text based on three broad categories- Positive, Neutral and Negative. It helps us getting to know the emotion displayed by a user through their choice of words in the text written by them.

This paper aims to check for the tweets of such people who used twitter to post about their experiences during Covid-19. It also thrives to check for the sentiments of the people through their tweets in order to acquire a generic sentiment analytical view of the people during Covid using Textblob technique. For the above mentioned procedure, we first collect data consisting of tweets related to Covid-19 and then we process them. After that we perform sentiment analysis of all the tweets available. We then filter the dataset based on the location to be India, and further perform sentiment analysis upon the filtered tweets. This helps us by providing a comparative perspective of the sentiments of the Indian audience with respect to rest of the world as a whole, including India.

Keywords: Sentiment Analysis · Covid-19 · Tweets · Twitter · Textblob

1 Introduction

The Corona virus played with the whole world and tossed all the running systems upside down. All living beings struggled during the pandemic. It has not only affected us physically and destructed our health, but it went inside our heads and played with our emotions as well. The whole humankind agonized through different emotions during these times. Some of such people used social media as a medium of sharing their thoughts and experiences. One of the widely used platforms was Twitter. Many people used twitter to share their sentiments about different topics related to the pandemic. In this paper these tweets are considered and analyzed based on Indian audience to extracts the sentiments experienced by them. There exists a lot of work related to the sentiment analysis of twitter tweets. In this paper, we attempt to categorize data based on location India and extract sentiments from those tweets which are classified into seven types - Positive, Negative, Neutral, Weakly Positive, Weakly Negative, Strongly Positive, and Strongly Negative.

M. Majumder et al. (Eds.): ICCTE 2023, CCIS 2376, pp. 173–183, 2025.
https://doi.org/10.1007/978-3-031-81935-3_15

The practice of recognizing text's sentiments through few words used in text is called sentiment analysis. Some organizations may use this technique to get an overview about brand reputation, product reviews or customer satisfaction level. In the skill of sentiment analysis, a text's polarity is calculated which tells us about text's sentiment [1]. Many researchers have worked upon tweets based on different categories and attempted identifying their sentiments in order to discover user opinions on those categories. Covid-19 being an alarming issue from November 2019 had made researches go through tweets posted about it and identify the emotions of people, and to infer how the population is reacting to different problems related to Covid. Many such researches recorded articles stating their experiments and observations related to this domain. The paper [3] was upon a rumor dataset, containing all the false news, spreading about the virus, giving us observations about how people reacted to these rumors. Another paper [6], worked upon tweets posted by the Indian population and classified it into three broad sentiment categories - Positive, Negative and Neutral. Several other papers worked upon finding the sentiments of the people based on tweets posted during Covid and classify them into Positive, Negative and Neutral. Rare articles were able to record more than the basic three sentiments. More classification to the sentiment can help us give a deeper view of the intensity of the sentiment such as weak or strong portrayed by the tweets.

This paper thrives to process Twitter tweets that are posted during the time of Covid and perform sentiment analysis upon those, for the purpose of getting to know the sentiments of people. Then we also try filter the data by finding the tweets twitted by the Indian audience and find their sentiments. So we get brief overview of sentiments of the Indian audience through their text. These sentiment analysis results helps us to

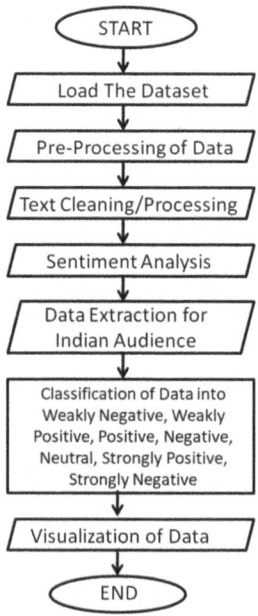

Fig. 1. Flow of Work

understand in concise about the number of twitter audience who posted tweets with not only positive, negative or neutral emotions but also weakly positive, weakly negative, strongly positive and strongly negative emotions, depending upon their word choices on their posts. These further added sentiments apart from positive, negative and neutral helps us in identifying the intensity of the emotion displayed by the text. Figure 1 displays the steps in the planned project.

1.1 Contribution of the Paper

This paper aims to contribute in the following ways-

a) This paper uses a dataset of tweets that are posted during the time of Covid and performs sentiment analysis upon it. It classifies the tweets into additional four sentiments other than the basic three sentiment classifications - Neutral, Negative and Positive. These additional sentiments are - Weakly Positive, Weakly Negative, Strongly Positive and Strongly Negative.

b) These sentiments give us an overview of the intensity of the emotions displayed by the tweets. This paper also tries to filter the data by finding the tweets twitted by the Indian audience and group their sentiments.

1.2 Structure of the Paper

This paper, primarily focuses on sentiment analysis as well as it discusses about comparative analysis of different sentiments portrayed by the Indian audience of twitter. Section 2 gives us the literature overview of some related work based on sentiment analysis, which lets us know about the interesting works done by researchers on the same field as well as the same domain of Covid-19. Section 3 gives us a case study based on the sentiment analysis of the Indian audience through their tweets in comparison with the rest of the tweets as whole from the selected dataset. Section 4 focuses on a brief discussion upon these tweets, and the algorithms used for the sentiment analysis. Finally, Sect. 5 gives us the conclusion for the whole project.

2 Literature Review

Covid-19 made us all use social media in an inappropriate amount, which created a huge amount of data on the Internet. In this paper we try to use and process such of 'Twitter' in order draw to inclusions about how people reacted or took the whole Covid-19 situation, through their tweets. We use sentiment analysis to extract the sentiments of the people through their written posts on Twitter. One of such similar experiment was performed and documented in the article "Sentiment Analysis on COVID-19 Twitter Data" [2], where tweets related to Covid-19 were considered, posted from November 2019 to May 2020. The tweets were categorized month-wise and state-wise into different datasets, and then using Text Blob API, sentiments were extracted from those tweets.

A similar intriguing article, "A COVID-19 Rumor Dataset" [3] discusses the widespread of false information, and rumors during the pandemic. Different datasets were acquired consisting of mis-informative tweets. In order to check the correctness of the tweets, the authors gathered online conversations from Twitter, which included real-time conversations with particular tags. They concentrated on a number of pertinent hash-tags and official accounts (such as ABC News, CNN, and BBC News) to track the updates of the trending subjects.

Another article, [4], discusses how in this study, they try to find the sentiments of the people tweeting about Covid in Twitter. The authors also try to identify the fear associated with the tweets. The global tweets of the year 2020 from Feb to Mar were considered for the analysis. The authors of this paper built a Transformation-based model of DistilBERT for the sentiment checking and represent them into one of three broad sentiment classifications- Positive, Negative and Neutral.

In a paper [5] the authors recorded a paper to determine the best ML model for forecasting COVID-19. In this paper they mentioned some of the ML algorithms as well as few introductory details about Covid-19. They described the ARIMA model for predicting counts of confirmed cases in India.

In the paper [6], it is mentioned how social media became a medium of expressing emotions during Covid-19. First, in the paper it is stated that they are dealing with tweets posted by the Indian population. Depending upon these tweets they attempt to incur their sentiments. Finally, they use methods like deep learning and lexicon-based techniques to categories sentiments gathered.

There is another paper [7] where the authors conduct a study in which they used LSTM RNN to retrieve sentiment classification and topic discovery on Covid-19. The authors consider the comments posted on reddit to understand people's sentiments about the Covid-19 outbreak. Methods like topic modeling are used to resurface various concerns posted by the people of Covid19.

In the paper, "Sentiment Analysis of Covid-19 Tweets" [8], the authors attempted to use an approach of ML, for classification of the Covid posts tweeted globally. Four machine learning algorithms were trained in order to be used in this experiment. The result of the experiment assists us in knowing the apt fitting technique for our job of finding the sentiments or emotions of the people through their tweets as well as spreading awareness about the pandemic.

The authors of [9] consider a dataset containing posts from five different social media platforms. The comments posted by people globally on these platforms are considered for retrieving the topics and the related sentiments of Covid-19. Furthermore, the authors try to differentiate the population into two groups depending upon their susceptibility to the diffusion of information from varied sources. Users on mainstream platforms are less likely to be exposed to the transmission of information from shady sources, and the method that information travels does not significantly differ depending on whether it comes from credible or dubious news sources. According to this paper, the unique characteristics of each social media platform's audience in addition to the interaction patterns of each play a critical part in the propagation of both accurate and false information.

In the paper, "Twitter Sentiment Analysis during COVID19 Outbreak" [10], from March 11 to 31, 2020, the author gathered Tweets on the corona-virus from more than 10 countries. He asserted that compared to other nations like Italy, Spain, and Belgium, individuals in France, Switzerland, and the Netherlands, displayed more mistrust and rage.

Another paper on Sentiment Analysis is [11]; the article examines the reactions of Indians to the announcements upon the lockdown. For the purpose of the study, authors assessed the Twitter tweets to understand Indians' attitudes regarding the shutdown. From March 25 to March 28, 2020, tweets with the two widely used hashtags #IndiaLockdown and #IndiafightsCorona were retrieved. A word-cloud that represents the feelings of the tweets was created after analysis using the R program. The Indians appeared to have faith in their government and were undoubtedly confident that it would carry out the lockdown smoothly, ensure that no residents would go without basic necessities during the lockdown, and make the necessary arrangements. Some tweets expressed complete surprise at the decision, but overall, it appeared that people were anticipating a lockdown. The only way to stop the virus from spreading was to practice hygiene measures like frequent hand washing with soap or alcohol-based sanitizer.

A recent research paper named, "Sentiment Analysis on COVID-19 Tweets: Machine Learning Approach", [12] shows the examining of tweets that surround a Covid incident. This research examines the opinions expressed on the ongoing COVID-19 outbreak in hundreds of tweets gathered from Kaggle. After using a data pretreatment strategy, the TF-IDF approach is used to extract uni-grams and bi-gram features. There are three types of supervised machine learning classifiers that are used: Random Forest (RF), Gaussian Naïve Bayes (GNB), and Bernoulli's Naïve Bayes (BNB).

The comparative study of the related articles mentioned above is given in the Table 1.

Table 1. Summary of Articles and Techniques Mentioned in Related Works.

Author(s) and Year	Method(s) Applied	Results	Limitations
Tanmay V. et al., (2020) [2]	Text Blob API was used for sentiment analysis	Tweets posted from November 2019 to May 2020 were considered. At the beginning a rise in the negativity in the sentiments of the people were seen. After April 2020, sentiments soon started drifting from negative to either positive or else neutral	While the pandemic continued till the year of 2021. This paper only recorded observations from Nov 2019 to May 2020
Mingxi C. et al., (2021) [3]	BERT and LSTM-based VAE (Variational Auto Encoder)	It discusses the widespread of false information during the pandemic. The sentiments of the rumor sentence were identified and categorized into five categories- positive, negative, neutral, very positive, and very negative	Classifying rumors based on its types or working upon a single category of rumor would provide precise conclusions of the experiment. Example- rumors based on how to cure Covid
ZuneraJalil et al., (2022) [4]	Transformation-based Multi-depth Distil BERT model	The paper finds the sentiments of the people tweeting about Covid in Twitter also try to identify the fear associated with the tweets. The global tweets of the year 2020 from Feb to Mar	This paper is specific and excludes emotions and mood detection

(continued)

Table 1. (*continued*)

Author(s) and Year	Method(s) Applied	Results	Limitations
Deepak P. et al., (2021) [5]	ARIMA Model	To determine the best machine learning model for forecasting and predicting COVID-19	The main motive of this paper was to find the best fitting model for sentiment analysis based on their accuracy value rather than gathering sentiments and conclusions from those sentiments
Bharati S. A. et al., (2023) [6]	Deep learning techniques VADER and NRCLex. Bi-LSTM and GRU	Sentiment analysis of Indian audience tweets on twitter platform using Deep learning and Lexicon based techniques	This study used many different techniques to find the algorithm giving the highest accuracy score for sentiment analysis
Rita Orji et al., (2020) [7]	LSTM- RNN, topic modeling	The sentiments of the comments on Reddit are classified into sentiments. The authors also analyze sentiment and opinion data from 10 sub-reddits to determine the polarity of the Covid remarks	Comments obtained only from Mar 19 to Jan 20 were used in this study
Uloko Emmanuel Junior et al., (2021) [8]	SVM, KNN, Random Forest and Naïve Bayes	The final result of this experiment assists us in knowing the best fitting algorithm for our job of finding the sentiments or emotions of the people through their tweets as well as spreading awareness about the pandemic	Interrogative phrases, queries, and sarcastic comments may produce unfavorable outcomes
Matteo Cinelli et al., (2020) [9]	NA	In this study, the authors compare user engagement across five different social media sites during the COVID-19 health emergency namely- Twitter, YouTube, Instagram, Gab and Reddit	The paper only deals with the tweets tweeted in the span of one month, which could create ambiguity regarding the observations obtained
Barkur G. et al., (2020) [10]	Data and content analysis	Indians are generally in favor of flattening the curve, although others are worried about day laborers' livelihoods	Eight different emotions were considered for this experiment, emotions like sarcasm and irony were not counted
Dubey (2020) [11]	Topic Modeling, R	The study suggests that while most individuals throughout the world choose a cheerful and upbeat attitude, there were still instances of fear, despair, and contempt displayed globally	The dataset used was not mentioned

(continued)

Table 1. (*continued*)

Author(s) and Year	Method(s) Applied	Results	Limitations
Janrhoni M. K. (2023) [12]	TF-IDF approach, Random Forest (RF), Gaussian Naïve Bayes (GNB)	the RF classifier outperformed the other two models on both feature extraction models, or the uni-gram and bi-gram feature extraction procedures	n-gram feature was not included into the experiment

3 Case Study: Indian Testimony

In this section, we discuss the steps mentioned in Fig. 1, and walk through each of those steps one by one.

3.1 Dataset Selection

The dataset used for this particular case study is 'Covid19_tweets' from Github. The dataset consists of thirteen columns.

3.2 Dataset Preparation and Data Selection

In order to proceed with our case study we first need to process our data for the better result generation from the selected dataset. For the above mentioned task we perform the following steps upon our data-

  i. We drop or eliminate any null data found in the dataset. Along with that any cell of the dataset containing 'Na' or 'NaN' values are converted to null values and are then removed. As shown in Fig. 2.

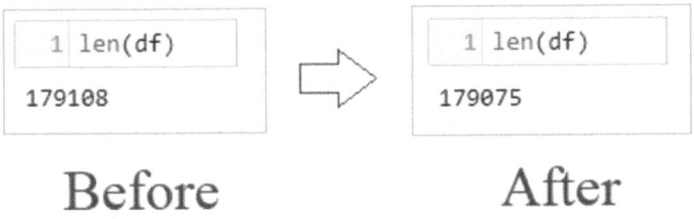

Before   **After**

Fig. 2. Comparison in the Length of the Dataset

  ii. The main text is cleaned from the dataset which contains the actual tweets upon which we would perform our further operations. The column is cleaned using certain commands.
  a. From the 'text' column the hashtags are extracted.
  b. User handles are removed from the 'text' and another column 'clean_tweet' is created.
  c. Multiple spaces are removed

   d. URLs are omitted

   e. Punctuations are eliminated

iii. There are many different columns present in our selected dataset which are redundant for our study. So, we select the necessary columns to eliminate any unnecessary ambiguity or representation errors.

3.3 Sentiment Analysis Using Textblob

In order to extract the sentiments of the global tweets available on the given dataset we use Textblob API to perform sentiment analysis, which in result categories tweets into 7 different categories.

a) Weakly Positive

b) Weakly Negative

c) Positive

d) Neutral

e) Negative

f) Strongly Positive

g) Strongly Negative

3.4 Processing Data with Special Filter

Our main aim in this case study is to be able to retrieve the sentiments experienced by the Indian audience through their tweets. Therefore we need to extract the tweets which are situated in India. For the above mentioned task we need to apply a filter to the dataset column as India. But the prior said filter would only include locations having the string 'India' in them.

a. One dataframe contains all the states and Union territories of India as well as the names of some common cities like Kolkata and Agartala.

b. Another dataframe contains the string India written in all the possible ways including its Hindi name 'Bharat'.

c. Now, we will check if our main dataset contains any of these values and store in a recent new dataframe.

   Now, we have a dataframe containing the tweets of only Indian audience as well as the sentiments corresponding to each of these tweets. We group these tweets to retrieve the sentiment measures of the Indian audience.

3.5 Visualization of Data

Figue 3 shows the sentiment proportion displayed by the (i) global tweet and (ii) Indian tweets respectively. The total number of global tweets is around 179075, whereas the total number of Indian tweets is 12139. In (i) the proportion scale is up to 100000, whereas in (ii) the proportion scale is up to 6000. According to both (i) and (ii), we can incur that 'Weakly Negative' is the most displayed sentiment displayed by people all over world as well as in India, and Neutral sentiment is the least displayed by peoples

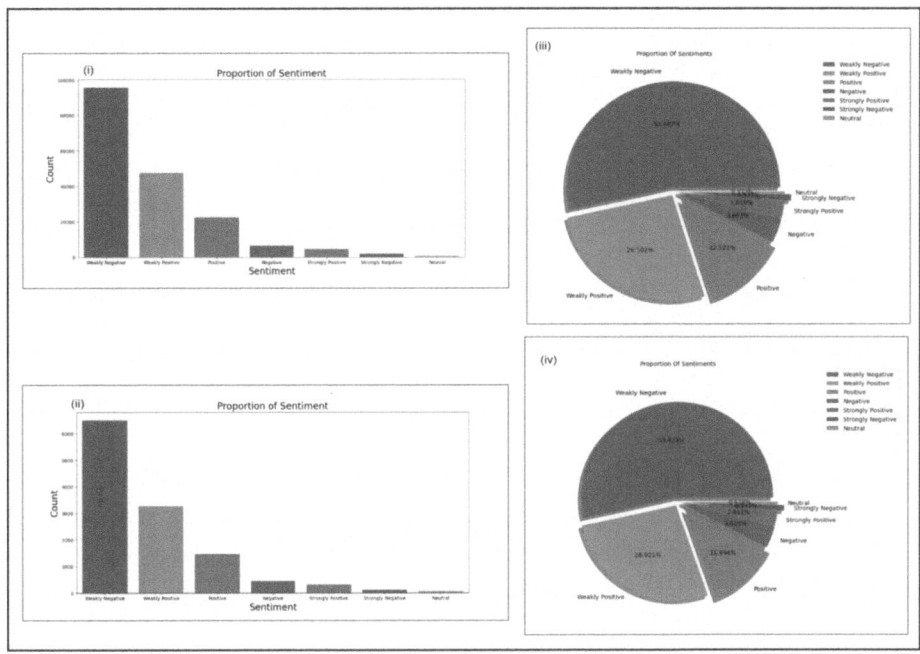

Fig. 3. Statistical Analysis

tweets. We can also conclude that only rare tweets portray 'Strong Negative' or 'Strong Positive' sentiments.

Figure (iii) and (iv) display the Pie-Chart for the give sentiment proportion in percentages. In (iii) the count of Weakly Negative sentiment in 100000 is represented in 53.387% and Neutral sentiment count 595 becomes 0.332%. In a similar manner, in figure (iv) the Weakly Negative sentiment count of 6485 becomes 53.423% in 12139 tweets as well as 58 Neutral tweets become 0.478%.

4   Result and Discussion

From the above experiment we can incur that 'Weakly Negative' tweets were a majority both in Global as well as Indian audience. On the other hand very few people decided to post Neutral tweets on Twitter displaying no such emotions. The sentiment counts of global as well Indian tweets are given in (Table 2).

In both the cases the sequence of the sentiments of containing the highest number of tweets to the lowest number remains the same, Weakly Negative and Neutral respectively.

Table 2. Overall Sentiment Counts

S No	Sentiments	Global Tweets	Percentage (%)	Indian Tweets	Percentage (%)
1	Weakly Negative	95586	53.387	6485	53.423
2	Weakly Positive	47450	26.502	3268	26.921
3	Positive	22420	12.522	1456	11.994
4	Negative	6558	3.663	440	3.625
5	Strongly Positive	4683	2.616	317	2.611
6	StronglyNegative	1750	0.977	115	0.941
7	Neutral	595	0.332	58	0.478

5 Conclusion

The importance of social media was prominent during the pandemic. Many people used social media as a way of expressing their feelings. Some used Twitter as one of such platforms. In this study, 179075 tweets were considered from Twitter. The sentiment counts of the global tweets were identified. Weakly Negative Tweets made the most of the counts among all the other sentiments, with a percentage of 53.387% of all the tweets, whereas Neutral tweets made up roughly up to 0.332% of all tweets. After Weakly Negative sentiment, the next sentiment having the highest count was seen to be Weakly Positive sentiment with a percentage of 26.502%. We then filtered out the tweets based on India, which made the number of tweets reduced to 12139. According to the polarity study of Indian audience, we get a similar report. Weakly Negative feelings dominated all the other sentiments with a percentage of 53.423%, while Neutral emotions were rarely identified among Indian tweets that had a percentage of 0.478%. This indicates that rare tweets were made of strong sentiments and mostly phlegmatic emotions were displayed by the people globally as well as in India. In this experiment, we were only able to select tweets based on India. For future studies, if the tweets could be state-wise classified it would generate observations according to state, and conclusions such as which state was most negatively or positively impacted. Also this research can be taken further to better observations by analyzing tweets not only based on sentiments but based on emotions like Fear, Anger, Hope, etc.

References

1. Ajibade, S., Tapales, C., Gido, N., Basillisco, G., Jesus, J., Oyebode, O.: An optimized data mining approach to analyze non-linear problem of online drug review of the Brazilian pharmaceutical industry, vol. 1, pp. 62–73 (2023). https://doi.org/10.17605/OSF.IO/GSF3X
2. Vijay, T., Chawla, A., Dhanka, B., Karmakar, P.: Sentiment analysis on COVID-19 twitter data. In: 2020 5th IEEE International Conference on Recent Advances and Innovations in Engineering (ICRAIE), Jaipur, India, pp. 1–7 (2020). https://doi.org/10.1109/ICRAIE51050.2020.9358301
3. Cheng, M., et al.: A COVID-19 rumor dataset. Front. Psychol. **31**(12), 644801 (2021). https://doi.org/10.3389/fpsyg.2021.644801. PMID: 34135812, PMCID:PMC8200409

4. Jalil, Z., et al.: COVID-19 related sentiment analysis using state-of-the-art machine learning and deep learning techniques. Front. Public Health **14**(9), 812735 (2022). https://doi.org/10. 3389/fpubh.2021.812735. PMID: 35096755, PMCID: PMC8795663

5. Painuli, D., Mishra, D., Bhardwaj, S., Aggarwal, M.: Forecast and prediction of COVID-19 using machine learning. In: Data Science for COVID-19, pp. 381–97 (2021). https://doi.org/ 10.1016/B978-0-12-824536-1.00027-7. Epub 21 May 2021, PMCID: PMC8138040

6. Ainapure, B.S., et al.: Sentiment analysis of COVID-19 tweets using deep learning and lexicon-based approaches. Sustainability **15**(3), 2573 (2023). https://doi.org/10.3390/su1503 2573

7. Jelodar, H., Wang, Y., Orji, R.: Deep sentiment classification and topic discovery on novel coronavirus or COVID-19 online discussions: NLP using LSTM recurrent neural network approach (2020). https://doi.org/10.1101/2020.04.22.054973

8. Azeez, N., Ogunlusi, V., Uloko, E.: Sentiment analysis of covid-19 tweets. Fudmam. J. Sci. **5**, 566–576 (2021). https://doi.org/10.33003/fjs-2021-0501-690

9. Cinelli, M., Quattrociocchi, W., Galeazzi, A., et al.: The COVID-19 social media infodemic. Sci. Rep. **10**, 16598 (2020). https://doi.org/10.1038/s41598-020-73510-5

10. Dubey, A.D.: Twitter sentiment analysis during COVID-19 Outbreak (2020). https://ssrn. com/abstract=3572023, https://doi.org/10.2139/ssrn.3572023

11. Barkur, G., Vibha, Kamath, G.B.: Sentiment analysis of nationwide lockdown due to COVID 19 outbreak: evidence from India. Asian J. Psychiatr. **51**, 102089 (2020). https://doi.org/10. 1016/j.ajp.2020.102089. Epub 12 Apr 2020, PMID: 32305035, PMCID: PMC7152888

12. Kikon, J., Kumar Bania, R.: Sentiment analysis on COVID-19 tweets: machine learning approach. In: Deka, J.K., Robi, P.S., Sharma, B. (eds.) EGTET 2022. LNEE, vol. 1061, pp. 339–348. Springer, Singapore (2023). https://doi.org/10.1007/978-981-99-4362-3_31

MythBuster: A Comparative Analysis of Few Machine Learning and Deep Learning Models for Fake News Detection

Barsha Pattanaik[1]([✉])(iD), Pratyush Mukherjee[2](iD), Sourav Mandal[1](iD),
Rohini Basak[2](iD), and Rudra M. Tripathy[1](iD)

[1] XIM University, Kakudia, Odisha, India
barsha@xustudent.edu.in, {sourav,rudramohan}@xim.edu.in
[2] Jadavpur University, Kolkata, India

Abstract. In the contemporary information age, most of our mornings begin with reading news and postings from various applications such as in short, Facebook, Twitter, and others, which differ from traditional printed media in that the news or posts' validity cannot be ensured. This situation necessitates the development of automatic ways for detecting and removing fraudulent content from the internet so that people are not misled by hoaxes propagated with the goal of profiting from the situation. To detect fake news, we examine various machine learning-based techniques and evaluate overall performance against fake news propaganda. We conducted a comparative analysis of the different deep learning (DL) and machine learning (ML) based systems we developed. Our proposed system gave 99.8% accuracy with the BERT (Bidirectional Encoder Representations from Transformers) word-embedding on the 'ISOT' fake news dataset.

1 Introduction

These days, the internet is becoming increasingly more available and has made our lives vastly more straightforward. Nonetheless, it accompanies its reasonable part of complexities since sites like Facebook and Twitter permit anyone to put data via virtual entertainment without confirmation. Subsequently, individuals can compose anything they wish to go from their viewpoints to untruths and falsehoods. Particularly considering the worldwide expression, this sort of falsehood and fabrication can make far-reaching frenzy and questions, which is extremely destructive for the public. Recently, sites like Google and Facebook have started to utilize some reality-checking instruments, for example, banishing scam destinations from publicizing stages and testing truth, actually taking a look at marks to have the option to recognize and signal phony news sites. Facebook additionally utilizes another framework for its clients to have the option to report any dubious stories that they could go over. Computational AI estimations have been exhibited to be precious in this space where the sheer volume of data overwhelms the human limit of assessment.

It's quite difficult to identify fake news on social media platforms nowadays. The limited amount of resources, such as datasets, is a significant disadvantage of detecting fake news on social media. According to Farokhian et al. [6], the main difficulty in these areas is to choose the proper features of the textual data. Deep neural networks have the benefit of enabling automated extraction of the most crucial news items during training. Furthermore, researchers like Ahmed et al. [2] mentioned that high-dimensional data sometimes develop problems in extracting features to identify fake information in social media. The huge amount of keywords, words, and phrases present in documents generate a lot of computational work in the learning process. Moreover, the accuracy and performance of classifiers were hampered by redundant and meaningless properties. Therefore, feature reduction should be used to reduce the size of the text feature and avoid having a big feature space.

This paper examines different ML and DL-based implementations utilized with the end goal of identifying fake news. The aim of the work is to think about the traditional and present-day draws near to see which works most successfully on true information. Further, this study will assist with utilizing the calculation via online social media sites by refreshing the information base with new news intermittently. We have likewise utilized the BERT profound learning approach, which achieves a promising precision of 99.8% on the ISOT dataset[1] along these lines outflanking any other learning models in the section.

In this paper, we have utilized different ML procedures to identify fake news in social media. So, We have utilized Text-classifiers ("Logistic Regression", "Naïve Bayes classifier", and "Passive Aggressive Classifier"), DL methods like LSTM ("Long Short-Term Memory"), and recent methods like BERT and analyzed their exhibition as examined in the following areas.

In this research paper, Sect. 2 depicts some examination works that have been done in the past on arranging counterfeit news; Sect. 3 we portrays the methodologies that we have used to take care of the issue of phony news recognition; Sect. 4 depicts the outcomes that we have acquired utilizing the strategies portrayed in Sect. 3; lastly, in Sect. 5 we have expressed as why our strategy is better when contrasted with the past works that have been done; likewise we have expressed an open examination regions that we might want to investigate from here on out.

2   Related Work

Recently, many researchers have developed various models to identify fake news by using different deep learning techniques Kaliyar et al. [9] designed a model consisting of both BERT and equal squares of single-layered profound CNN (Convolutional Neural Network) having various channels and portion estimates and accomplished a precision of 98.90% on the Kaggle datasets based on 2016 U.S. General Presidential Election. Ruchansky et al. [17] proposed a crossbreed technique joining the text, its client reaction, and the client advancing article on

---

[1] https://www.uvic.ca/ecs/ece/isot/datasets/fake-news/index.php.

a dataset containing posts from Twitter and Weibo and accomplished precision of 89.2% and 95.3% on Twitter and Weibo information individually. OBrien et al. [12] accomplished an exactness of 93.5% by applying profound learning techniques to arrange counterfeit news in clever subjects just from language designs. Singh et al. [20] fostered a text-handling-based AI model and got a precision of 87.00% on a Kaggle dataset.

Ahmed et al. [2] focused on detecting online fake reviews. To extract features from the data, the authors used term frequency and "Term Frequency-Inverse Document Frequency" (TF-IDF) techniques. They fed them onto different classifiers and obtained an accuracy of 92% using TF-IDF features and LSVM classifier. Ghosh and Shah [7] looked at how social media affects political choices and how false news affects international relations.

Rath and Basak [16] developed a textual entailment recognizer and trained the recognizer on the "Stanford Natural Language Inference dataset" (SNLI) for detecting fake news. In the comparison of their method with others, they reported more than 90% accuracy on the standard FakeNewsAMT dataset. Drif et al. [5] combined LSTM (to learn long-distance dependencies) and CNN models (to extract coarse-grained local features) and got an accuracy of 62.3% on the LIAR dataset and 72.5% on news articles.

Okano et al. [14] used "Hierarchical Attention Network" (HAN) for detecting fake news in "Brazilian Portuguese fake news parallel corpus" and obtained an accuracy of 97% for entire text and 90% for truncated texts. Sharma and Kalra [18] used the XG Boost algorithm in combination with state-of-the-art BERT to categorise fake and true news in the FakeNewsNet [19] dataset and thereby obtained accuracy of 98.53%.

Farokhian et al. [6] proposed a BERT-based fake news detection model that consisted of two parallel BERT networks to encode headlines and news bodies. To assess the model's effectiveness, the authors used Fakenewsnet [19] datasets and showed that their model gave 85.4% accuracy. Qasim et al. [15] proposed transfer learning models for text classification. To gauge the model's effectiveness, the authors used a range of datasets related to COVID-19 [1,13], and their model showed 99% accuracy.

3 Proposed Method

In this paper, different Machine Learning procedures are utilized for grouping fake and genuine news. Here, we attempted to distinguish counterfeit news utilizing Text-classifiers (Logistic Regression, Passive Aggressive Classifier, and Naïve Bayes classifier), Deep Learning procedures like LSTM, and current word embedding techniques like BERT and analyzed their exhibition as examined in the ensuing areas. The work process of the cycle is portrayed in Fig. 1.

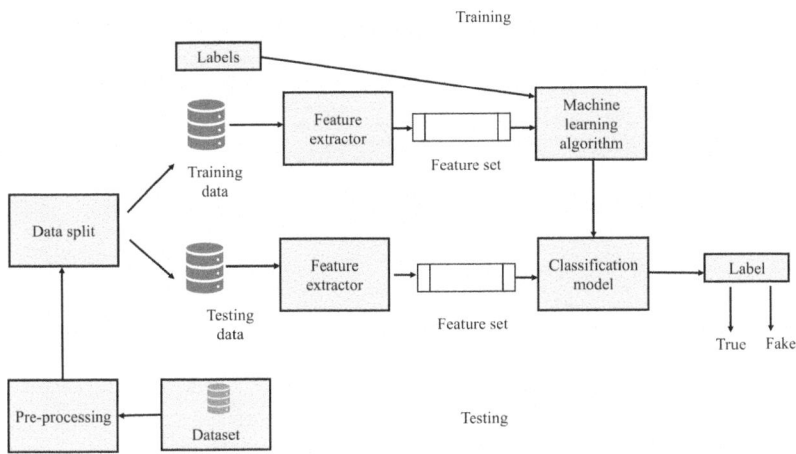

Fig. 1. Process flow for detecting fake news with machine learning

3.1   Dataset

We have utilized the standard ISOT dataset [2] here for an exploratory reason. There are 45,000 news stories in the ISOT dataset. There are more than 23502 fake news gathered from various sources hailed by truth actually looking at the association "Politifact[2] However, there are more than 21,417 genuine or confirmed news gathered from confided-in sources, for example, 'Reuters'. All the news in the dataset is primarily gathered from 2016 to 2017 during the U.S. Official decisions when there was an enormous blast of phony news in web-based entertainment like Facebook, Twitter, and so forth. The dataset comprises 4 segments, specifically the title, text, subject, and the date of distribution of the article. The news stories are predominantly political, world news, and government news. Figure 2 and Fig. 3 address sample datasets.

3.2   Pre-processing

Here, we first eliminate the irrelevant information from the news text and dispose of the news texts that become invalid after the previous interaction. Then, at that point, a class field is added where 0 is relegated to FAKE information, and 1 is allocated to REAL information. Then, we combine the title and the news to shape the refreshed text content. After playing out this, the consolidated dataset is made by joining the phony and genuine news datasets. After that, the news is

---

[2] www.politifact.com".

title	text	subject	date
Trump on Twitter (Dec 28) - Global Warming		politicsNews	December 29, 2017
	The following statementsÂ were posted to the verified Twitter accounts of U.S. President Donald Trump, @realDonaldTrump and @POTUS. The opinions expressed are his own.Â Reuters has not edited the statements or confirmed their accuracy. @realDonaldTrump : - Together, we are MAKING AMERICA GREAT AGAIN! bit.ly/2InpKaq [1814 EST] - In the East, it could be the COLDEST New Yearâ€™s Eve on record. Perhaps we could use a little bit of that good old Global Warming that our Country, but not other countries, was going to pay TRILLIONS OF DOLLARS to protect against. Bundle up! [1901 EST] -- Source link: (bit.ly/2jBh4LU) (bit.ly/2jpEXYR)		

Fig. 2. True news sample from the dataset

title	text	subject	date
NEIL CAVUTO Gives A HUGE Reality Check To College Activist Who Wants Free College [Video]	This is awesome! Cavuto rips into this activist big time!	left-news	Nov 12, 2015
NO JOKE! THE LYINâ€™ HILLARY DOLL IS HERE!â€¦ 18 Lies Included! [Video]	HERE S SOMEONE GOING THROUGH ALL 18 LIES WOW!	politics	Oct 6, 2016

Fig. 3. Fake news sample from the dataset

changed over to lowercase, and the stop words and unique characters are taken out utilizing "Natural Language Toolkit" (NLTK) [3] and normal articulations. Then the following steps are performed.

3.3 Text Classification Algorithms Used

Our model proposes tf-idf vectorizer and count-vectorizer for extricating highlights from the text in the dataset, after which those elements are taken care of onto various classifiers. Since this is a text-arrangement issue, we carry out the accompanying three classification algorithms to order fake and genuine information:

- Naïve Bayes: Naive Bayes, a conventional text-classification algorithm built on the Bayes theorem, is one of the most widely used methods for text classification issues. If we consider collected data as D and the class of data as (Ci), it can be fake or real. The probability of data D in class Ci can be calculated as follows [3]: $P(Ci/D) = P(D/Ci) *P(Ci) /P(D)$
- Logistic Regression: This Machine learning algorithm has borrowed the Logistic Regression technique from the field of statistics. It is the most popular method for binary classification problems (problems which has two class values).
- Passive Aggressive: This Machine learning model can be of two types - batch learning and online machine learning algorithms. All training datasets are utilized simultaneously in batch learning., whereas the data input for online machine learning algorithms arrives progressively in sequence. The passive-aggressive method is employed for the huge amount of data as an "online learning algorithm." For example, Twitter is a social media website where new data is added constantly; therefore, the data is huge. So, using the "Online Learning Algorithm" is ideal.

We have extracted text features using count vectorizer and tf-idf vectorizer and used them to classify the fake news using the "Naïve Bayes Classifier", "Passive Aggressive Classifier", and "Logistic Regression Techniques". Ultimately, we have compared the accuracies of all the six ways of classification. The process is illustrated in Fig. 4.

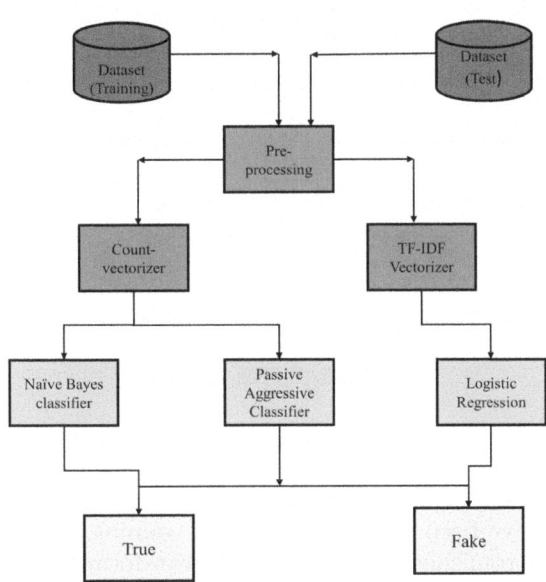

Fig. 4. Workflow of the ML-based classifiers

LSTM is a kind of RNN (Recurrent Neural Network) that contains sequential/successive data. It is additionally fit for learning the request reliance in issues of arrangement forecast. This conduct is useful in tackling complex issues like speech recognition or machine translation [4]. The architecture of LSTM [10] is given in the below Fig. 5.

3.4 Proposed Neural Network-Based Approach Using BERT

BERT helps in learning a deep contextualized representation of words. Although called a bi-directional network, its transformer encoding learns the word context at once instead of two passes. It provides us with pre-trained word vectors, which really helps in two major tasks in the implementation phase. The first one is the feature extraction from the dataset using a BERT tokenizer and encoder, and the second one is fine-tuning the existent model for sequence classification. The extension makes use of 110 million parameters, 12 layers, and a BERT Base with 12 attention heads.

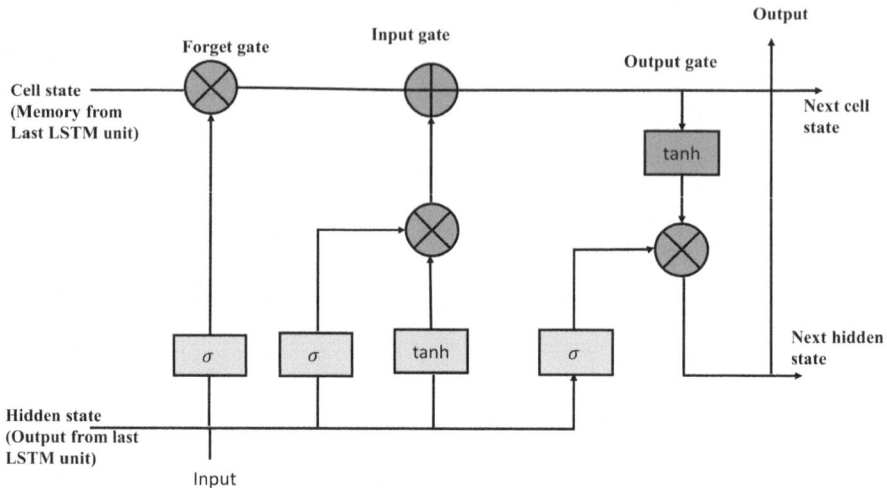

Fig. 5. LSTM unit

The main advantage of BERT is its pre-trained model for sequence classification, whose bottom layers and weights are already trained [4]. Otherwise, much more time would have been spent building and training a neural network from scratch to produce promising results. PyTorch-Transformers have been used here for the implementation of BERT. The maximum length of a document in BERT is 512 tokens. We tested from the highest possible length to lower; as high as 512 causes run time error due to the large space it takes to run at this size. Therefore, we finally selected 70 tokens as the length of each article and the size of the batches to be 16 based on performance as the standard. We used Adam optimizer as the loss function. To optimize the training, we used a learning rate scheduler to change the hyper-parameters and reduce the learning rate as epochs increase. We used 'utils.clip_grad_norm' from the PyTorch exploding gradient problem, where a large error gradient accumulates, leading to an unstable model that cannot learn. The below Fig. 6 depicts the flow of our model.

– Basic Configuration
  The basic configurations used in the model are as follows: • Validation split = 0.3 of train data-set (part of training data-set is used for validation) • Maximum sequence length = 100 • Activation function = sigmoid • Loss function = binary_crossentropy • Word embedding = (Word2Vec and BERT) • Learning rate = 0.1 • Epochs = 60.

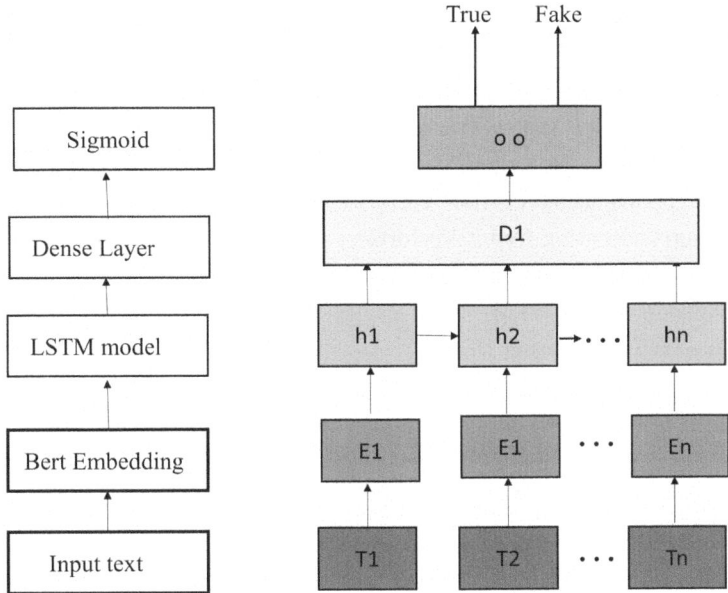

Fig. 6. BERT-LSTM model for fake news detection

4   Experiments and Results

Since a binary classification problem has been solved here, there can be four different outcomes:

- The actual is REAL, and the expected prediction is REAL: this is known as "True Positive" (TP).
- The actual is Fake, and the prediction is REAL: this is known as "False Positive" (FP).
- The actual news is FAKE, and the prediction is FAKE: this is known as "True Negative" (TN).
- The actual is REAL, and the prediction is FAKE: this is known as "False Negative" (FN).

The confusion matrix is created by counting how many times each of these outcomes occurs. To compare the models, we compute some performance metrics using confusion matrix, which helps us to analyze the performance of different models.

Table 1. Performance comparison of various ML and DL-based models

Models	Accuracy	Precision	Recall	F1 Score
Naïve Bayes using Count Vectorizer	94.3%	94%	94%	94%
Naïve Bayes using TF-IDF Vectorizer	91.8%	93%	91%	92%
Passive Aggressive using Count Vectorizer	91.3%	92%	91%	91%
Passive Aggressive using TF-IDF Vectorizer	83.7%	87%	84%	83%
Logistic regression using Count Vectorizer	92.1%	92%	92%	92%
Logistic regression using TF-IDF Vectorizer	66.2%	79%	66%	61%
Neural Network using LSTM	98.8%	99%	99%	99%
Our Proposed System using BERT	99.8%	99.8%	99.8%	99.8%

4.1 Performance Evaluation and Confusion Matrices

The confusion matrix is a tabular representation that evaluates the degree to which the categorization model properly assigns instances to various categories [8]. To assess the efficacy of the models, the confusion matrix is used for this research work. We tested our models' performance with standard classification metrics such as 'Accuracy', 'Precision', 'Recall', and 'F1 score' (see Table 1). The definitions of all these metrics are available in any standard textbooks like [8].

All the performance results obtained by the ML and DL-based algorithms, as mentioned previously, are depicted in Table 1 (Table 2).

Table 2. Comparison with similar systems on ISOT dataset

Systems	Technology/methodology used	Accuracy (%)
Ahmed et al. [2]	Machine learning algorithms, LSVM	92%
Nasir et al. [11]	Hybrid CNN-RNN	99
Our Proposed System using BERT	LSTM (or BiLSTM) using BERT	99.8

Figure 7 8 9 10 11 12 13 and Fig. 14 depicted the confusion matrices for all our applied methods.

4.2 Critical Discussions and Result Analysis

The ISOT dataset was used in this study, and the model we proposed outperformed two other comparable systems that had generated respectable results on the identical ISOT dataset. The performance of our suggested solution is most obviously due to the utilization of BERT. Table 2 displays the outcome. In this study, we demonstrate that our model provided 99.8% accuracy when the BERT model was applied.

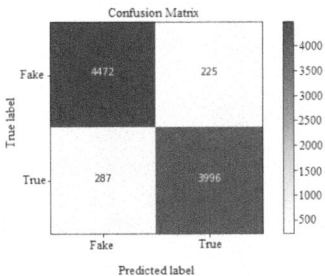

Fig. 7. Classifier using Naive Bayes and Count Vectorizer

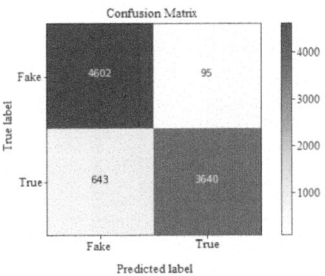

Fig. 8. Classifier using Naive Bayes and TF-IDF Vectorizer

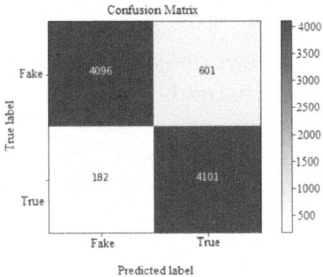

Fig. 9. Classifier using Passive Aggressive and Count Vectorizer

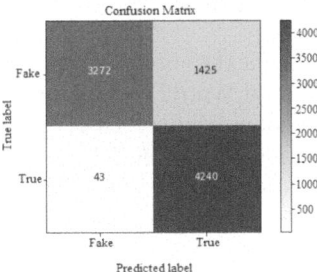

Fig. 10. Classifier using Passive Aggressive Classifier using TF-IDF Vectorizer

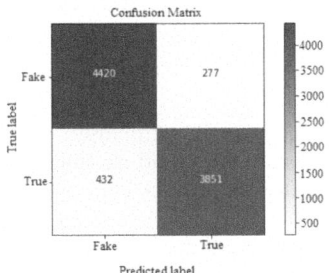

Fig. 11. Classifier using Logistic Regression and Count Vectorizer

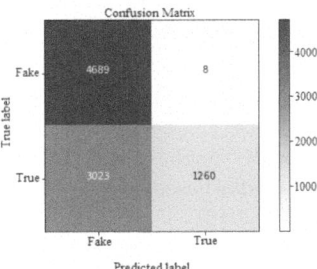

Fig. 12. Classifier using Logistic Regression and TF-IDF Vectorizer

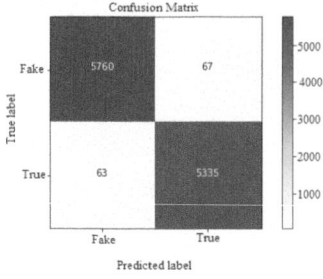

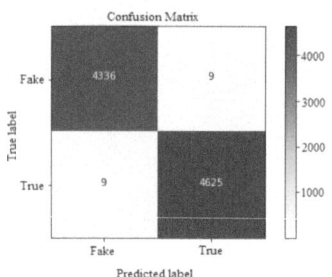

Fig. 13. Classifier using LSTM   **Fig. 14.** Classifier using BERT

5   Conclusion

This study used a common dataset for automatic fake news detection work. Here, We have used Word2Vec for implementing LSTM and PyTorch-Transformers for the implementation of BERT. The LSTM model yields an accuracy of 98.8% whereas the BERT model results in an accuracy of 99.8%, which outperforms the work done by [9], having an accuracy of 98.9%. Thus, our work outperforms all the fake news detection methods by giving a promising accuracy very close to 100%.

However, we have used a pre-defined dataset for our work, whereas in the real world every second millions of gigabytes of data are generated every second. Therefore, in the future, we would like to test the algorithms on real-time data, which are absolutely new and of different kinds, and test and develop methods to detect the presence of such hoaxes. We would also like to build a model to detect and eliminate fake news at a very early stage before reaching the masses and causing havoc on societal issues.

References

1. Ahmad, S., Asghar, M.Z., Alotaibi, F.M., Awan, I.: Detection and classification of social media-based extremist affiliations using sentiment analysis techniques. HCIS **9**, 1–23 (2019)
2. Ahmed, H., Traoré, I., Saad, S.: Detection of online fake news using n-gram analysis and machine learning techniques. In: Intelligent, Secure, and Dependable Systems in Distributed and Cloud Environments - First International Conference, ISDDC 2017, Vancouver, BC, Canada, 26–28 October 2017, Proceedings, volume 10618 of Lecture Notes in Computer Science, pp. 127–138. Springer (2017)
3. Bird, S., Klein, E., Loper, E.: Natural Language Processing with Python. O'Reilly, Sebastopol (2009)
4. Devlin, J., Chang, M.-W., Lee, K., and Toutanova, K.: BERT: pre-training of deep bidirectional transformers for language understanding. In: Proceedings of the 2019 Conference of the North American Chapter of the Association for Computational Linguistics: Human Language Technologies, Volume 1 (Long and Short Papers), pp. 4171–4186, Minneapolis, Minnesota. Association for Computational Linguistics (2019)

5. Drif, A., Hamida, Z.F., Giordano, S.: Fake news detection method based on text-features. In: France, International Academy, Research, and Industry Association (IARIA), pp. 27–32 (2019)

6. Farokhian, M., Rafe, V., Veisi, H.: Fake news detection using parallel BERT deep neural networks (2022). CoRR, abs/2204.04793

7. Ghosh, S., Shah, C.: Towards automatic fake news classification. Proc. Assoc. Inf. Sci. Technol. **55**, 805–807 (2018)

8. Han, J., Kamber, M.: Data Mining: Concepts and Techniques, Second Edition. The Morgan Kaufmann series in data management systems. Elsevier (2006)

9. Kaliyar, R.K., Goswami, A., Narang, P.: Fakebert: fake news detection in social media with a bert-based deep learning approach. Multim. Tools Appl. **80**(8), 11765–11788 (2021)

10. Le, X.-H., Ho, H.V., Lee, G., Jung, S.: Application of long short-term memory (lstm) neural network for flood forecasting. Water **11**(7), 1387 (2019)

11. Nasir, J.A., Khan, O.S., Varlamis, I.: Fake news detection: a hybrid CNN-RNN based deep learning approach. Int. J. Inf. Manag. Data Insights **1**(1), 100007 (2021)

12. O'Brien, N., Latessa, S., Evangelopoulos, G., Boix, X.: The language of fake news: opening the black-box of deep learning based detectors (2018)

13. Ohashi, S., Kajiwara, T., Chu, C., Takemura, N., Nakashima, Y., Nagahara, H.: Idsou at wnut-2020 task 2: identification of informative covid-19 english tweets. In: Proceedings of the Sixth Workshop on Noisy User-generated Text (W-NUT 2020), pp. 428–433 (2020)

14. Okano, E.Y., Liu, Z., Ji, D., Ruiz, E.E.S.: Fake news detection on fake.br using hierarchical attention networks. In: Computational Processing of the Portuguese Language - 14th International Conference, PROPOR 2020, Evora, Portugal, 2–4 March 2020, Proceedings, volume 12037 of Lecture Notes in Computer Science, pp. 143–152. Springer (2020)

15. Qasim, R., Bangyal, W.H., Alqarni, M.A., Ali Almazroi, A., et al.: A fine-tuned bert-based transfer learning approach for text classification. J. Healthc. Eng. **2022** (2022)

16. Rath, P.K., Basak, R.: Automatic detection of fake news using textual entailment recognition. In: 2020 IEEE 17th India Council International Conference (INDI-CON), pp. 1–6. IEEE (2020)

17. Ruchansky, N., Seo, S., Liu, Y.: CSI: a hybrid deep model for fake news detection. In: Proceedings of the 2017 ACM on Conference on Information and Knowledge Management, CIKM 2017, Singapore, 06–10 November 2017, pp. 797–806. ACM (2017)

18. Sharma, S., Kalra, V.: A deep learning based approach for fake news detection. Int. J. Sci. Res. Sci. Eng. Technol. (2021)

19. Shu, K., Mahudeswaran, D., Wang, S., Lee, D., Liu, H.: Fakenewsnet: a data repository with news content, social context, and spatiotemporal information for studying fake news on social media. Big Data **8**(3), 171–188 (2020)

20. Singh, V., Dasgupta, R., Sonagra, D., Raman, K., Ghosh, I.: Automated fake news detection using linguistic analysis and machine learning. In: International Conference on Social Computing, Behavioral-Cultural Modeling, & Prediction and Behavior Representation in Modeling and Simulation (SBP-BRiMS), pp. 1–3 (2017)

Unsupervised Approach for Word Sense Disambiguation in Bengali

Ratul Das[1]([⊠]) [iD], Alok Ranjan Pal[2] [iD], and Diganta Saha[1] [iD]

[1] Jadavpur University, Kolkata 700032, West Bengal, India
ratul2006das@gmail.com
[2] College of Engineering and Management, Kolaghat 721171, West Bengal, India

Abstract. Word sense disambiguation is the process of identifying a word sense based on its context. This article presents a novel approach for word sense disambiguation for Bengali text using unsupervised technique. The effect of clustering algorithms on 100 popular polysemous Bengali words are demonstrated. The features on which the clustering is applied are sentence embeddings and target word embeddings obtained from fastText and BERT model. Sentence embeddings are low dimensional real valued vector representation of a whole sentence. The novelty of the approach lies in the fact that instead of using hand crafted features the authors rely on the use of embeddings to form clusters. Sentences with similar semantics tend to be grouped into similar clusters. Hence the sentences in a cluster appear in similar context. Thus, it can be interpreted that a cluster effectively is associated with one particular meaning of a polysemous/homonymous word. Each of the cluster identifies a coarse grained sense distinction of the word based on its context. The evaluation of the clusters for a particular target word are done based on the random score. The results obtained are very promising. Out of a maximum possible adjusted random index of 1 the average value obtained for 100 words is 0.82 using a specific BERT implementation.

Keywords: Word Sense Disambiguation · Unsupervised approach · fastText · Semantic Similarity Measure · Word Embeddings · Sentence Embeddings

1 Introduction

In natural languages the relation between the meaning of a word and the word itself is not very straight forward. It is the context in which the word appears, determines the meaning of the word. This is the main idea behind distributional hypothesis [1]. Ambiguity in word senses occurs due to two important linguistic properties of words homonymy and polysemy [2]. Both refer to words having multiple meanings. But there is a difference between the two. Polysemy refers to the coexistence of many possible meanings for a word or phrase. Homonymy refers to the existence of two or more words having the same spelling or pronunciation but different meanings and origins. Homonymy gives coarse grained distinction in meanings and hence easier to detect. On the other hand, polysemy presents fine grained distinction which is a harder problem to solve. Here is an example to show the difference between them: Let's us take the target word চাল (chāl) (Table 1):

M. Majumder et al. (Eds.): ICCTE 2023, CCIS 2376, pp. 196–206, 2025.
https://doi.org/10.1007/978-3-031-81935-3_17

Table 1. Example sentences illustrating the different meanings of the word চাল (chāl).

Id	Sentence	Meaning
1	একজন খেলোয়াড় নিজের চাল দেবার পর একটি বোতাম টিপে অথবা একটি হাতলে চাপ দিয়ে প্রতিপক্ষের ঘড়ি সচল করে দেন (Ekjan kheloār nijer chāl debār por ekti botām tipe athabā ekti hātale chāp diye pratipaksher ghori sachal kore den: A player moves his opponent's clock by pressing a button or pressing a hand after making his move)	দাবা লুডো পাশা প্রভৃতি খেলায় ঘুঁটির দান (daba ludo pasha probhiti khelai ghuntir dan: turn of dice in games like chess, ludo, dice, etc.)
2	ভুটানের প্রধান খাবার হলো লাল চাল বাজরা ভুট্টা ইত্যাদি (Bhhutāner pradhān khābār holo lāl chāl bājrā bhuttā ityādi: The main food of Bhutan is red rice, buckwheat, corn etc)	ধানের খোসা ছাড়ালে যে খাদ্যশস্য পাওয়া যায় (dhaner khosha charale je khadyasashya paoa jai; Peel a squash, grate it and squeeze the juice)
3	এমন চালে খুশিই হয় কারখানা কর্তৃপক্ষ পুরস্কৃত হয় শ্যামলেন্দু (Eman chāle khushii hoy kārkhānār katripaksya puraskritya hoy Shyamalendu: The factory authorities are happy with such diplomacy and Shyamlendu is rewarded)	ফন্দি, কৌশল (Phandi, koushal: Tricks, tricks)
4	গ্রামের ঘরের টিনের চালের কোণে বাসা বাঁধে (Grāmer gharer tiner chāler kone bāsā bāndhe: Nestled in the corner of the tin roof of the village house)	বাঁশ টিন খড় তৃণ ইত্যাদির তৈরি কাঁচা ঘরের আচ্ছাদন বা ছাদ (bansh tin khar trina ithadir tairi kancha ghorer achhadan ba chad: Raw house cover or roof made of bamboo tin straw grass etc.)
5	প্রথম যৌবনের উচ্ছলতার প্রকাশ পেতো তার কথাবার্তা ও চাল চলনে (Prothom youboner uchhlatār prokāsh peto tār kathābārtā o chāl chalane: The eloquence of his early youth was reflected in his speech and mannerisms)	রীতিনীতি; স্বভাবচরিত্র; আচারব্যবহার (ritiniti; sbhabcharitra; acharbyabahar: Custom; Character Behavior)

The sentence/sentences (1, 3, 5), 2, 4 illustrates homonymy and 1, 3 and 5 illustrates polysemy. The clear sense demarcation in homonymy leads to a completely different set of context words.

Disambiguating word senses can be performed by using three different approaches. They are (a) knowledge based methodology, (b) Supervised methodology and (c) unsupervised methodology. The supervised methodologies give the highest accuracy rate. However, training data preparation for supervised approach involves extensive human labor and budget. Especially for low resources languages like Bengali development of a sense tagged corpus like Senseval [3] has not been undertaken.

The current research targets the problem of coarse-grained WSD using unsupervised (clustering) based approach.

The novelty of the approach lies in the use of the sentence embeddings in the clustering process. Traditionally researchers had been targeting the unsupervised approach by using hand-crafted features. Recent researchers have started using word embeddings to cluster sentences. The current approach extends the Transfer Learning paradigm in the field of unsupervised algorithms. Normally in Transfer learning for NLP large scale language models/word embeddings are transferred to supervised learning to fine tune the supervised model. In this approach the word embeddings are used to cluster sentences with the intention behind one sense per cluster. Since the current research uses word embeddings it handles several drawbacks of the traditional tf-idf models, e.g., handling of the stop-words and lemmatization. The research as a baseline uses subword embeddings from fasttext which can efficiently handle unknown word forms in most cases. The use of BERT models further improves the accuracy of the model. This is because BERT is context sensitive and the sentences if they have similar context, they can easily be clustered. The data used for clustering comprises of 100 polysemous (includes homonymy) words and on average 3–4 senses per word. Clustering approach for Bengali has never been attempted with 100 target words.

This research focuses on a comparative study of the unsupervised approach using different word/sentence embedding techniques.

The embeddings used for comparison can be broadly classified into two groups:

- Fixed embeddings using wor2vec/fasttext [4].
- Context sensitive embeddings using transformer based masked language models like BERT [5].

These groups mentioned above depend on the way the embeddings are built. Another way to categorize the groups is based on whether the embeddings are obtained from the whole sentence/set of context words/specific target word.

A comparative analysis of the different embeddings' generation technique and its effectiveness in the creation of clusters is closely studied. The Adjusted Rand Score [6] is used to measure the quality of the generated clusters by comparing them to the ground truth or reference clusters, while considering the possibility of randomness.

The fasttext word embeddings applied upon the whole sentence is treated as the baseline for evaluating the effectiveness of the clustering.

The paper is organized as follows: Sect. 2 delves into other works carried out in this area. Section 3 touches upon the resources and the following Sect. 4 presents a detailed discussion on the proposed approach. Section 5 presents the results and Sect. 6 provides a conclusion with a hint towards the future scope of work.

2   Related Work

In this section the use of unsupervised approaches towards word sense disambiguation is explored. Some notable directions and approaches that were being explored in this section.

Graph-based Approaches: Unsupervised methods often leverage thesauri or knowledge graphs to create word sense representations. These representations can be used to compute word similarities or to build sense clusters. Graph-based algorithms, such as Random Walks or PageRank [7], can then be applied to determine dominant senses based on the graph structure.

Distributional Semantics: Distributional methods capture word meanings based on the distributional patterns of words in a large corpus.

Topic Modeling: Topic modeling, such as Latent Dirichlet Allocation (LDA) [8], can be adapted to WSD by treating word senses as topics and assigning words to different senses based on their topic distributions in the context. Chaplot and Salakhutdinov 2018 [9] addressed the challenge of unsupervised Word Sense Disambiguation (WSD) by utilizing a topic modeling approach, allowing for linear scalability with the context size. Unlike traditional WSD systems that use small context windows, the proposed method employed a variant of Latent Dirichlet Allocation, leveraging synset proportions and WordNet information, enabling the entire document to be used as context.

Spectral Clustering: Spectral clustering techniques can be used to identify clusters of similar word senses in a graph representation of word occurrences. Goyal and Hovy in 2014 [10] were able to report a paired F-Score accuracy of 61.5% on the Semeval-2010 task using Spectral clustering.

Bootstrapping and Co-training: These techniques involve iteratively refining the sense disambiguation process by leveraging unlabeled data. They can start with a small seed set of disambiguated instances and gradually expand the set as more instances are correctly disambiguated.

Cross-Lingual and Multilingual Approaches: Unsupervised methods can be extended to work across multiple languages, leveraging language similarities to perform WSD. B. Moradi et al. in 2019 [11] by leveraging a trained English word2vec model and a bilingual dictionary were able to disambiguate Persian words. The approach identified the correct sense of a polysemous word in one language by comparing its translations against embeddings of surrounding translated words in another language. The method showcased strong accuracy (almost 90%) in experiments using a manually created test dataset. But it is to be noted that the number of target words used in the experiment was only 4.

The more recent research focuses on the use of Deep Learning based techniques and Large Language Models to address the problem of WSD.

Recurrent networks: A. Saeed et al. in 2021 [12] demonstrated the use of Recurrent Networks and its variations like LSTM's, GRU, BiLSTM and Ensemble learning on two bench marked Urdu corpora and reports an accuracy of 72.63%.

Prompt Based Approach: Q. He et al. in 2023 [13] demonstrated the use of prompt-based contextual word representation (PCWR), leveraging pre-trained models to enhance WSD performance, particularly in low-resource scenarios in both zero-shot and few-shot cases.

A lot of research is performed on Bengali WSD using Bengali WordNet as a Knowledge base or supervised methods. Since the current research focuses on unsupervised approach it is worth mentioning research performed by A. R Pal in 2019 [14]. The authors claimed an accuracy of 61% on 12 target words. But their research had dependencies on lemmatization methods, stop word identification and quality of WordNet. The current research proposes a novel approach that is not at all dependent on the quality of available resource like WordNet or lexicon in Bengali. The use of BERT model also ensures that the approach doesn't need any sophisticated stemming/lemmatization or stop words removal. Thus, the approach doesn't depend on the quality of these preprocessing activities.

3   Resources

The dataset used in the current research is obtained from Kaggle [15]. This is a huge dataset containing sentences from 100 polysemous words. The number of polysemous words having 3 senses is 55 and the rest 45 words have 4 senses.

Structure of each file is as shown below (Fig. 1):

<word-জল>< sense1-এক প্রকার পানীয় বিশেষ >

পানি বা জল হলো একটি যৌগ পদার্থ , যার রাসায়নিক সংকেত হল H2O ।

এক অণু জল দুটি হাইড্রোজেন পরমাণু এবং একটি অক্সিজেন পরমাণুর সমযোজী বন্ধনে গঠিত ।

সাধারণত পৃথিবীতে জল তরল অবস্থায় থাকলেও এটি কঠিন (বরফ) এবং বাষ্পীয় অবস্থাতেও (জলীয় বাষ্প) পাওয়া যায় ।

Fig. 1. Polysemous File for the target word জল (jal: water).

Each file contains headers and then paragraphs. The header contains the target word and the particular sense which is catered by the file. Paragraphs are separated by an empty line. Each paragraph is composed of a group of coherent sentences. Each sentence is separated by a newline.

Table 2. Example word জল (Jala: water) and চাল (chāl) from the dataset.

Id	Word	Sense Id	Sense
W1	জল (Jala: water)	W1S1	এক প্রকার পানীয় বিশেষ (Ēka prakāra pānīẏa biśēṣa: One type of drink)
		W1S2	অত্যাধিক পরিশ্রম (Atyādhika pariśrama: Exhausted)
		W1S3	বৃষ্টি (Bṛṣṭi: Rain)
		W1S4	অশ্রু (Aśru: Tear)
W2	চাল (Cāla)	W2S1	খাদ্যদ্রব্য বিশেষ (Khādyadrabhya bisesh: a food)
		W2S2	ঘরের ছাউনি (gharer chāuni: Thatched roof)
		W2S3	ছল চাতুরি (chol chāturi: Trick)
		W2S4	দাবার দান (Dābār dān: A move in chess)

Here is a high level description of two such polysemous word (Table 2):
The summary of the dataset is presented below:

- Total number of target polysemous words:100
- Total number of sentences: 6012
- Total number of words: 52789
- Avg number of words per sentence: 8.8

4   Proposed Approach

The high level flow diagram of the approach is presented in Fig. 2. The data is first stored in a dataFrame as per the structure given in Table 3.

Table 3. dataFrame structure.

Attribute name	Value
target word	অজ
sentence_id	ID2_1
sentence	অজ পাড়াগায়ে এমন বউ কজন আছে যিনি কি না সংবাদে...
gold label	sense3
bag of words	[বউ, কজন, এগিয়ে, এলেন, সংবাদের, সূত্র, পাড়াগা...]
bag of words context	[পাড়াগায়ে, বউ]
bag of words context with stop-words	[পাড়াগায়ে, এমন]

In the next step fastText is used to create sentence embeddings for the sentence and average word embeddings for the words in the bag of words respectively.

The BERT model is then used to create sentence-embedding and the target word specific embedding. The generation of BERT embeddings are done using pretrained BERT models obtained from Hugging Face.

The current research uses 3 different BERT models to compare their performance on the clustering task. The comparison results are discussed in the Sect. 6. The pretrained BERT models used in this research are:

- sagorsarker/bangla-bert-base [16]: Bangla-Bert-Base is a pretrained language model of Bengali language using mask language modeling. The model has bert-base-uncased model architecture which comprises of 12-layer, 768-hidden, 12-heads, 110M parameters.
- bert-base-multilingual-cased [17]: This model has been pretrained 104 languages and has similar parameters like the above.
- l3cube-pune/bengali-bert [18]: This model is trained on publicly available Bengali monolingual datasets.

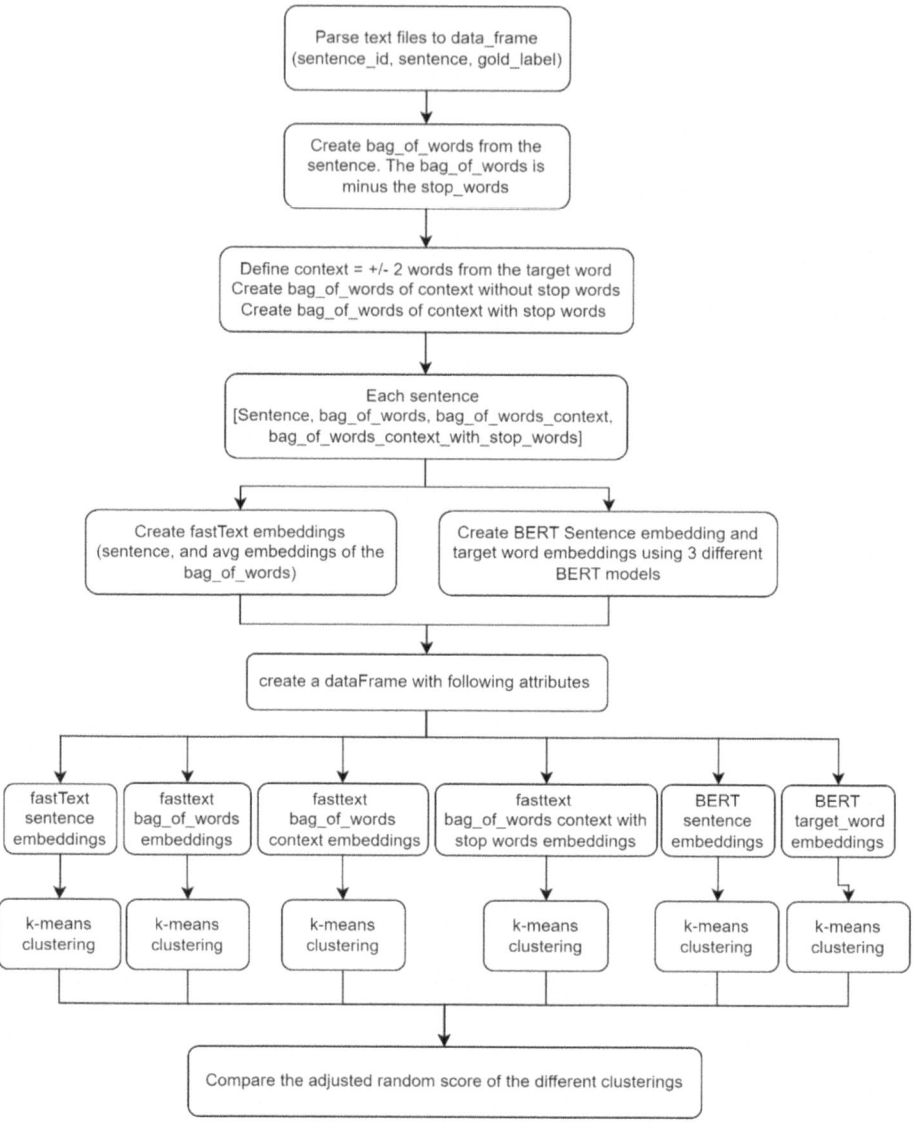

Fig. 2. High level flow chart for the proposed approach.

The embedding generation from BERT is a multi-step process. The process is highlighted in Fig. 3.

In the initial step the sentence or phrase is fed into a BERT Tokenizer. The tokenizer breaks the sentence into tokens that are understood by the BERT Model.

Normally it's a token per word but sometimes words are broken into multiple tokens or subwords. The input embeddings against each token are fed into the BERT model. The BERT model runs a forward pass and generates enhanced context sensitive output

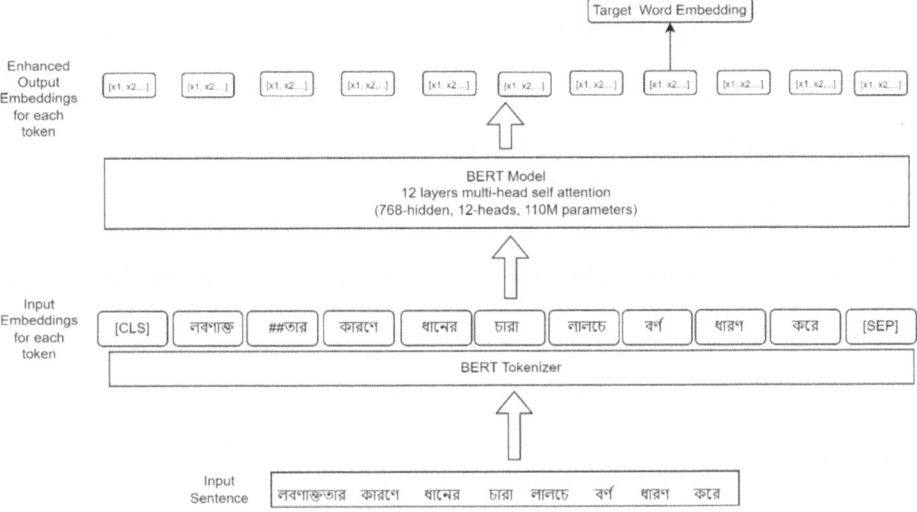

Fig. 3. Word Embedding generation using BERT.

embeddings. The current paper picks up the particular word embeddings against the target word which needs to be disambiguated.

Thus, for each sentence containing the target word we have a 768 dimensional real value embedding. These embeddings are now fed into the k-means clustering engine.

One of the hyper-parameters for the clustering algorithm is the number of clusters. Since the dataset already has a maximum of 4 senses per word, the number of clusters is set to 4. But this can be an important parameter which can be dynamically set up. Section 7 presents a detailed discussion on this.

5 Result Analysis

To measure the effectiveness of the clustering a metric known as Adjusted Random Score or Adjusted Random Index [19] is used. This is given by the following formula:

$$ARI = \frac{RI - Expected\_RI}{(max(RI) - Expected\_RI)} \qquad (1)$$

where,

$$RI = \frac{number\ of\ agreeing\ pairs}{number\ of\ pairs} \qquad (2)$$

The results of ARS or ARI are discussed in Table 4. Out of the 4 different fastText embeddings the whole sentence of fasttext embeddings gives the best result. Then the clustering experiment was repeated with sentence and target word embedding using 3 variants of BERT model (refer Sect. 5). The observations show that the target word embeddings result for each of the BERT models outperform the results obtained from sentence embeddings using the same BERT model. Also, the l3cube Pune Bangla BERT model target word embeddings clearly shows a much higher Adjusted Random Score = 0.82. The highest possible value is 1.0 which is interpreted as all agreeing clustering result with the gold labels. Note that all Adjusted random score values are the average of the Adjusted random score for all the 100 target words.

Table 4. Results of Clustering using k-means Adjusted random score.

Embeddings Name	Adjusted random score
Avg fastText sentence embeddings	0.55
Avg fasttext bag of words embeddings	0.50
Avg fasttext bag of words context embeddings	0.46
Avg fasttext bag of words context with stop words embeddings	0.49
Avg Bangla BERT base sentence embeddings	0.42
Avg Bangla BERT base target word embeddings	0.58
Avg BERT base multilingual cased sentence embeddings	0.45
Avg BERT base multilingual cased target word embeddings	0.57
Avg l3cube Pune Bangla BERT sentence embeddings	0.60
Avg l3cube Pune Bangla BERT target word embeddings	0.82

The scope of the paper is limited to only finding the cluster and not labelling it. Hence reporting a confusion metric or accuracy is not effective. Another metric which is presented in the paper to show the clustering effectiveness is silhouette score (Table 5).

Table 5. Results of Clustering using k-means.

Embeddings Name	Silhouette score
Avg fastText sentence embeddings	0.33
Avg fasttext bag of words embeddings	0.39
Avg fasttext bag of words context embeddings	0.45
Avg fasttext bag of words context with stop words embeddings	0.43
Avg Bangla BERT base sentence embeddings	0.31
Avg Bangla BERT base target word embeddings	0.36
Avg BERT base multilingual cased sentence embeddings	0.32
Avg BERT base multilingual cased target word embeddings	0.33
Avg l3cube Pune Bangla BERT sentence embeddings	0.34
Avg l3cube Pune Bangla BERT target word embeddings	0.50

The average silhouette score for the 100 target words doesn't show very promising results. This may be because the clusters overlap, and the inter-cluster distance is not much in most cases.

6 Limitations

Unsupervised approaches in solving WSD is very valuable in scenarios where labeled training data is expensive to obtain. However, in this section a discussion on some of the limitations of the proposed approach is presented.

- Unsupervised methods depend on the inherent patterns in the data. They don't leverage examples for training. Hence certain subtleties of the language are not captured in this approach. This leads to compromised accuracy in comparison to supervised or fine-tuned approaches.
- Unsupervised methods fail to distinguish between fine-grained senses or in cases when the context is somewhat similar.
- They are good in detecting homonymy but not so good in detecting polysemy. This is because polysemous words have in most cases very fine grained sense distinction and hence context are similar with minor differences.
- The current method creates sense clusters but labeling the cluster with the appropriate sense would require a rich WordNet.
- Natural Language is highly creative hence the word embeddings used in this approach should be trained from a large corpus containing examples from different domains. If there is biasness when creating the word embeddings, that would propagate to the clustering process.

7 Conclusion and Future Work

This approach of clustering using k-means to solve the problem of WSD has several benefits. The method is only dependent on the availability of sentences containing the target words. There is no requirement of sense tagged corpus and good dictionary or WordNet to achieve the sense tagging. If required, the tagging can be done for the whole cluster with very minimal manual effort.

The number of clusters used for the experiment, in the paper, was fixed at 4 because the data used for the experiment was having 3–4 senses per word. However, deciding the number of clusters is an important hyper parameter and requires careful study. Two of the popular approaches Elbow method and Silhouette Method [20] can be used to determine the number of clusters.

The current paper uses one way to perform the clustering which is k-means. The experiments can be repeated with other clustering methods like spectral clustering, hierarchical clustering, DBSCAN etc. The effect of the clustering method on the adjusted random score can be an interesting study.

The current clustering method gives good results when the sentences have clearly segregated context. This can happen when the target words exhibit homonymy. For polysemy the embeddings are not very apart. Hence determining the candidate cluster is a challenge.

For polysemy detection the experiments can be extended with a fine-tuned classifier on top of a large language model [21].

References

1. Harris, Z.S.: Distributional structure. WORD **10**, 146–162 (1954)
2. Dash, N.S.: Polysemy and homonymy: a conceptual labyrinth. In: IndoWordNet Workshop 2012. Indian Institute of Technology, Kharagpur, India (2012)
3. Senseval 3, Evaluation exercises for Word Sense Disambiguation Organized by ACL-SIGLEX. https://web.eecs.umich.edu/~mihalcea/senseval/senseval3/data.html. Accessed 01 July 2023
4. Bojanowski, P., Grave, E., Joulin, A., Mikolov, T.: Enriching word vectors with subword information. Trans. Assoc. Comput. Linguist. **5**, 135–146 (2017)
5. Devlin, J., Chang, M.-W., Lee, K., Toutanova, K.: BERT: pre-training of Deep Bidirectional Transformers for Language Understanding (2018)
6. sklearn.metrics.adjusted_rand_score: Rand index adjusted for chance. https://scikit-learn.org/stable/modules/generated/sklearn.metrics.adjusted_rand_score.html. Accessed 01 July 2023
7. Agirre, E., de Lacalle, O.L., Soroa, A.: Random walks for knowledge-based word sense disambiguation. Comput. Linguist. **40**, 57–84 (2014)
8. Blei, D.M., Ng, A.Y., Jordan, M.I.: Latent Dirichlet allocation. J. Mach. Learn. Res. **3**, 993–1022 (2003)
9. Chaplot, D.S., Salakhutdinov, R.: Knowledge-based word sense disambiguation using topic models. In: Proceedings of the AAAI Conference on Artificial Intelligence, vol. 32, no. 1 (2018)
10. Goyal, K., Hovy, E.: Unsupervised word sense induction using distributional statistics. In: Proceedings of COLING 2014, the 25th International Conference on Computational Linguistics: Technical Paper 13 (2014)
11. Moradi, B., Ansari, E., Zabokrtsky, Z.: Unsupervised word sense disambiguation using word embeddings. In: 2019 25th Conference of Open Innovations Association (FRUCT) (2019)
12. Saeed, A., Nawab, R.M., Stevenson, M.: Investigating the feasibility of deep learning methods for Urdu word sense disambiguation. ACM Trans. Asian Low-Resour. Lang. Inf. Process. **21**, 1–16 (2021)
13. He, Q., Zhang, J., Huang, X.: Improved word sense disambiguation via prompt-based contextual word representation. In: 023 International Joint Conference on Neural Networks (IJCNN) (2023)
14. Pal, A.R., Saha, D.: Word sense disambiguation in Bengali language using unsupervised methodology with modifications. Sādhanā **44** (2019)
15. Das Dawn, D., Khan, A., Shaikh, S.H., Pal, R.K.: A dataset for evaluating Bengali word sense disambiguation techniques. J. Ambient. Intell. Humaniz. Comput. **14**, 4057–4086 (2022)
16. sagorbrur GitHub - sagorbrur/bangla-bert: Bangla-Bert is a pretrained BERT model for Bengali language, GitHub. https://github.com/sagorbrur/bangla-bert. Accessed 01 July 2023
17. Devlin, J., Chang, M.-W., Lee, K., Toutanova, K.: BERT: pre-training of deep bidirectional transformers for language understanding. In: Proceedings of NAACL-HLT 2019. Association for Computational Linguistics (2018)
18. Joshi, R.: L3Cube-HindBERT and DevBERT: pre-trained bert transformer models for Devanagari based Hindi and Marathi languages (2022)
19. Warrens, M.J., van der Hoef, H.: Understanding the adjusted Rand index and other partition comparison indices based on counting object pairs. J. Classif. **39**, 487–509 (2022)
20. https://towardsdatascience.com/how-many-clusters-6b3f220f0ef5. Accessed 01 July 2023
21. Ding, N., et al.: Parameter-efficient fine-tuning of large-scale pre-trained language models. Nat. Mach. Intell. **5**, 220–235 (2023)

A Hybrid Method for Bengali Word Segmentation from Handwritten Copies of School Students

Moumita Moitra[1], Souvik Ganguly[2], and Sujan Kumar Saha[1(✉)]

[1] CSE Department, National Institute of Technology Durgapur, Durgapur, India
mm.23cs1108@phd.nitdgp.ac.in , sksaha.cse@nitdgp.ac.in
[2] Dr. B. C. Roy Engineering College, Durgapur, India

Abstract. This paper presents a novel approach for word segmentation from handwritten Bengali document images. We have employed a modified version of the Scale Space method, where it is combined with the Shuffled Frog-Leaping algorithm (SFLA). Our approach overcomes certain limitations of the standard Scale Space method in choosing the right scaling parameters. The incorporation of the SFLA with the Scale-Space method for handwritten word segmentation allows adaptive parameter tuning and, in this way, optimizes the right scale for the segmentation process. This method is employed to segment words from handwritten Bengali answer sheets gathered from schools to develop a handwritten character recognition (HCR) system. The proposed method is compared with a few existing methods, and the experimental results show that the proposed method is superior to others.

Keywords: Word Segmentation · Handwritten character recognition · Handwritten Bengali script

1 Introduction

The process of converting handwritten text into machine-readable form is known as handwritten character recognition (HCR). When the input to the HCR system is an image of a page containing several lines of text, it is essential to extract the individual words for better processing. Handwritten word segmentation is the process of extracting the words from handwritten images. Although segmentation primarily focuses on separating words, it also refers to the separation of lines or characters in some domains.

In the case of printed documents or images of printed documents, the space between any two consecutive words or characters and the space between two lines almost always remains consistent. However, in handwritten documents, the space varies since each person's writing style is unique. When addressing handwritten word segmentation, intricacies arise due to the presence of slant and skew at the word level and the non-uniform inter-word and intra-word gap. Therefore,

M. Majumder et al. (Eds.): ICCTE 2023, CCIS 2376, pp. 207–218, 2025.
https://doi.org/10.1007/978-3-031-81935-3_18

the segmentation task is comparatively difficult when the input is a handwritten document image. Also, the presence of noise and other language-specific features like matra, zone, compound characters, etc., make the segmentation of handwritten documents more challenging.

A possible approach for word segmentation maybe to first segment the text into lines, then segment those lines into words. However, this two-step approach might not always be the best for handwritten recognition. The difficulties faced in the word segmentation step could become even more pronounced due to the complexities added during the text-line extraction process. Alternatively, if we directly extract words from the document image, this may effectively help to overcome the aforementioned errors.

In our study, a hybrid method has been considered for handwritten word segmentation. In this method, we have used SFLA to optimize the performance of the traditional Scale Space method. Scale Space method has been used in the literature for handwritten word segmentation in English and some other languages. The success of the Scale-Space method depends on the accurate selection of parameters for scaling. The primary objective of integrating SFLA is to automate the parameter tuning process, thereby achieving an optimal scale for the segmentation task. SFLA is a population-based cooperative search metaphor inspired by natural memetics, which helps in the dynamic adjustment of the scale parameters. With the SFLA, we have systematically searched and adapted the scaling parameters over iterative computations. The algorithm helped in identifying the optimal parameter values that yield the highest segmentation accuracy and enhanced the overall performance of the Scale-Space method.

To assess the efficiency of the proposed technique, we have also analyzed the performance of three other methods: the vanilla Scale Space method, the Connected Component Analysis method, and the Stroke Width Analysis method. All four methods are implemented and tested with our test dataset to measure the relative performance. The test dataset contains images from an open online Bengali HCR dataset and our in-house dataset that contains images of handwritten answer books of grade six Bengali medium students. In our experiments, we found that the proposed technique performs better than others.

2   Related Work

Bengali, a major language in the Indian subcontinent and the seventh most spoken native language globally, has a complex script that combines consonants with inherent vowels and uses diacritics for other vowels. Its script includes compound characters and conjuncts, making word boundary identification challenging. This complexity has led to numerous research efforts in handwriting text recognition and document segmentation for various Indian scripts [1–3]. Some of the popular text segmentation methods are the Projection Profile method, Connected Component Analysis (CCA), Scale Space method, Stroke Width Analysis, and machine learning and deep learning-based methods. There are some studies in the literature that focus solely on segmenting handwritten Bengali lines and words.

U. Pal et al. [4] applied the Projection Profile method for text line segmentation from Bengali handwritten documents. R. MamathaH et al. [5] proposed a segmentation technique for extracting individual text-lines from handwritten document images using morphological operation and run-length smearing algorithm (RLSA). A word in Bengali script can be divided into three horizontally adjacent zones called the bottom zone, middle zone, and upper zone. These zones were utilized in this article [9] to identify words that were contained within a single line. R. Pramanik et al. [6] proposes a method for recognizing handwritten Bengali and Devanagari words. It detects and corrects skew within words, estimates the headline, and segments words into meaningful pseudo characters. [12] used the scale space approach for word segmentation from handwritten historical documents. [13] proposed a tri-level handwritten text segmentation technique where they have used the scale space method for word-level segmentation of handwritten documents. To our knowledge, this is the only work that recognizes Bengali words from images. All these methods have some drawbacks in the case of oblique texts and critical overlapping situations. Also, most of these methods were applied on datasets where the space between lines and words was enough clear and noticeable. After observing the efficacy of all existing methods, we discerned that there is still significant scope for the enhancement of the word segmentation process from handwritten document images.

3   Segmentation Methods Implemented

In the previous section, we found that in the literature, several techniques have been applied for the segmentation of handwritten words. As per our study of our domain of interest and theoretical understanding, we have chosen three methods that we implemented first. These are discussed below.

3.1   Connected Component Analysis (CCA)

CCA, also known as Connected Component Labelling, is a popular technique that finds a wide range of applications in the domain of computer vision and pattern recognition. In handwritten word segmentation, CCA identifies and extracts distinct word components in a given document image [7]. CCA works by grouping connected pixels with the same pixel intensity value [8]. Based on the connectivity, CCA divides the image into individual components. For handwritten text, CCA can help determine separate words, characters, or even parts of characters by identifying these connected components. In the case of handwritten text, this approach can be effective if the units (characters or words) are well-separated and do not overlap. This can especially work well when the writing is cursive and a word is written in a single stroke. However, it might not work well in overlapping handwriting where individual letters and words are not neatly separated. Therefore, the performance of CCA might be influenced by factors like writing style, ink spread, or noise in the image.

3.2 Stroke Width Analysis (SWA)

This method estimates Stroke Width (SW) by detecting and measuring sequences of continuous black pixels in each image row. The mode of these lengths provides a representative SW value, guiding vertical and horizontal smoothing to identify word-level components.

Vertical Smoothing: While traversing the image vertically, if a sequence of black pixels (valued 0) is smaller than or equal to the stroke width (SW), they are considered as noise. Those pixels were replaced by white (value 1) pixels which represent the background of the image. Otherwise, the pixels are considered to be part of the foreground which is represented by black pixels.

Horizontal Smoothing: While scanning the image from left to right, if a series of white pixels (valued at 1) is shorter than five times the stroke width (SW), they're viewed as part of the text. These white pixels are then converted to black (valued at 0), symbolizing the main content. If this isn't the case, these pixels are seen as "noise" and are changed to white, representing the background, following the same equation.

Post the initial steps, a morphological opening operation merges characters into single components [10]. After that, this method divides the image into word-level segments, each ideally representing a distinct word, with segmentation guided by the spaces between words and Stroke Width (SW).

3.3 Scale Space Method

The Scale-Space technique, effective for segmenting handwritten texts, involves preprocessing the document image for noise removal and clarity enhancement, followed by applying scale-space theory which emphasizes the importance of scale in identifying significant objects and features. The concept applies a convolution of a function f(x, y) with a two-dimensional Gaussian kernel to generate a sequence of smoothed signals L(x, y; t). These signals, forming a linear scale-space representation, offer a series of progressively blurred versions of the original two-dimensional image, thus revealing information at various scales [11].

$$G(x, y; t) = \frac{1}{2\pi t} e^{-\left(x^2 + y^2\right)/2t} \tag{1}$$

In the above equation, the scaling parameter t indicates the variance for the Gaussian filter at the scale level where t $\geq$ 0.

$$L(x, y; t) = G(x, y; t) * f(x, y) \tag{2}$$

In Scale-Space representation, an image is transformed across various scales by a convolution operation, as shown in Eq. 2. Here, L(x, y; t) denotes the scale-space image, 'f(x, y)' is the original image, and '*' indicates the convolution operation. As we increase the scale level 't', it progressively reduces the image's fine details as a result of the smoothing effect of the Gaussian kernel. This phenomenon is encapsulated in Eq. 2, where the convolution operation (f * G) modifies the shape of the image, smoothing it across the increasing scale levels [12]. Subsequently, specific features like lines and curves are extracted at each scale to identify word boundaries in Bengali handwritten words, focusing on scaling level, Gaussian filter size, and sigma. The key challenge is selecting the correct scale, requiring precise algorithm tuning.

4   Proposed Method

Here, we propose a word segmentation algorithm that identifies the words from a document image. In this study, we are working with Bengali handwritten answer sheet images collected from a school. Most of the pages were written on the ruled pages, and the images were noisy. We first applied a few preprocessing steps to clean the images, such as image binarization and noise reduction. Then, we applied our proposed algorithm, which combines the shuffled frog leaping algorithm (SFLA) algorithm and the Scale-Space method. Finally, we compare the proposed technique with the traditional scale-space method to assess its effectiveness.

Now, we discussed tuning the hyper-parameters of the scale-space technique using SFLA for handwritten Bengali word segmentation. The workflow is summarized in Fig. 1.

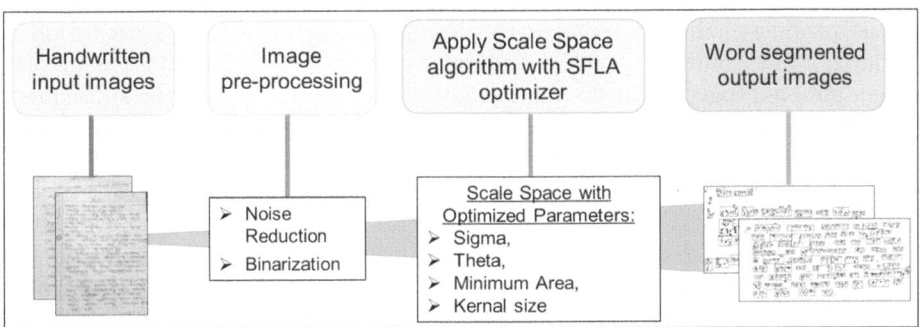

Fig. 1. Workflow of the proposed method

SFLA, a population-based heuristic search technique, combines aspects of the Memetic Algorithm (MA) [15] and Particle Swarm Optimization (PSO) [16] to solve complex optimization problems. Introduced by Yusuf and Lansey in 2003, it uses a population of potential solutions (frogs) in a D-dimensional space to conduct local and global searches, ranking frogs based on an objective function [17]. The M/N number of frogs is distributed to each memeplex if there are M number of frogs and N number of memeplexes. According to this, the foremost memeplex has the frog with the highest fitness value demarcated by an objective function, and the second memeplex possesses the next-best frog. As long as M^{th}

frog does not obtain the Mth memeplex, this process will continue. Iteration-wise, following this order, $(M+1)^{th}$ frog with the loftiest fitness value is again allocated to the first memeplex. Once the population search mechanism is over, the best frog is addressed as X_{NBest} whereas the worst one is addressed as X_{NWorst}. In the entire population, the frog with the global best fitness value is opted as the global best frog and is identified as X_G. In every memeplex, the worst frog is moving in the direction of the best or ideal one. If it is determined after each iteration that the new position is superior to the old one, the position is improved. The movement of X_{NWorst} has been obtained by Eqs. 3 to 5.

$$[M^T]_N = rand(X_{NBest} - X_{NWorst}) \tag{3}$$

In Eq. 3, $[M^T]_N$ identifies the movement of the frogs with T^{th} iterations within the N memeplex with worst fitness value. Hence, rand () is a random number generation method that generates the numbers between 0 and 1.

$$M_{min} \leq [M^T]_N \leq M_{max} \tag{4}$$

From Eq. 4, M_{min} and M_{max} both determine that the maximum number of alterations are permitted for the worst frog in the memeplex.

$$tX_{new} = X_{NWorst} + [M^T]_N \tag{5}$$

In Eq. 5, X_{NWorst} denote the positions of the worst. If a better solution X_{new} is produced, then it replaces the X_{NWorst}. Accordingly, if no solution is attained, a new solution is picked randomly to replace X_{NWorst}. The memeplex is then jumbled together to share information and reallocated for the subsequent search process after being refreshed and reordered within local search durations.

This population-based meta-heuristic search method is utilized to tune the hyperparameters of the Scale-Space technique for word segmentation. To dissect an image's content at various granularities, scale-space space theory, a concept in image processing and computer vision, calls for portraying the image at several scales. Scale space carries its cue from the human optic system, which is capable of perceiving objects and patterns in a range of scales, from minute fragments to comprehensive structures. A convolution of f(x, y) with the two-dimensional Gaussian kernel is used to yield a lineage of emanated signals L(x, y; t) for the linear scale space miniature of consecutive signals with arbitrarily sized signals for a two-dimensional image [18].

5   Experimental Setup

5.1   Dataset Collection

As we could not find any work or dataset on handwritten character recognition from Bengali student answer books, there needs to be a standardized dataset composed of student answer sheets. In the absence of a suitable dataset, we collected some handwritten Bengali images to test the proposed system. We have collected a few images from the open online BN-HTR dataset [13,14], and some images are taken from an in-house dataset. The in-house dataset is collected from a Bengali medium school where the images of answer books of students in grades six and seven were scanned using a mobile scanner. The images of the whole pages were used as input to our system. Although the online dataset images are cleaner, the images in the in-house dataset introduced an array of challenges that added complexity to the segmentation task, including issues like skew, curves, and closely spaced or touching lines of text; unequal intra-word and inter-word space; noise including poor paper quality, and scanning distortion. The test dataset contains eight images, four of which are taken from the in-house dataset, and the remaining four images are taken from the BN-HTR dataset. Figure 2 shows a few representative images from both sources.

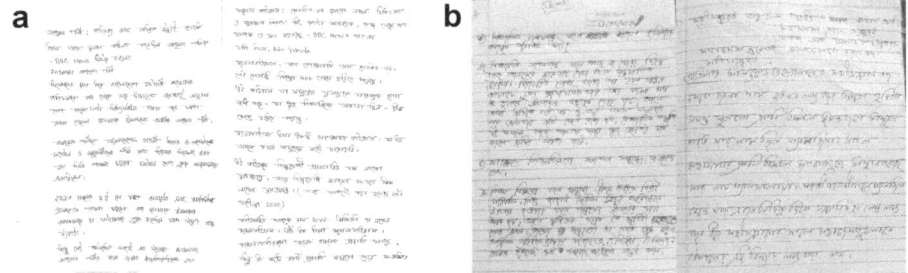

Fig. 2. Representative images of the dataset used in our study (a) online collected handwritten dataset, (b) handwritten images collected from the school.

5.2   System Evaluation Measure

Word segmentation in handwritten Bengali documents is challenging due to complex handwriting styles, unique script features like compound characters, Matra, and Diacritical marks. In our study, we addressed these challenges by identifying each word within a bounding box.

When we mark a machine-generated segmentation as correct or not, we might have considered two classes - correct and wrong. However, when we studied various output files, we found incorrect segmentation can be categorized into three classes: over-segmentation, under-segmentation, and partial-segmentation. And perfect-segmentation is the correctly segmented class where a single bounding box identifies the whole word. In the case of over-segmentation, a single word is inaccurately identified in multiple parts. This may occur often in Bengali text segmentation due to inconsistencies in handwriting, such as variations in space between letters within a word or Matra connecting some part and the rest not. On the other hand, under-segmentation is a situation where more than one word is mistakenly identified as a single word. This happens typically due to an

214   M. Moitra et al.

unequal gap between words. Another problem occurred: partial segmentation, where a word is segmented, but some parts are incorrectly left out or included in another word.

6   Result and Discussion

To assess the effectiveness of the proposed method, we implemented some existing techniques and tested them using the same test data. The existing techniques are Connected Component Analysis, Stroke Width Analysis, and Scale space method, and the proposed one is the proposed SFLA with Scale Space technique. The segmentation process was executed on the test set images, followed by a detailed evaluation based on the factors outlined earlier. In Fig. 3, we have shown the outputs of the four different methods applied to an in-house image containing student answers. The figure shows that our proposed method, SFLA with Scale Space, demonstrated the best performance.

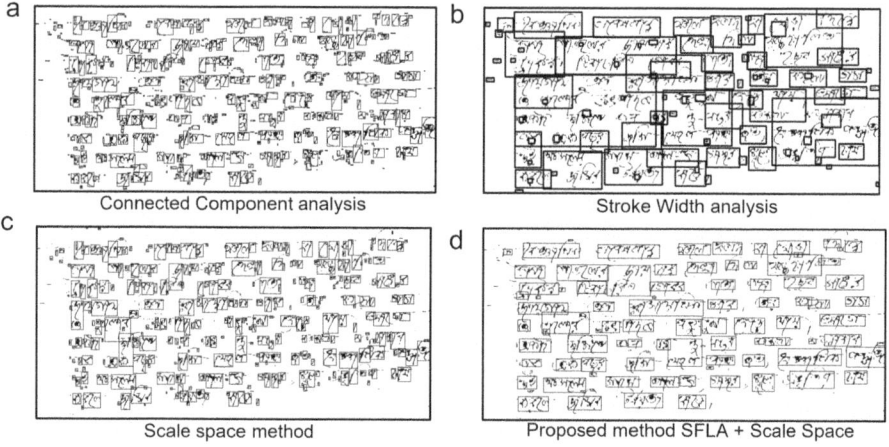

Fig. 3. Segmented outputs: (a–c) Images segmented using Connected Component Analysis, Stroke Width Analysis, and Scale Space method. (d) Image segmented with our proposed method.

Similarly, we apply all four techniques to the online images. The output of all four techniques applied on an online dataset image is shown in Fig. 4.

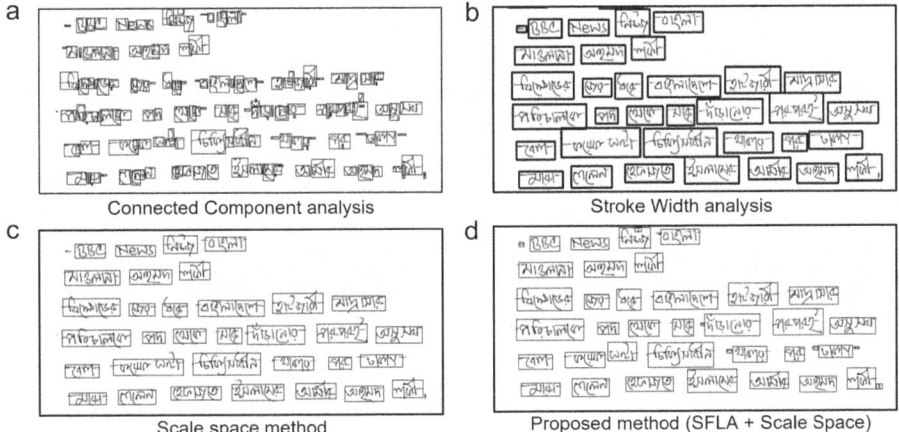

Fig. 4. Compared the resulting output images after word-level segmentation on hand-written downloaded image, (a, b, c) Segmented output images using Connected Component Analysis, Stroke Width Analysis, and Scale space method. (c) Image output after segmentation using our proposed method

Now, we perform a quantitative analysis. This analysis considers the result of all eight images. For this analysis, we manually counted the segmented boundaries from all the output images and compared the values as per the metrics specified in Sect. 5.2. Our in-house test dataset, comprising 4 images with 241 words, yielded a 56.84% accuracy rate using the proposed technique: 137 words were perfectly segmented, while there were 61 over-segmented, 16 under-segmented, and 27 partially segmented words. This accuracy was calculated based on perfect segmentation only. The results of applying three other techniques to the same dataset are summarized in Table 1, showing a comparative analysis.

Table 1. Quantitative analysis of word-level segmentation performed on handwritten in-house dataset.

Segmentation Methods	Total Words	Perfectly segmented	Over segmented	Under segmented	Partial segmented	Accuracy based on perfectly segmented words
CCA	241	94	89	20	38	39%
SWA	241	25	211	5	0	10.37%
Scale Space	241	62	58	119	0	25.72%
SFLA + Scale Space	**241**	**137**	61	16	27	**56.84%**

In the case of the downloaded handwritten image dataset, we have a total of 236 words in 4 pages. The proposed technique performed quite well on these images. 221 words were perfectly-segmented, and 15 words were over-segmented. There were no under-segmented and partial-segmented images were presented. So, the proposed model has achieved 93% accuracy on downloaded handwritten datasets. Similarly, we manually count and compute these values when other techniques were applied to the dataset. Table 2 presents the values.

Table 2. Quantitative analysis of word level segmentation performed on handwritten online dataset.

Segmentation Methods	Total Words	Perfectly segmented	Over segmented	Under segmented	Partial segmented	Accuracy based on perfectly segmented words
CCA	236	151	85	0	0	63.98%
SWA	236	115	121	0	0	48.72%
Scale Space	236	69	167	0	0	29.23%
SFLA + Scale Space	**236**	**221**	15	0	0	**93.64%**

The values in Table 1 and Table 2 prove the superiority of the proposed technique over the existing techniques.

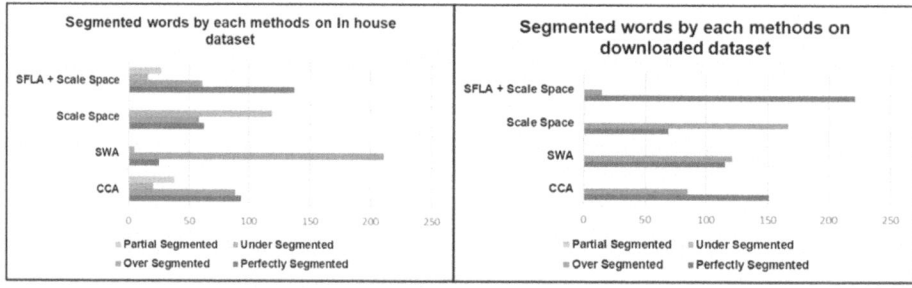

Fig. 5. Bar plot representing the performance of each method on in house dataset and downloaded dataset.

A bar plot (Fig. 5) with detailed error analysis highlighted that our method increased the number of perfectly segmented words by 18%, reduced over-segmentation by 12%, under-segmentation by 2%, and partial-segmentation by 5% compared to the next best-performing baseline method in segmenting in-house dataset. In the case of the downloaded dataset, proposed method increased the number of perfectly segmented words by 30% and reduced over-segmentation by 30%. These figures not only demonstrate the enhanced precision of our technique but also underscore its robustness in handling the complexities of handwritten text segmentation.

In our experiments, we observed that both the proposed and existing techniques perform better on the images collected from the BN-HTR dataset. However, their performance became poor when applied to the in-house images of student test copies. Segmentation of the student handwritten images is complex due to several handwriting inconsistencies, such as unequal intra-word and inter-word space, image noise, poor paper quality, and scanning distortion. We need more sophisticated techniques to perform word segmentation from those images.

7 Conclusion

In this paper, we propose a word segmentation methodology for unconstrained Bangla handwritten document images. The main novelty of the proposed approach is merging the Scale Space method with the Shuffled Frog-Leaping algorithm. This fusion not only addresses the challenges of the traditional Scale Space technique regarding parameter selection but also offers adaptive tuning. The proposed method is found to be superior when tested on our test dataset.

However, there are scopes for further improvement. We have received a greater number of over-segmented words that need further splitting to get an individual word. This problem specifically arises due to the absence of the required gap between two words or the discontinuation of 'Matra' inside a single word. These challenges might be mitigated by using machine learning and deep learning-based models, which can adapt different handwriting styles by learning directly from data. Deep learning, especially convolutional neural networks (CNN's), can automatically learn hierarchical representations from raw pixel values; we shall explore this possibility in the future.

Acknowledgment. This work is partially supported by SERB-GoI (EEQ/2021/ 000687).

References

1. Jindal, A., Ghosh, R.: Word and character segmentation in ancient handwritten documents in Devanagari and Maithili scripts using horizontal zoning. Expert Syst. Appl. **225**, 120127 (2023)
2. Priyadarshi, A., Saha, S.K.: The first named entity recognizer in Maithili: resource creation and system development. IFS. **41**, 1083–1095 (2021)
3. Inunganbi, S., Choudhary, P., Manglem, K.: Meitei Mayek handwritten dataset: compilation, segmentation, and character recognition. Vis. Comput. **37**, 291–305 (2021)
4. Pal, U., Datta, S.: Segmentation of Bangla unconstrained handwritten text. In: 2003 Proceedings of the Seventh International Conference on Document Analysis and Recognition, pp. 1128–1132. IEEE Computer Society, Edinburgh (2003)
5. Mamatha, H.R., Srikantamurthy, K.: Morphological operations and projection profiles based segmentation of handwritten Kannada document. IJAIS **4**, 13–19 (2012)

6. Pramanik, R., Bag, S.: Segmentation-based recognition system for handwritten Bangla and Devanagari words using conventional classification and transfer learning. IET Image Process. **14**, 959–972 (2020)

7. Khandelwal, A., Choudhury, P., Sarkar, R., Basu, S., Nasipuri, M., Das, N.: Text line segmentation for unconstrained handwritten document images using neighborhood connected component analysis. In: Chaudhury, S., Mitra, S., Murthy, C.A., Sastry, P.S., Pal, S.K. (eds.) Pattern Recognition and Machine Intelligence, pp. 369–374. Springer, Heidelberg (2009)

8. Wang, Y., Wang, W., Li, Z., Han, Y., Wang, X.: Research on text line segmentation of historical tibetan documents based on the connected component analysis. In: Lai, J.-H., et al. (eds.) Pattern Recognition and Computer Vision, pp. 74–87. Springer, Cham (2018)

9. Basu, S., Chaudhuri, C., Kundu, M., Nasipuri, M., Basu, D.K.: Text line extraction from multi-skewed handwritten documents. Pattern Recogn. **40**, 1825–1839 (2007)

10. Shivakumara, P., Jain, T., Pal, U., Surana, N., Antonacopoulos, A., Lu, T.: Text line segmentation from struck-out handwritten document images. Expert Syst. Appl. **210**, 118266 (2022)

11. Manmatha, R., Srimal, N.: Scale space technique for word segmentation in handwritten documents. In: Nielsen, M., Johansen, P., Olsen, O.F., Weickert, J. (eds.) Scale-Space Theories in Computer Vision, pp. 22–33. Springer, Heidelberg (1999)

12. Capobianco, G., et al.: Image convolution: a linear programming approach for filters design. Soft. Comput. **25**, 8941–8956 (2021)

13. Manmatha, R., Rothfeder, J.L.: A scale space approach for automatically segmenting words from historical handwritten documents. IEEE Trans. Pattern Anal. Machine Intell. **27**, 1212–1225 (2005)

14. Rahman, M.A.: BN-HTRd: a benchmark dataset for document level offline bangla handwritten text recognition (HTR) (2021). https://data.mendeley.com/datasets/743k6dm543/1

15. Nguyen, Q.H., Ong, Y.S., Krasnogor, N.: A study on the design issues of Memetic Algorithm. In: 2007 IEEE Congress on Evolutionary Computation, pp. 2390–2397 (2007)

16. Wang, D., Tan, D., Liu, L.: Particle swarm optimization algorithm: an overview. Soft. Comput. **22**, 387–408 (2018)

17. Eusuff, M., Lansey, K., Pasha, F.: Shuffled frog-leaping algorithm: a memetic metaheuristic for discrete optimization. Eng. Optim. **38**, 129–154 (2006)

18. Lee, J.-S.: Digital image smoothing and the sigma filter (1983). https://doi.org/10.1016/0734-189X-90047-6

Securing Social Spaces: Harnessing Deep Learning to Eradicate Cyberbullying

Rohan Biswas$^{(\boxtimes)}$ ⓘ, Kasturi Ganguly ⓘ, Arijit Das ⓘ, and Diganta Saha ⓘ

Department of Computer Science and Engineering, Jadavpur University, Kolkata,
West Bengal, India
`biswasrohan7@gmail.com, arijit.das@ieee.org`

Abstract. In today's digital world, cyberbullying is a serious problem that can harm the mental and physical health of people who use social media. This paper explains just how serious cyberbullying is and how it really affects individuals exposed to it. It also stresses how important it is to find better ways to detect cyberbullying so that online spaces can be safer. Plus, it talks about how making more accurate tools to spot cyberbullying will be really helpful in the future. Our paper introduces a deep learning-based approach, primarily employing BERT and BiLSTM architectures, to effectively address cyberbullying. This approach is designed to analyse large volumes of posts and predict potential instances of cyberbullying in online spaces. Our results demonstrate the superiority of the hateBERT model, an extension of BERT focused on hate speech detection, among the five models, achieving an accuracy rate of 89.16%. This research is a significant contribution to "Computational Intelligence for Social Transformation", promising a safer and more inclusive digital landscape.

Keywords: Cyberbullying Prevention · NLP · Deep Learning · BERT · hateBERT · Social transformation

1 Introduction

Digital communication tools, such as social media, websites, email, text messages, or other online platforms provide individuals with the opportunity to openly and publicly share their thoughts and emotions with others. In spite of the considerable advantages offered by digital communication tools, their misuse leads to significant and adverse consequences. Harmful behaviors such as cyberbullying, cyber-aggression, hate speech, offensive language, and various forms of negativity have been growing more prevalent in these forms of communication. Cyberbullying [26] can cause emotional distress, low self-esteem, isolation, and even lead to more severe consequences such as self-harm and suicidal thoughts.

Due to the immediate requirement and the absence of suitable systems to manage the challenge of hate speech online, numerous researchers are driven to automate

---

A. Das and D. Saha—Senior Member, IEEE.

the process of identifying offensive content. The Natural Language Processing (NLP) [27] community plays a significant role in this. BERT, which stands for Bidirectional Encoder Representations from Transformers, is a groundbreaking natural language processing (NLP) model introduced by researchers at Google AI in 2018. HateBERT and RoBERTa are advanced variations of the original BERT model. HateBERT is specialized in detecting hate speech and offensive language. It is trained with data sourced from communities with banned content, enhancing its ability to identify harmful language patterns where the BERT model is also proficient in understanding language context but not mainly focused on hate speech detection. RoBERTa achieves better language representation through improved context understanding by using larger training datasets and more iterations during pretraining.

Cyberbully detection is challenging due to the complex, context-dependent nature of online interactions. The diverse forms of communication, evolving language, subtle indicators, and cultural nuances make it hard for automated systems to accurately identify instances of cyberbullying. Additionally, imbalanced data, user privacy concerns, and the need to balance precision with recall further complicate the task. The dynamic and ever-changing online landscape, coupled with the potential for malicious users to craft content that evades detection, amplifies the difficulty of developing effective and reliable cyberbully detection methods.

BERT models tackle these difficulties by leveraging their deep contextual understanding of language. BERT's bidirectional architecture captures intricate language nuances and context, helping to identify subtle indicators of cyberbullying. HateBERT specifically enhances its training with data from banned online communities, making it more adept at recognizing offensive language. These models' pretraining and fine-tuning processes enable them to learn patterns indicative of cyberbullying, thus addressing the complexity of the task and contributing to more accurate and efficient detection in diverse online environments.

Automated cyberbully detection functions like a sensor by continuously monitoring online content, detecting potentially harmful language, and alerting users or moderators when such content is identified. When the system detects content that meets predefined criteria for cyberbullying, it triggers alerts or actions, such as notifying the platform's moderators, flagging the content for review, or even issuing warnings to users. This "sensor-like" approach allows platforms to proactively address harmful behavior and create safer online environments.

The novelties of this work are – a) Proposing a generalized state-of-the-art methodology for detecting cyberbullying comments in any languages, b) The proposed method works for non-dictionary, code-mixed, colloquial text containing misspelled and incomplete slang words as well which is generally very difficult to detect with classical NLP technique, c) The proposed system has achieved almost 90% accuracy in best case which is better than the benchmark system [9] run over same standard dataset. The proposed model can be very useful for intelligent crowdsensing in social media for detecting cyberbullying.

2 Literature Review

Researchers have proposed various techniques for cyberbullying detection across different languages and platforms. Ahmed et al. [1] introduced a hybrid neural network for Bengali, achieving 87.91% binary and 85% multiclass accuracy. Ranasinghe et al. [2] used cross-lingual embeddings for offense detection in multiple languages. Islam et al. [3] improved Bengali sentiment analysis with multi-lingual BERT. Balakrishnan et al. [4] explored Twitter psychological features for effective detection. Talpur et al. [5] used PMI-based methodologies for identifying cyberbullying occurrences on the Twitter platform. In another work by Muneer et al. [6] achieved 90.57% accuracy with Logistic Regression. Islam et al. [7] combined NLP and ML for abusive message detection. Samghabadi et al. [8] used BERT for aggression identification. Iwendi et al. [9] found BLSTM effective for insult detection. Traditional methods were outperformed by deep learning for cyberbullying in social networks [10]. Research [11] reviewed prediction models and ML techniques for cyberbullying detection. Hani et al. [12] achieved 92.8% accuracy with NN. Emon et al. [13] improved Bengali sentiment analysis using RNN. Chakraborty et al. [14] used SVM for Bengali abusive language detection. Tarwani et al. [15] achieved 80.26% accuracy for cyberbullying in Hinglish. Murnion et al. [16] collected in-game chat data to address cyberbullying. Al-Ajlan et al. [17] introduced CNN-CB with 95% accuracy. Research [18] addressed Arabic cyberbullying detection. Zhang et al. [19] proposed PCNN for detecting misspelled cyberbullying. Al-Garadi et al. [20] achieved high accuracy in Twitter cyberbullying detection using unique features.

3 Methodologies

In this section, we present the implementation details of our experimental setup. Here 3 transformer models and 1 deep learning model are used. The implemented approach that is shown in Fig. 1.

At first two datasets are merged to get the desired dataset. Then the dataset is preprocessed and produces tokenized data. That tokenized data is splitted into training, validation and test sets. Word embeddings are generated for each set. Then the model is trained and it generates accuracy and evaluation metrics using the test set.

3.1 Algorithm: The Proposed Classification Algorithm of Cyberbullying Detection

Input:
Merged dataset <- By merging 2 datasets("train.csv" and "test_with_solutions.csv") [21].

Output: Accuracy, loss, classification reports.

1. Start
2. Load the merged dataset and apply data preprocessing to clean the text data by following steps:
   a. Removing escape characters (e.g., backslashes).
   b. Removing numbers.

    c. Removing URLs and mentions.

    d. Expanding contractions.

    e. Removing punctuation and converting text to lowercase.

    f. Removing single alphabet characters.

3. Store the cleaned data into a separate file
4. Split the preprocessed dataset into training, validation, and test sets

    a. Step 1: 60% Training, 40% Test-Valid split

    b. Step 2: 50% Test-Valid split into Validation and Test

5. Remove Unnecessary Columns < - Remove "Date" column
6. Training, Testing and Validation datasets are tokenized:

    a. Cleaned comments tokenized into 'input_ids', 'token_type_ids', 'attention_mask'

7. Create TensorFlow datasets for each set (training, validation, and test) by converting tokenized data into TensorFlow-compatible format.
8. Prepare the datasets to fit in the BERT model

    a. target tokens: 'input_ids', 'token_type_ids', 'attention_mask'

    b. prediction: 'Insult' (0 = neutral comment, 1 = insulting comment)

9. Load Pre-trained model for sequence classification
10. Compile Model

    a. Hyperparameters:

        (1) Set optimizer = Adam

        (2) learning rate = 5e−5

        (3) Set loss function = SparseCategoricalCrossentropy

        (4) Set evaluation metrics = SparseCategoricalAccuracy

11. Train Model <- training dataset, validation dataset

    a. Hyperparameter: Desired number of epochs

12. Evaluate Model and calculate loss and accuracy <- Test dataset
13. Compute Classification Report <- precision, recall, f1-score for Neutral and Insulting classes
14. Stop

Here is a description about the models which are used for the experiment.

3.2 BERT

Built upon the transformer architecture, Bidirectional Encoder Representation from Transformers (BERT) [1] leverages an extensive corpus of unlabeled text, including Wikipedia (2500 million words) and Book Corpus (800 million words), for its pre-training. The significant achievement of BERT predominantly arises from its pre-training phase, during which it learns from an extensive collection of texts. The BERT model assimilates information from both the preceding and subsequent parts of a sentence context.

3.3 hateBERT

HateBERT [22] is a model built upon the foundation of the BERT architecture, which is specifically designed to address issues related to hate speech detection and offensive

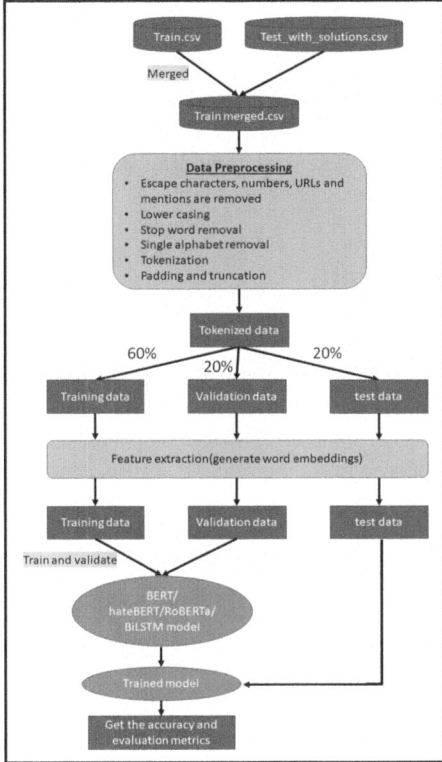

Fig. 1. Implemented approach

language identification in text. HateBERT is an enhanced version of the English BERT base uncased model. It's created by extending the training of the original model using over a million posts sourced from Reddit communities that have been banned due to their offensive content. By leveraging the inherent contextual understanding capabilities of BERT, HateBERT aims to accurately identify and classify instances of hate speech, offensive content, and potentially harmful language within text data.

3.4 RoBERTa

RoBERTa [23], short for "A Robustly Optimized BERT Pretraining Approach", is a derivative of BERT model that retains the same architecture as BERT but introduces some modifications in the training procedure. It employs a larger batch size and trains on more data for a longer duration. This extended training duration helps RoBERTa to thoroughly capture context and nuances in the text. Unlike BERT, which uses masked language modeling as one of its pretraining tasks, RoBERTa trains solely using a variant of the masked language modeling task.

3.5 BiLSTM

Bidirectional Long Short-Term Memory (BiLSTM) [25] is designed to capture sequential dependencies in data, particularly in sequences of variable length, such as sentences or time series data. BiLSTM networks offer a valuable tool for processing sequential data, especially in NLP tasks where understanding the context from both directions is crucial for accurate predictions.

4 Experiments and Results

The dataset [21] contains labeled comments with two attributes: the time the comment was made and the content of the comment in English language, sometimes with formatting. The dataset consists of two files, "train.csv" and "test_with_solutions.csv". The "train.csv" file contains a total of 3947 tweets, with class labels 0 or 1 indicating Neutral or Insulting, respectively. It includes 1049 Insulting tweets and 2898 Neutral tweets. The "test_with_solutions.csv" file contains 2647 tweets, comprising 693 Insulting tweets and 1954 Neutral tweets. These two files are combined into one file named "train_merged.csv".

This dataset is split into train, test, and validation sets using a 60:20:20 ratio to assess the model's performance comprehensively.

The results of all different experiments that we performed are presented in Table 1.

In this research, hateBERT outperforms other 4 models considering evaluation metrics for detection of cyberbullying.

The hatebert model shows the accuracy of 89.16% where the RoBERTa model shows 85.59% accuracy.

Clearly RoBERTa shows the second best accuracy among these models. For the BiLSTM model, two different cases i.e. without any embedding and with FastText embedding are shown and from the accuracy it is clear that embedded BiLSTM works better.

Table 1 presents the precision, recall and F1-measure values for neutral and insulting classes and macro average F1 and accuracy for all 4 models and baseline paper [9]'s result is also defined in the table.

Table 1. Classification report and accuracy for all models

Model	Labels	Precision	Recall	F1 Score	Macro Average F1 Score	Accuracy
BERT	Neutral	0.88	0.91	0.89	0.78	83.78%
	Insulting	0.72	0.65	0.68		
HateBERT	Neutral	0.94	0.91	0.93	0.86	89.16%
	Insulting	0.76	0.83	0.79		
RoBERTa	Neutral	0.86	0.96	0.91	0.80	85.59%
	Insulting	0.84	0.57	0.68		
BiLSTM (without Embedding)	Neutral	0.84	0.94	0.89	0.76	82.49%
	Insulting	0.76	0.53	0.63		
BiLSTM (with FastText Embedding)	Neutral	0.87	0.90	0.88	0.78	83.32%
	Insulting	0.72	0.65	0.68		
Baseline Paper [9] Results using BiLSTM	Neutral	0.86	0.91	0.88	0.76	82.18%
	Insulting	0.71	0.60	0.65		

For better understanding, Fig. 2 shows the comparison of Precision, Recall and F1-Measure for all 4 models.

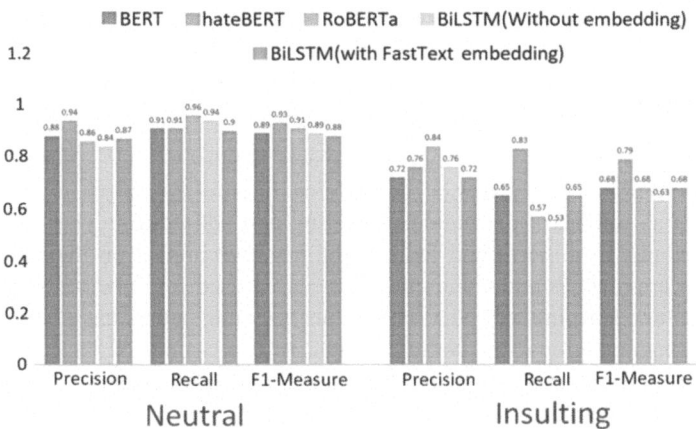

Fig. 2. Classification report

Figure 3 shows the comparative discussion of 4 different models' accuracy and Fig. 4 shows the comparison between our best result and baseline paper [9] result.

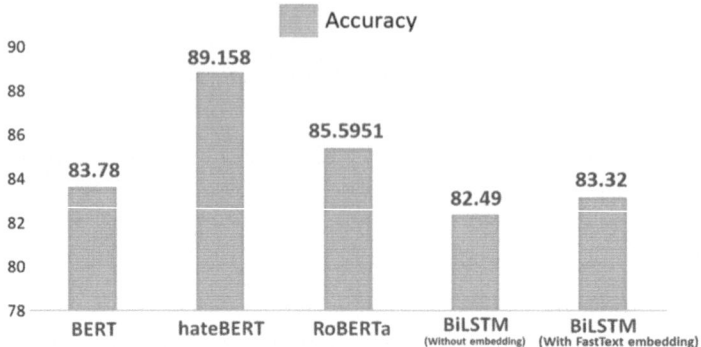

Fig. 3. Classifiers accuracy

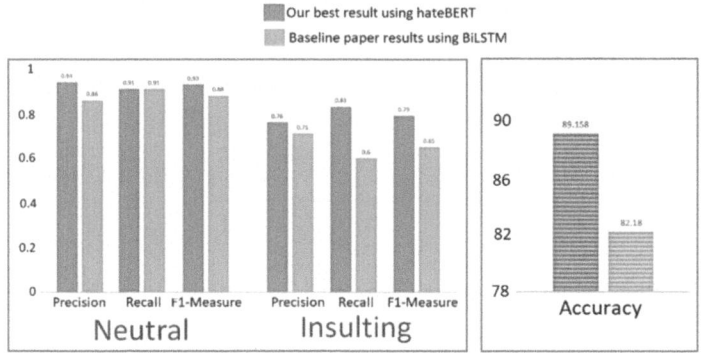

Fig. 4. Comparison with Baseline paper [9]

5   Conclusion

Cyberbullying can have serious negative impacts on the victims' mental and emotional well-being. In some cases, cyberbullying has even resulted in tragic consequences, such as self-harm or suicide. Detecting and addressing cyberbullying helps protect individuals from harm and reduces the negative impact on their mental well-being.Various approaches and techniques that can be used for cyberbullying detection including machine learning and NLP. Numerous research studies have been conducted on cyberbullying detection using Natural Language Processing (NLP) techniques [24].

In this study, three transformer models, namely BERT, hateBERT, RoBERTa and BiLSTM models are used for experiments. Data preprocessing steps that are applied include: text cleaning, stop word and single alphabet removal, tokenization, Padding and truncation. hateBERT model achieved the highest accuracy and F1-Measure compared to other models. hateBERT model showed 89.158% accuracy where BERT, RoBERTa and BiLSTM could achieve 83.78%, 85.5951% and 83.32% accuracy respectively.

In the future, we shall explore advanced techniques like oversampling, undersampling, data augmentation, Class Weighting, Ensemble Methods to handle the class imbalance in the dataset and improve the model's performance on the minority class. We can use hybrid models like combination of BERT's contextual embeddings with Bidirectional LSTM (BLSTM). Hyperparameters can be optimized. We can also develop a real-time implementation of the model for dynamic monitoring and immediate response to cyberbullying instances, providing a proactive approach to online safety.

Acknowledgement. We extend our appreciation to our supervisors and mentors Dr. Arijit Das and Professor Dr. Diganta Saha for their invaluable guidance. We are deeply grateful to my friends Rupam Saha and Imanur Rahaman for their invaluable assistance in the research work. We would also like to acknowledge the support of Jadavpur University and our families unwavering support. We would like to thank Gautam Srivastava for sharing the dataset [21].

Competing Interests. The authors declare that they have no competing interests.

References

1. Ahmed, M.F., Mahmud, Z., Biash, Z.T., Ryen, A.A.N., Hossain, A., Ashraf, F.B.: Cyberbullying detection using deep neural network from social media comments in Bangla language. arXiv preprint arXiv:2106.04506 (2021)
2. Ranasinghe, T., Zampieri, M.: Multilingual offensive language identification for low-resource languages. Trans. Asian Low-Resour. Lang. Inf. Process. **21**(1), 1–13 (2021)
3. Islam, K.I., Islam, M.S., Amin, M.R.: Sentiment analysis in Bengali via transfer learning using multi-lingual BERT. In: 2020 23rd International Conference on Computer and Information Technology (ICCIT), pp. 1–5. IEEE (2020)
4. Balakrishnan, V., Khan, S., Arabnia, H.R.: Improving cyberbullying detection using Twitter users' psychological features and machine learning. Comput. Secur. **90**, 101710 (2020)
5. Talpur, B.A., O'Sullivan, D.: Cyberbullying severity detection: a machine learning approach. PLoS ONE **15**(10), e0240924 (2020)
6. Muneer, A., Fati, S.M.: A comparative analysis of machine learning techniques for cyberbullying detection on twitter. Future Internet **12**(11), 187 (2020)
7. Islam, M.M., Uddin, M.A., Islam, L., Akter, A., Sharmin, S., Acharjee, U.K.: Cyberbullying detection on social networks using machine learning approaches. In: 2020 IEEE Asia-Pacific Conference on Computer Science and Data Engineering (CSDE), pp. 1–6. IEEE (2020)
8. Samghabadi, N.S., Patwa, P., Pykl, S., Mukherjee, P., Das, A., Solorio, T.: Aggression and misogyny detection using BERT: a multi-task approach. In: Proceedings of the Second Workshop on Trolling, Aggression and Cyberbullying, pp. 126–131 (2020)
9. Iwendi, C., Srivastava, G., Khan, S., Maddikunta, P.K.R.: Cyberbullying detection solutions based on deep learning architectures. Multimed. Syst. 1–14 (2020)
10. Dadvar, M., Eckert, K.: Cyberbullying detection in social networks using deep learning based models. In: Big Data Analytics and Knowledge Discovery: 22nd International Conference, DaWaK 2020, Bratislava, Slovakia, 14–17 September 2020, pp. 245–255. Springer (2020)
11. Al-Garadi, M.A., et al.: Predicting cyberbullying on social media in the big data era using machine learning algorithms: review of literature and open challenges. IEEE Access **7**, 70701–70718 (2019)
12. Hani, J., Mohamed, N., Ahmed, M., Emad, Z., Amer, E., Ammar, M.: Social media cyberbullying detection using machine learning. Int. J. Adv. Comput. Sci. Appl. **10**(5) (2019)
13. Emon, E.A., Rahman, S., Banarjee, J., Das, A.K., Mittra, T.: A deep learning approach to detect abusive Bengali text. In: 2019 7th International Conference on Smart Computing & Communications (ICSCC), pp. 1–5. IEEE (2019)

14. Chakraborty, P., Seddiqui, M.H.: Threat and abusive language detection on social media in Bengali language. In: 2019 1st International Conference on Advances in Science, Engineering and Robotics Technology (ICASERT), pp. 1–6. IEEE (2019)
15. Tarwani, S., Jethanandani, M., Kant, V.: Cyberbullying detection in Hindi-English code-mixed language using sentiment classification. In: Advances in Computing and Data Sciences: Third International Conference, ICACDS 2019, Ghaziabad, India, 12–13 April 2019, Revised Selected Papers, Part II, pp. 543–551. Springer, Singapore (2019)
16. Murnion, S., Buchanan, W.J., Smales, A., Russell, G.: Machine learning and semantic analysis of in-game chat for cyberbullying. Comput. Secur. **76**, 197–213 (2018)
17. Al-Ajlan, M.A., Ykhlef, M.: Deep learning algorithm for cyberbullying detection. Int. J. Adv. Comput. Sci. Appl. **9**(9) (2018)
18. Haidar, B., Chamoun, M., Serrhrouchni, A.: A multilingual system for cyberbullying detection: Arabic content detection using machine learning. Adv. Sci. Technol. Eng. Syst. J. **2**(6), 275–284 (2017)
19. Zhang, X., et al.: Cyberbullying detection with a pronunciation based convolutional neural network. In: 2016 15th IEEE international conference on machine learning and applications (ICMLA), pp. 740–745. IEEE (2016)
20. Al-Garadi, M.A., Varathan, K.D., Ravana, S.D.: Cybercrime detection in online communications: the experimental case of cyberbullying detection in the Twitter network. Comput. Hum. Behav. **63**, 433–443 (2016)
21. Kaggle. Detecting Insults in Social Commentary (n.d.). https://www.kaggle.com/competitions/detecting-insults-in-social-commentary/data. Accessed 20 July 2023
22. Caselli, T., Basile, V., Mitrovic, J., and Granitzer, M.: HateBERT: retraining BERT for abusive language detection in English. CoRR, abs/2010.12472 (2020)
23. Briskilal, J., Subalalitha, C.N.: An ensemble model for classifying idioms and literal texts using BERT and RoBERTa. Inf. Process. Manage. **59**(1), 102756 (2022)
24. Sun, S., Luo, C., Chen, J.: A review of natural language processing techniques for opinion mining systems. Inf. Fusion **36**, 10–25 (2017)
25. Hameed, Z., Garcia-Zapirain, B.: Sentiment classification using a single-layered BiLSTM model. IEEE Access **8**, 73992–74001 (2020)
26. Slonje, R., Smith, P.K., Frisén, A.: The nature of cyberbullying, and strategies for prevention. Comput. Hum. Behav. **29**(1), 26–32 (2013)
27. Cambria, E., White, B.: Jumping NLP curves: a review of natural language processing research. IEEE Comput. Intell. Mag. **9**(2), 48–57 (2014)

Exploring Intonation Patterns in Nepali Speech: A Phonetic and Linguistic Analysis for Text-to-Speech System

Pratika Rai[1]([✉]), Suraj Gurung[2], Sharad Sinha[1], Ranjit Subba[3], and Roshan Kumar Prasad[1]

[1] Department of Computer Science and Technology, University of North Bengal, Siliguri, Darjeeling, India
`rs_pratika@nbu.ac.in`
[2] Department of Nepali, University of North Bengal, Siliguri, Darjeeling, India
[3] Department of Computer Science and Applications, Hijli College, Kharagpur, India

Abstract. Phonetic, prosodic and linguistic analysis of intonation occurring in Nepali speech have been done in this paper. Different types of intonation patterns depending on various factors and types of sentences were identified and studied. In addition to pattern identification, a discussion has also been done on the significance of incorporating intonation while developing a Text-to-Speech system. Further, analysis has also been done on how intonation patterns are affected by different parts of speech. The investigation shows how nouns, verbs, adjectives, adverbs and other linguistic elements affect the rise and fall of the pitch. The dataset used for the analysis includes audio recordings obtained from All India Radio, Kurseong.

Keywords: TTS · Phonetics · Intonation · Speech synthesis · Fundamental Frequency (F0) · Semantic ambiguity · Phoneme · MFCC

1 Introduction

Intonation refers to a combination of acoustic parameters, encompassing duration, intensity and pitch which serve to convey the meaning of the discourse [1]. Among these factors, pitch, primarily determined by the fundamental frequency (F0) of the speech, plays a pivotal role. F0 is measured in Hertz or semitones [2]. Fundamental frequency is only present in voiced segments and so it can be measured. In voiceless portions, the lack of F0 leads to a void in pitch measurement.

Intonation currently stands as the central point of focus in speech synthesis research and is swiftly gaining prominence within the realm of speech recognition as well [3].

Phonemic analysis in intonation can be extremely beneficial for building a TTS system for Nepali or any language. Research on intonation in Nepali has not been extensive and well documented compared to more widely studied languages.

M. Majumder et al. (Eds.): ICCTE 2023, CCIS 2376, pp. 229–239, 2025.
https://doi.org/10.1007/978-3-031-81935-3_20

1.1   Need for Exploring Intonation for TTS

Intonation is one of the crucial aspects of speech communication. While Developing a Text-to-Speech System, incorporating intonation patterns is important in order to synthesize a speech that closely resembles human speech, leading to a more natural, comprehensible and pleasant listening experience [4]. Without proper intonation, the speech can sound robotic and monotonous, which makes it harder for the listeners to engage with the content as well as to comprehend the content. In humans, intonation is a primary carrier of emotional and linguistic meaning in speech. Including such a feature in a tts system allows it to accurately convey the emotional nuances present in the input text, making the synthesized speech more expressive and contextually appropriate. another aid that intonation provides is distinction between different elements in a sentence, such as differentiating between nouns and verbs or marking emphasis on specific words. This can improve the clarity and comprehensibility of the synthesized speech, especially when dealing with sentences that have potential semantic ambiguity.

1.2   Literature Review

Recent research on intonations provides valuable insights into language-specific patterns, cross-linguistic variations and the application of advanced technologies in studying prosody. Allwood et al., 2019 [5], delved into the intonation patterns in Nepali feedback units. The study revealed the use of rising intonation in feedback-eliciting units contrasting with falling intonation in other feedback-giving units. Banerjee et al., [6] conducted a comprehensive analysis of Hindi intonation to create a semi-automatically labelled prosody database. This resource serves as valuable training data for Automatic Speech Recognition (ASR) and Text-to-Speech systems. Further, Rao et al., 2009 [7] explored the application of neural network models to capture intonation patterns in Indian languages, focusing on Hindi, Telugu and Tamil. Their study involved analyzing recorded news data using both neural network and Support Vector Machines (SVM) models to predict fundamental frequency (F0) of syllables.

2   Intonation in Nepali

Intonation patterns in Nepali language, like any other language, can convey various meanings, emotions and pragmatic functions. Intonation occurs primarily at the sentence level although it can affect individual words within a sentence. Applying intonation in different words in a sentence changes the meaning of the sentence altogether. On the other hand, phonemes can also contribute to intonation in Nepali as in case of other languages. Phonemes, being the smallest distinctive units of sound in a language, can play a role in shaping the intonation patterns in sentences. Each of these aspects has been discussed in detail in the sections to follow.

2.1   Sentence Level Intonation

Intonation can indicate whether a sentence is a question, statement, command or an exclamation. Different intonation patterns are associated with different types of sentences. Changing the intonation of a sentence while using the same exact words can completely change its meaning. For example, if we consider the Nepali sentence: तिमीले खाना खायौ [timilekʰanakʰajʌu], this sentence can have different meanings depending on the change in intonation applied at various places. A few of such instances have been discussed below. The sound utterances were recorded and their pitch were studied.

तिमीले खाना खायौ। [timile kʰana kʰajʌu] There is no significant rise or fall in this statement which shows that there is a flat pattern, hence it is a neutral statement. In Fig. 1, it can be observed that there is no rise or fall.

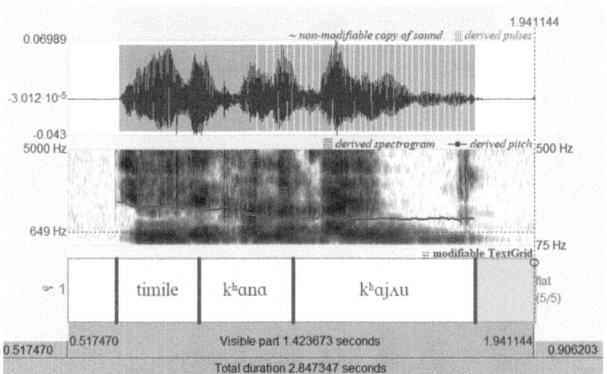

Fig. 1. Neutral statement

तिमीले खाना खायौ? [timile kʰana kʰajʌu?] In the same statement, if a rising intonation is applied at the end of the sentence, it becomes an interrogative sentence as shown in Fig. 2.

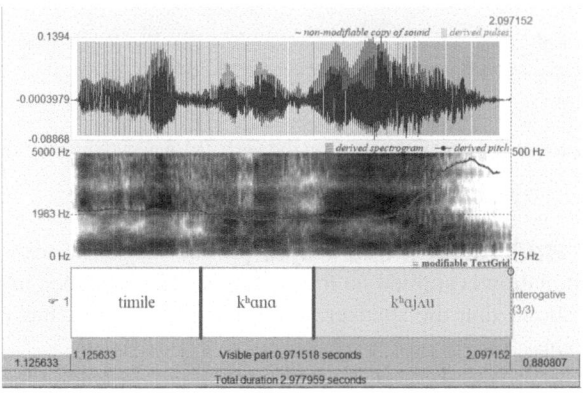

Fig. 2. Interrogative statement

तिमीले खाना खायौ [timile kʰana kʰajʌu!] The Fig. 3a shows an instance of applying intonation at different places where there is a sharp rise followed by a sharp fall in the first segment of the last word, which shows that the speaker is surprised. The rising intonation in the word खायौ [kʰajʌu] shows that the speaker is surprised and that the person he/she is addressing has eaten, where the stress is on verb.

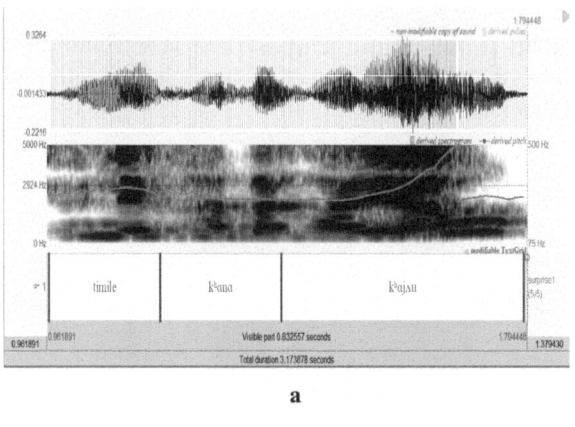

a

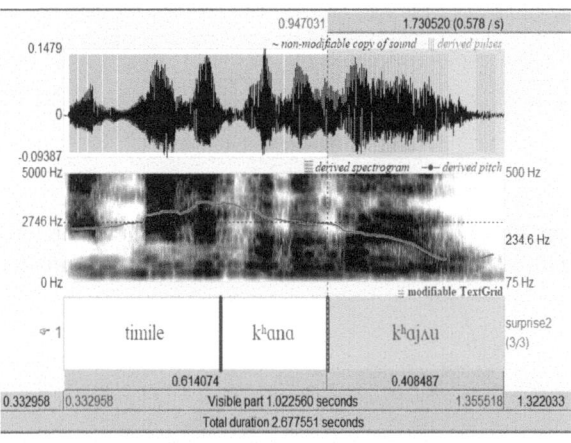

b

Fig. 3. (a) Exclamatory sentence with stress on the verb. (b) An exclamatory sentence with stress on the noun

तिमीले खाना खायौ [timile kʰana kʰajʌu!]In contrast to the example in Fig. 3a, here, the rising intonation is applied on the first word itself. As seen in Fig. 3b, this rise in the first word shows that the speaker is surprised that he/she has eaten, where the stress is on the noun.

2.2 Analysis of Intonation Patterns

Based on the analysis of intonation pattern, variations in different types of sentences and inferences drawn thereof, intonation in Nepali has been classified into following categories:

Falling Intonation: It is observed when the pitch falls as the end of a sentence and it indicates that the sentence is coming to an end. It indicates certainty or completion of the sentence. This type of intonation can be observed in declarative statements, as well as imperative and exclamatory sentences. Figure 4 shows the pitch of the recorded audio which is a declarative sentence, so a falling intonation can be seen.

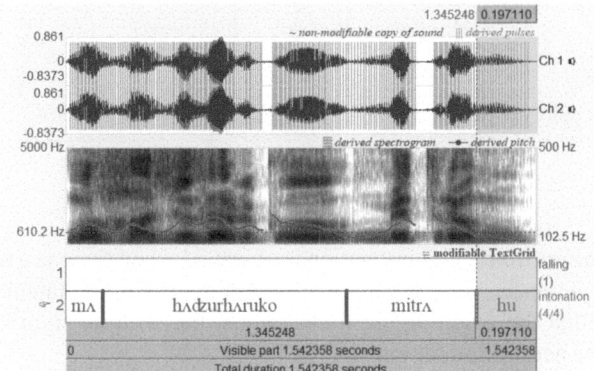

Fig. 4. Falling intonation

Rising Intonation: This type of pattern is often associated with yes-no questions in Nepali. The pitch rises at the end of the sentence, creating a questioning tone. It can be observed in Fig. 5 that there is a rise in pitch at the end of a sentence which shows that it is a question.

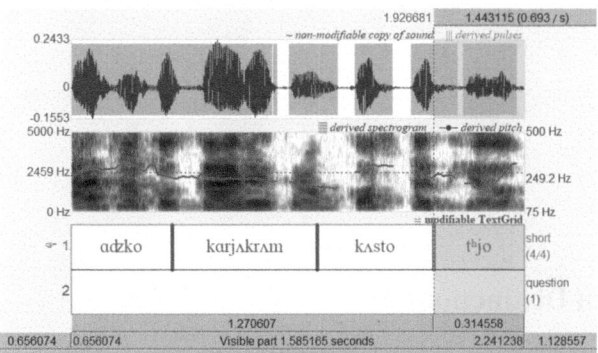

Fig. 5. Rising intonation

Rising-Falling Intonation: This intonation pattern combines a rise followed by a fall in pitch. It is used in more complex questions of sentences that seek elaboration or clarification. In Fig. 6 the question demands an elaborate answer and so a rising-falling intonation can be observed.

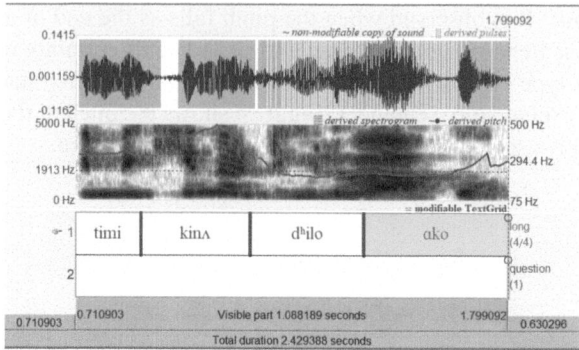

Fig. 6. Rising falling intonation

Flat (Neutral) Intonation: Flat intonation is observed when there is a relatively steady pitch across the sentence, often used for informative or neutral statement. In Fig. 7 it can be observed that the whole figure looks almost flat with no significant rise or fall signifying a flat or neutral intonation.

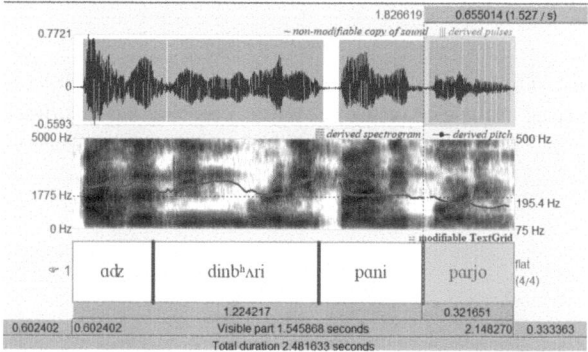

Fig. 7. Flat intonation

3   Verb-Noun Distinction

The roles of words within a sentence can influence their pitch placements, impacting the overall intonation patterns. In Nepali, although intonation alone might not be the sole factor in distinguishing verbs and nouns, it can contribute to the overall context. In most cases, it was observed that pitch placement can help signal whether a word is functioning as a verb or a noun. Verbs seem to have a more stable pitch pattern, not showing a significant rise or fall at the end. Whereas nouns seem to exhibit a rise or fall at the end, indicating that the word is functioning as a noun. It was also observed that the pitch subtly changed when a subject is followed by a verb, indicating that the word is functioning as a verb in agreement with the subject. Figure 8 shows that the recorded sentence is आमा बजार जानुभयो। [ama bʌdʒar dʒanubʰʌjo]. The segment 'a' indicates a noun and a rise in pitch can be seen. The segment 'b' indicates the verb with no significant rise or fall in pitch.

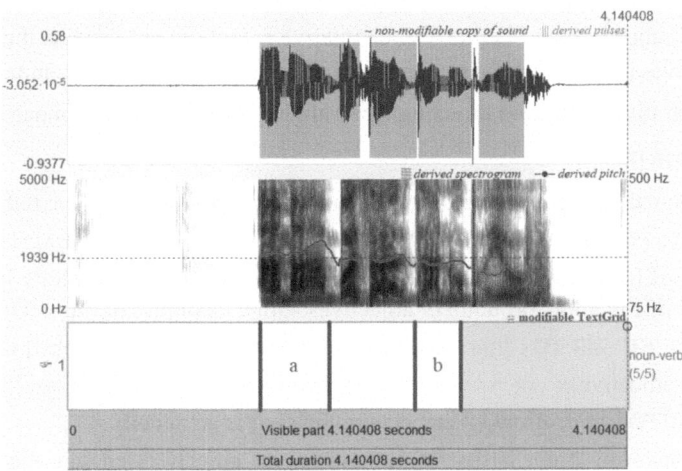

Fig. 8. Distinction between noun and verb

4 Intonation Caused by Parts of Speech

Intonation can be influenced by various aspects of language, including the parts of speech being used [8].

4.1 Adjectives and Adverbs

Adjectives, being an essential part of speech, can influence the intonation patterns in sentences. They are used to emphasize certain qualities of the noun/pronoun they modify and also highlight a specific quality. Based on the analysis done on the dataset, it was observed that, a rising or falling intonation might be used to draw attention to it.

Let us consider some examples:

उनी **असल** मानिस हुन्। [uni ʌsʌl mɑnis hun]

उनी **साह्रै असल** मानिस हुन्। [uni sahrʌi ʌsʌl mɑnis hun]

In the second sentence, the emphatic form of simple qualitative adjective like **साह्रै**

असल [sahrʌi ʌsʌl] has played a vital role in changing the intonation where both [ʌsʌl]
(Qualitative adjective) and [sɑhrʌi] (Quantitative adjective) are used as the main factors. Adjectives used in comparisons can also affect intonation. While comparing things
using comparatives, the intonation might rise slightly to indicate the comparison (eg. धेरै

[dʰerʌi] , कम्ती [kʌɑti]).

Apart from this, quantitative adjective like [sɑhrʌi] is also considered as adverb
when we say, त्यो कुरा सुनेर उ साह्रै रिसायो। [tjo kura sunerʌ u sɑhrʌi risajo]. Here
the position of [sɑhrʌi] is just in front of the verb [risajo], so it is directly linked with
verb instead of any noun, pronoun or adjectives. Other examples of this occurrence are:

हरिले **धेरै** कुरा जान्दछ। [hʌri le **dʰerʌi** kura dʒandʌcʰʌ] here, [dʰerʌi] is indefinite
quantitative adjective.

हरि **धेरै** हाँस्छ। [hʌri **dʰerʌi** hascʰʌ] here, [dʰerʌi] is an adverb.

Another noteworthy point is that some quantitative adjectives change the rhythm of
speech when they are used as emphatic forms like:

सबै मानिस प्रकृतिप्रेमी हुन्छन्। [sʌbʌi mɑnis prʌkriti premi huncʰʌn]

सप्पै मानिस प्रकृतिप्रेमी हुन्छन्। [sʌppʌi mɑnis prʌkriti premi huncʰʌn]

मैले उसको कुरामा **उति** ध्यान दिइनँ। [mʌile usko kurama **uti** dhjan dijinʌ]

मैले उसको कुरामा **उत्ति** ध्यान दिइनँ। [mʌile usko kurama **utti** dhjan dijinʌ]

आज **निकै** जाडो छ। [adʒ **nikʌi** dʒaɖo cʰʌ]

आज **निक्कै** जाडो छ। [adʒ **nikkʌi** dʒaɖo cʰʌ]

4.2 Particles

Particles also seem to influence intonation patterns in Nepali Particles are often used
to convey various nuances and emotions in speech. Some amount of emphasis is given
using particles like न[nʌ]in between adjectives, which look like negation, tend to change
intonation patterns in the speech.

For example:

अनुहार पनि **कालो**, मन पनि **उस्तै** कालो रहेछ। [ʌnuhar pʌni **kalo**, mʌn pʌni **ustʌi**
kalo rʌhecʰʌ]

अनुहार पनि **कालो न कालो**, मन पनि उस्तै कालो रहेछ। [ʌnuhar pʌni **kalo nʌ kalo**,
mʌn pʌni **ustʌi** kalo rʌhecʰʌ]

The presence of न [nʌ] in the sentence show a rising intonation, which in turn shows that the sentence is stressing on the point.

4.3 Compound Verbs

In Nepali, compound verbs can convey meanings and modify the overall tone of a sentence, which can impact the intonation during speech. Depending on the context, the emphasis on the compound action may lead to variation in intonation. A rising or falling intonation could be used to highlight the complexity or significance of the action. A rising intonation seemed to suggest curiosity or seeking clarification about the details of the action, while a falling intonation seemed to imply confidence in conveying the intended meaning. A rising intonation implied an ongoing process, whereas a falling intonation indicated its completion.

Consider the following examples:

श्यामले त्यो पुस्तक **पढ्यो**। [sjamle tjo pustʌk **pʌɖʰjo**]

श्यामले त्यो पुस्तक **पढिसक्यो**। [sjamle tjo pustʌk **pʌɖʰisʌkjo**]

मलाई तिमीले त्यो कुरा **भन्दा** हुन्छ। [mʌlai timile tjo kura **bʰʌnda** huncʌ]

मलाई तिमीले त्यो कुरा **भनिहाल्दा** हुन्छ। [mʌlai timile tjo kura **bʰʌnihalda** huncʌ] (compound form of non-finite verb)

मलाई तिमीले त्यो कुरा **भनिदिइहाल्दा** हुन्छ। [mʌlai timile tjo kura **bʰʌnidihalda** huncʌ] (more complex compound form of non-finite verb)

Intonation can be affected in some complex sentences when perfect participles like भनेको [bʰʌneko], गरेको [gʌreko], खाएको [kʰaeko] and particles like रे[re] and त[tʌ] come together. Examples of such cases are listed below:

रिना बजार गइन भनेर सुजितले **भनेको रे**। [rina bʌdʒar gʌjin bhʌnerʌ sudʒit le **bʰʌneko re**]

रिना बजार गइन् भनेर सुजितले **भनेको रे** हौ। [rina bʌdʒar gʌjin bhʌnerʌ sudʒit le **bʰʌneko re hʌu**]

रिना बजार गइन् भनेर सुजितले पो **भनेको रे** त। [rina bʌdʒar gʌjin bhʌnerʌ sudʒit le **bʰʌneko retʌ**]

आज बजार जानु भनेको, गाडी नै पो पाइनँ। [ajʌ bʌdʒar dʒanu bʰʌneko, gaɖinʌi po painʌ]

आज बजार जानु **भनेको त**, गाडी नै पो पाइनँ। [ajʌ bʌdʒar dʒanu bʰʌneko tʌ, gaɖinʌi po painʌ]

हिज सुन्तला खाएको थिएँ, ज्वरो पो आयो। [hidʒʌ suntʌla kʰaeko, dʒoro po ajo]

हिज सुन्तला **खाएको त**, ज्वरो पो आयो। [hidʒʌ suntʌla **kʰaekotʌ**, dʒoro po ajo]

5   Training Model for a TTS System

As discussed, incorporating various language features and phenomena is of utmost importance in order to develop a TTS system that can generate a more natural sounding synthetic speech. The steps involved in doing have been proposed as follows:

5.1   Data Collection

A large dataset of spoken language samples with transcriptions will be required. The dataset should consist of a wide range of speakers, dialects, speaking styles including both male and female voices. The target dataset for this purpose would be from recorded discussions and interviews for radio broadcast.

5.2   Annotation

Annotation of the collected speech data will be done with labels and markers that indicate intonation patterns, such as pitch accents and phrase boundaries.

5.3   Feature Extraction

The relevant acoustic and linguistic features need to be extracted. Acoustic features include fundamental frequency (F0), duration, energy, spectral feature. Whereas linguistic feature would include syntactic structure, word stress, syllable level stress, lexical information etc., the study needs to be done on the variation of these features. Mel-frequency Cepstral Coefficients (MFCC), is a feature extraction technique widely used in speech and audio processing. It represents the spectral characteristics of sound in a way that is well suited for various machine learning tasks such as speaker recognition and Text-to-Speech systems. The MFCC are based on the known variation of the human ear's critical bandwidth frequencies with filters spaced linearly at low frequencies and logarithmically at high frequencies used to capture the important characteristics of speech. The signal is divided into overlapping frames to compute MFCC coefficients [9].

5.4   Model Selection

Machine Learning techniques will be applied to learn the relationship between the linguistic feature and the corresponding waveforms. Models like Hidden Markov Models (HMMs), Gaussian Mixture Models (GMMs), Support Vector Regression (SVR), and neural network models (RNNs) will be employed. Neural networks have the generalization ability to predict intonation patterns reasonably well for the patterns which are not present in the learning phase [10].

5.5 Training

The selected model will be trained using annotated speech data and extracted features. The desired result would be the model to learn the relationship between the input text or linguistic features and the corresponding intonation patterns. The dataset needs to be divided into training and test subsets. The training date is used to train the model and testing data is used to evaluate the performance of the model. The model is expected to predict the F0 based on the linguistics and contextual features extracted from the text. To optimize the performance, hyperparameter tuning will be needed.

6 Conclusion

Prosody plays a vital role within a language. The significance of prosody in comprehension and production becomes particularly evident in languages like Hindi and Nepali, where word order can be relatively flexible. This flexibility expands the possibilities for conveying information structure [11]. Intonation being one of the prosodic components of speech needs to be explored in order to understand the meaning and function of suprasegmentals within a sentence. Incorporating intonation into a TTS system involves understanding linguistic and prosodic rules governing intonation patterns in different languages and contexts. This can be achieved through advanced linguistic modelling and training data that captures a wide range of intonation variations. As TTS technology advances, efforts to improve intonation modelling will continue to be an essential aspect of creating high-quality and human-like synthetic speech. Apart from the intonation patterns discussed in this paper, further classification needs to be done on the basis of intensity of pitch variation. For instance, the rising or falling tone could be further subdivided into sharp, mid-rising or mid falling variations.

References

1. Bolinger, D., Bolinger, D.L.M.: Intonation and Its Parts: Melody in Spoken English. Stanford University Press (1986)
2. Levis, J.: Suprasegmentals: Intonation. The Encyclopedia of Applied Linguistics (2012)
3. Botinis, A.: Intonation: Analysis, Modelling and Technology. Springer (2012)
4. Martin, P.: Deep learning and intonation in Text to Speech systems (2019)
5. Allwood, J., Regmi, B.N.: Intonation patterns in Nepali feedback units. In: Proceedings of Oriental COCOSDA 2010, Paper 61, Kathmandu (2010)
6. Banerjee, E., Kr Ojha, A., Nath Jha, G.: Prosody labelled dataset for Hindi using Semi-automated approach. ArXiv, pp. 14–19 (2021). https://doi.org/10.48550/arXiv.2112.05973
7. Rao, K.S., Yegnanarayana, B.: Intonation modeling for Indian languages. Comput. Speech Lang. 23(2), 240–256 (2009). https://doi.org/10.1016/j.csl.2008.06.005
8. Clark, T.W.: Introduction to Nepali: A First-Year Language Course (1977)
9. Kurzekar, P.K., Deshmukh, R.R., Waghmare, V.B., Shrishrimal, P.P.: A comparative study of feature extraction techniques for speech recognition system. Int. J. Innov. Res. Sci. Eng. Technol. 03(12), 18006–18016 (2014)
10. Reddy, V.R., Rao, K.S.: Two-stage intonation modeling using feedforward neural networks for syllable based text-to-speech synthesis. Comput. Speech Lang. 27(5), 1105–1126 (2013). https://doi.org/10.1016/j.csl.2013.02.003
11. Patil, U., Kentner, G., Gollrad, A., Vasishth, S.: Focus, word order and intonation in Hindi (2008)

Sentiment Analysis on Airline Customer Review Using Language Model and Capsule Network

Nilanjana Das[1], Rakesh Dutta[2(✉)], Uttam Kumar Mondal[1],
and Jyotsna Kumar Mandal[3]

[1] Department of Computer Science, Vidyasagar University,
Midnapore, West Bengal, India
[2] Department of Computer Science and Application, Hijli College,
Kharagpur, West Bengal, India
rakeshhijli@gmail.com
[3] Department of Computer Science and Engineering, University of Kalyani,
Kalyani, Nadia, West Bengal, India

Abstract. In the past three decades, the competitive airline industry has experienced rapid growth. However, choosing a particular flight for travelling remains a crucial task and requires detail analysis and prior experience. These could be leveraged by intellectually processing existing passengers' reviews or feedbacks. Traditional methods of collecting and analyzing such feedbacks or reviews are time consuming and repetitive. To address this challenge, sentiment analysis can be an effective tool. Twitter data has emerged as a valuable resource for collecting such reviews or feedbacks and conducting sentiment analysis. This research paper introduces a new deep learning model that combines BERT (a language model), self-attention based Bidirectional Gated Recurrent Unit with Capsule network, and Softmax classifier to assess airline passenger review tweets on six different US airlines, with three sentiment classes: positive, neutral, and negative. The performance of the sentiment analysis system is evaluated with various contemporary methods. This assessment demonstrates that the suggested model exhibits comparable performance to state-of-the-art methods.

Keywords: Sentiment Analysis · BERT · Bi-GRU · Attention Mechanism · Captual network

1 Introduction

In the era of artificial intelligence, sentiment analysis has emerged as a critical aspect, especially when it comes to passenger feedback in the airline industry. The market of airline has experienced rapid growth and intense competition in recent decades [1]. This study aims to predict public reactions to airlines and assess customer satisfaction with pricing and services provided. Sentiment analysis includes the extraction of sentiments/opinion to determine attitudes and

M. Majumder et al. (Eds.): ICCTE 2023, CCIS 2376, pp. 240–250, 2025.
https://doi.org/10.1007/978-3-031-81935-3_21

emotions expressed in textual data [2]. Nowadays, consumers heavily trust on reviews and personal suggestions when making decisions. For airline companies, customer feedback plays a crucial part in improving the quality of services and facilities offered. Traditionally, sentiment analysis in the airline industry has been carried out through customer satisfaction surveys and forms. However, these processes can be time-taking, labor-intensive, and costly to analyze. Additionally, the information collected through questionnaires may not always be accurate or compatible with the desired analysis [3].

Twitter, with its vast user base of over 100 million people, representing a significant fraction of the global population, serves as a valuable online microblogging platform. Its users express their genuine thoughts, opinions, and feedback, making it a reliable source of knowledge for research purposes [4]. In the context of this study, a substantial number of positive evaluations can greatly influence both customers and the airline industry. This phenomenon is observed in scenarios involving customer questionnaires, where items with a higher number of positive evaluations tend to be preferred by people. As a result, Twitter becomes an even more reliable data source as users share their feelings and feedback, making it suitable for investigation [5]. Conversely, items with negative assessments may dissuade customers from choosing them, which would be problematic for the company. Hence, sentiment analysis plays a critical role in understanding the customer-airline relationship. The primary objective of the article is to assist the airline sector with a complete understanding of customer sentiments and address their needs effectively.

In this study, a novel framework named B-ABiGRUCaps is introduced with the objective of improving the model's capacity to gather significant data. The framework incorporates multiple components to accomplish the goal. Initially, the BERT model is employed to enhance the expression capability of word vector information through word embedded representation. Second, the BiGRU model is employed to collect contextual data. A self-attention mechanism has been developed to combat the issue of redundant information processing. This mechanism selectively directs its attention to the features captured within the hidden layer of BiGRU. Lastly, the introduction of CapsNet is suggested to classify entity categories, enabling the generation of more comprehensive information for desired classes. This proposed model is specifically applied to perform sentiment analysis on airline datasets obtained from KAGGLE [6]. This study makes the following contributions:

1. Language model (BERT) to generate word-level vectors, effectively enhances the capacity of neural network model to capture essential information.
2. BiGRU is employed to acquire contextual information from both preceding and succeeding.
3. A self-attention mechanism is introduced to enhance feature representation obtained by the BiGRU hidden layer.
4. CapsNet is used to capture hierarchical relationships and spatial information from the context for improving classification accuracy.

The other parts of the article are as follows: Sect. 2 discusses a summary of related works. In Sect. 3, we explain about the Gated Recurrent Unit (GRU) framework employed for classification. Our proposed approach is presented in Sect. 4. Description of experimental results appears in Sect. 5. The article concludes with future scope in Sect. 6.

2   Related Work

The classification task has gained significant attention across diverse fields, including applications in cancer detection and diagnosis [7,8], article classification [9], analyzing data in agricultural and environmental biology [10], sentiment classification [11], fake news classification [12], and more. Furthermore, the airline industry has evolved its research methods by incorporating consumer feedback. For instance, in the article [13,14] researchers used a supervised learning model to assign sentiment labels to a large dataset of unlabeled product reviews on Amazon, offering valuable insights into consumer preferences and opinions. The preprocessing phase involved two key steps: tokenization and the removal of stop words. Following this, feature extraction was carried out using chi-square and TF-IDF methods on a bag-of-words. Subsequently, the performance metrics, including precision, recall, F-measure, and accuracy, were computed for various machine learning techniques applied in these studies. A newly Adaboost approach was introduced for conducting sentiment analysis on tweets dataset related to US airlines [15]. The dataset used in this research spans from 2014 to 2017 and was sourced from Skytrax. To ensure data quality, a thorough data preparation phase was carried out, eliminating unwanted information. The performance of the Adaboost method was benchmarked against various other machine learning approaches. In addition, statistical procedures, including correlation and regression analysis, were employed through data mining techniques. The description of experimental outcomes demonstrated that Adaboost performed superiorly to the other existing methods, showcasing its exceptional performance in the study. Kumawat et al. focused on sentiment analysis using language models [2]. They employed advanced models like BERT, Roberta, and Electra to examine sentiment in the context of US airlines. These techniques involved extracting data from an extensive pool of unsupervised data to encompass various textual characteristics. The study involved the creation of different learning curves for the aforementioned models. Notably, BERT exhibited higher accuracy compared to previous models, marking a notable improvement in the performance of sentiment analysis. Rane et al. introduced a Sentiment Classification process for analyzing US airline services using Twitter data [16]. This study exclusively relies on commonly employed machine learning approaches for text classification. The authors adopted doc2vec technique, a embedding of phrase technique, to produce a vector representation from a sentence, forming a key component of their methodology. Hasib et al. proposed a newly designed sentiment analysis process using deep learning approach on Twitter Data for US Airline Service [1]. They used CNN (Convolutional Neural Network) model. Their approach involved translating their observations into metadata and subsequently applying TF-IDF to feed the data into the four layers of the DNN (Deep Neural Network).

The study included a comparative analysis of machine learning algorithms and neural network approaches based on their respective accuracy, forming a significant part of the research investigation. Saad et al. focused on opinion extraction from a Twitter dataset on airline reviews, the proposed model was structured into three distinct levels: pre-processing, classification, and validation [4]. The pre-processing included removal of stop words, punctuation, and stemming. Subsequently, during the classification, a range of machine learning approaches were employed to categorize the data effectively. For validation, the Twitter dataset was divided into a 70% training set and a 30% testing set, with the additional use of a 10-fold cross-validation technique to ensure robustness and accuracy in the results. Safrin et al. studied sentiment analysis from consumer feedback in the form of textual expressions, including thumbs up or down, emojis, and various symbols [17]. Subsequently, feature modeling was applied, followed by performance evaluation. The study's performance was assessed using metrics such as accuracy, precision, and recall determining its effectiveness and reliability. Rahat et al. conducted a comparative analysis between Support Vector Machine and Naïve Bayes methods for sentiment analysis using Twitter review [18]. The outcomes of both methods were compared, revealing that the SVM outperformed Naïve Bayes in terms of efficiency and effectiveness. Hasib et al. conducted experiment to evaluate customer sentiment based on online Airline Review dataset of Bangladesh using six classifiers [19]. In their predictive analysis, the authors employed a combination of three deep learning approaches (CNN, LSTM, and BERT) and three machine learning approaches (Random Forest, Decision Tree, and XGBoost). This diverse approach allowed for a comprehensive assessment of sentiment prediction in the study.

3   Gated Recurrent Unit (GRU)

The Gated Recurrent Units incorporate the recurrent neural network mechanism [20]. These GRUs, despite having fewer parameters, bear a resemblance to the Long Short-Term Memory (LSTM) architecture. LSTM, another RNN variant, is capable of retaining information over time intervals. The LSTM includes a cell, an input gate, an output gate, and a forget gate. Notably, LSTM networks have effectively addressed issues like the vanishing gradient and exploding problems. To address the vanishing gradient issue in GRUs, two gates, namely the Update Gate and Reset Gate, are employed. Figure 1 illustrates a comprehensive version of a GRU.

The related calculations are given in the following:

$$r_t = \sigma \left(w_r \cdot [h_{t-1} \oplus x_t] + b_r \right). \tag{1}$$

$$z_t = \sigma \left(w_z \cdot [h_{t-1} \oplus x_t] + b_z \right). \tag{2}$$

$$\widetilde{h}_t = tanh \left(w_h \cdot [r_t \otimes h_{t-1} \oplus x_t] + b_h \right). \tag{3}$$

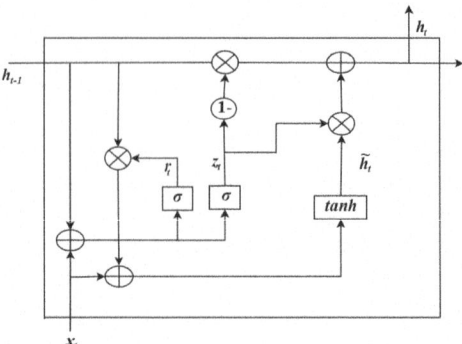

Fig. 1. The basic architecture of GRU Cell.

$$h_t = (1 - z_t) * h_{t-1} + z_t * \widetilde{h}_t. \tag{4}$$

In these mathematical expressions, x_t represents the input vector, h_t represents the output vector, z_t signifies the vector associated with the update gate, and r_t represents the vector corresponding to the reset gate. The parameters W and b refer to the weight and bias matrices that undergo learning during the training process. Additionally, the symbols σ and $tanh$ represent the Sigmoid and activation functions, respectively.

4   Proposed Method

This section presents a summary of the dataset, describes the data pre-processing steps, and introduces the proposed sentiment analysis model known as "B-ABiGRUCaps" for evaluating sentiments from airline customer review tweets. The basic structure of the proposed classification model is illustrated in Fig. 2.

4.1   Dataset

In this study, the dataset (Twitter US Airline Sentiment) comprises a diverse collection of tweets sourced from Kaggle. The dataset encompasses a total of 14,640 tweets. These tweets were specifically gathered from six prominent US airlines, namely United, Southwest, Delta, US Airways, and Virgin America. The sentiment distribution of the tweets is presented in Table 1 below.

4.2   Data Preprocessing

This represents a fundamental step of data mining approach employed to convert data into a comprehensible format. To prepare the tweets for subsequent analysis, they undergo a series of preprocessing stages aimed at generating cleaned and

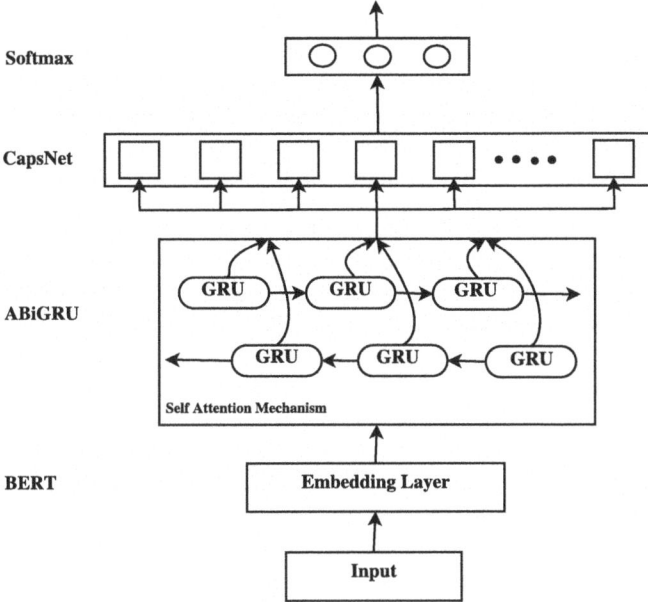

Fig. 2. The overall structure of our proposed B-ABiGRUCaps model.

Table 1. Sentiment distribution of tweets.

Sentiment	Tweet Count
Positive	2363
Negative	9178
Neutral	3099

usable tweet data. The specific pre-processing procedures encompass tasks such as converting text to lowercase, eliminating punctuation marks, removing special characters, and tokenizing the text.

The B-ABiGRUCaps approach comprises of four parts: input layer, BiGRU with self-attention layer (ABiGRU), CapsNet layer, and Softmax layer. The subsequent sections provide a comprehensive overview of the model's component structure.

Embedding Layer: The BERT relies on Bi-directional Transformers as its encoder, which effectively combines contextual data from both preceding and succeeding sides of the present sentence and its tokens. During the sentence vector training phase, the encoder adopts a different strategy compared to traditional left-to-right or right-to-left context encoding. Instead, it involves the random concealment or substitution of certain sentences based on a predefined proportion, followed by predicting the original sentences based on the surrounding context.

Each sentence vector is represented through three key components: character embedding, sentence embedding and position embedding. These components together contribute to the sentence vector representation, which is subsequently input into a BiGRU featuring a Self-Attention Mechanism. This architectural design enriches the ability of the model to capture and utilize context, creating it a powerful approach for various natural language understanding tasks.

BiGRU with Self-Attention Mechanism Layer: Intuitively, sentiment analysis of tweets involves processing the sequential information within sentences. The BiGRU has been used in sequence modelling to extract contextual information from sequences of sentence vectors. BiGRU comprises both a forward and a backward GRU unit, allowing it to incorporate both preceding and subsequent contextual information from the sentence vectors. A self-attention mechanism has been used after BiGRU to reduce the huge volume of useless information in the context. This mechanism assigns weight to essential information, thereby enhancing the understanding of sequence semantics. Specifically, the Scaled Dot-Product Attention function has been applied within the proposed network [21]. The model follows a many-to-one architecture, where the words in the present sentence are treated as pairs of Key (K), Value (V), and Query (Q) in relation to the output words in the target sentences. The computations, as detailed in Eqs. (5, 6, and 7), involve generating an output matrix as shown in Eq. (5). Estimating the attention score involves measuring the similarities between Key and Query. The Scaled Dot-Product function includes a division by the square root of the dimension of the key vectors ($\sqrt{d_k}$) to prevent the inner product from becoming excessively large [22]. The calculation of the self-attention level is as follows:

$$Attention\,(Q, K, V) = softmax\left(\frac{QK^T}{\sqrt{d_k}}\right)V. \tag{5}$$

$$Where \quad Q\epsilon\mathbb{R}^{n \times d_k}, K\epsilon\mathbb{R}^{m \times d_k}, V\epsilon\mathbb{R}^{m \times d}. \tag{6}$$

$$h \quad Sequence \quad Q : n * d_k \xrightarrow{attention\,layer} n * d_v. \tag{7}$$

Capsule Neural Network (CapsNet): The CapsNet is a supervised technique dependent on Artificial Neural Networks (ANN) [23]. Capsule essentially constitutes a group of neurons with activity vectors that serve as immediate descriptors for a specific type of entity [24]. These vectors convey both the probability associated with the entity and its instantaneous parameters via their length and direction, respectively. To transform shorter vectors, close to 0 and longer vectors toward 1, a non-linear squashing function is used. This function is a mathematical function that partitions the input into small intervals. As indicated in Eqs. (8, 9, 10), the capsule input (v_j) is associated with the output vector (s_j). The overall capsule input s_j is obtained through a weighted summation $\widehat{U}_{i|j}$. The vectors denoted as $\widehat{U}_{i|j}$, commonly known as "prediction vectors", are produced by taking the output u_i from a capsule in the lower layer and performing a multiplication operation using the weight matrix W_{ij}. Additionally,

c_{ij} represents coupling coefficients, and these coefficients are determined through an iterative dynamic routing process [23]. The variable u_i represents the output, while W_{ij} stands for the weight matrix.

$$v_j = \frac{\|s_j\|^2}{1 + \|s_j\|^2} \frac{s_j}{\|s_j\|}. \tag{8}$$

$$s_j = \sum_{i=1} c_{ij} \widehat{u}_{i|j}. \tag{9}$$

$$\widehat{u}_{i|j} = W_{ij} u_i. \tag{10}$$

In Eq. (9), the term c_{ij} represents coupling coefficients, which are precisely determined through an iterative dynamic routing process. These coupling coefficients signify the relationships between capsule i and all the capsules situated in the higher layer. Importantly, they adhere to the constraint of summing up to 1 and are computed using a routing softmax mechanism [23]. We use the softmax layer to process class label.

5   Result and Discussion

In our experiment, we have employed precision (P), recall (R), and $F1$ score as the metrics for evaluation. Regarding the configuration of model parameters, the character embedding size is defined as 768, hidden state dimension is configured to be 100, learning rate is set to be 0.001, and rate of dropout is 0.5.

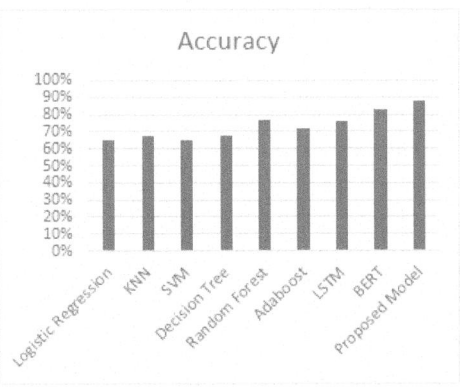

Fig. 3. The comparison of different contemporary approaches with the proposed model.

Numerous researchers have conducted studies focused on sentiment analysis using this dataset, employing a diverse range of machine learning algorithms, including NB, SVM, RF, and deep learning methods such as LSTM and BERT.

Subsequently, we have performed a comparative analysis of the proposed model with contemporary approaches on the same dataset. The model exhibits an impressive performance by achieving an accuracy rate of 88%, P score of 0.90, R score of 0.85, and $F1$-score of 0.87. In Fig. 3, we have depicted the comparison of different recent approaches used by AKSH PATEL et al. [25], using the same dataset, with the proposed model in team of accuracy. Furthermore, we have conducted tests on the results produced by the proposed model alongside other state-of-the-art approaches and the outcomes are presented in Table 2.

Table 2. Performance Comparison with existing methods.

Model Name	Precision	Recall	F1-measure	Accuracy
Logistic Regression	0.65	0.57	0.48	65%
KNN	0.61	0.59	0.44	67%
SVM	0.65	0.56	0.49	65%
Decision Tree	0.57	0.56	0.65	67%
Random Forest	0.78	0.75	0.73	77%
Adaboost	0.69	0.56	0.61	72%
LSTM	0.72	0.74	0.72	76%
BERT	0.81	0.86	0.83	83%
Proposed Model	**0.90**	0.85	**0.87**	**88%**

From Fig. 3 and Table 2, it is evident that the suggested model outperforms other contemporary approaches.

6   Conclusion

In this work we have focussed on analyzing the sentiment of passenger reviews on popular US airlines from Twitter data. This article has introduced the B-ABiGRUCaps method for sentiment classification. The B-ABiGRUCaps model consists of four parts: input layer, self-attention based BiGRU layer (ABiGRU), CapsNet layer, and Softmax layer. The proposed B-ABiGRUCaps method has obtained 88% accuracy, which is better than traditional machine learning and deep learning models. This research might be expanded in the future to include multilingual data. The findings of this study will have practical implications for both airline industries and passengers. Specifically, the results may be used to develop a recommendation system that offers personalised suggestions to passengers based on their individual preferences, travel objectives, and financial constraints, while also considering criteria such as the desired destination.

References

1. Hasib, K.M., Habib, M.A., Towhid, N.A., Showrov, M.I.H.: A novel deep learning based sentiment analysis of twitter data for us airline service. In: 2021 International Conference on Information and Communication Technology for Sustainable Development (ICICT4SD), pp. 450–455. IEEE (2021)

2. Kumawat, S., Yadav, I., Pahal, N., Goel, D.: Sentiment analysis using language models: a study. In: 2021 11th International Conference on Cloud Computing, Data Science & Engineering (Confluence), pp. 984–988. IEEE (2021)

3. Kumar, S., Zymbler, M.: A machine learning approach to analyze customer satisfaction from airline tweets. J. Big Data **6**(1), 1–16 (2019)

4. Saad, A.I.: Opinion mining on us airline twitter data using machine learning techniques. In: 2020 16th International Computer Engineering Conference (ICENCO), pp. 59–63. IEEE (2020)

5. Kamal, S., Dey, N., Ashour, A.S., Ripon, S., Balas, V.E., Kaysar, M.S.: FbMapping: an automated system for monitoring Facebook data. Neural Netw. World **27**(1), 27–57 (2017)

6. Twitter us airline sentiment. https://www.kaggle.com/datasets/crowdflower/twitter-airline-sentiment. Accessed 09 July 2023

7. Oza, P., Sharma, P., Patel, S., Kumar, P.: Deep convolutional neural networks for computer-aided breast cancer diagnostic: a survey. Neural Comput. Appl. **34**(3), 1815–1836 (2022)

8. Oza, P., Sharma, P., Patel, S., Adedoyin, F., Bruno, A.: Image augmentation techniques for mammogram analysis. J. Imaging **8**(5), 141 (2022)

9. Dutta, R., Majumder, M.: Attention-based bidirectional LSTM with embedding technique for classification of COVID-19 articles. Intell. Decis. Technol. **16**(1), 205–215 (2022)

10. Thomas, J., Sharma, N.C., Kumar, P., Chauhan, A., Chauhan, P.: Effect of biostimulant and biofertilizers on soil biochemical properties and plant growth of apple (malus x domestica borkh.) nursery. J. Environ. Biol. **43**(2), 276–283 (2022)

11. Dutta, R., Das, N., Majumder, M., Jana, B.: Aspect based sentiment analysis using multi-criteria decision-making and deep learning under COVID-19 pandemic in india. CAAI Trans. Intell. Technol. **8**(1), 219–234 (2023)

12. Dutta, R., Adhikary, D.R.D., Majumder, M.: A deep learning model for classification of covid-19 related fake news. In: Electronic Systems and Intelligent Computing: Proceedings of ESIC 2021, pp. 449–456. Springer (2022)

13. Güner, L., Coyne, E., Smit, J.: Sentiment analysis for amazon.com reviews. Big Data Media Technol. (DM2583) KTH Roy. Inst. Technol. **9** (2019)

14. Ul Haque, T., Saber, N.N., Shah, F.M.: Sentiment analysis on large scale amazon product reviews. In: 2018 IEEE International Conference on Innovative Research and Development (ICIRD), pp. 1–6. IEEE (2018)

15. Prabhakar, E., Santhosh, M., Hari Krishnan, A., Kumar, T., Sudhakar, R.: Sentiment analysis of us airline twitter data using new adaboost approach. Int. J. Eng. Res. Technol. (IJERT) **7**(1), 1–6 (2019)

16. Rane, A., Kumar, A.: Sentiment classification system of twitter data for us airline service analysis. In: 2018 IEEE 42nd Annual Computer Software and Applications Conference (COMPSAC), vol. 1, pp. 769–773. IEEE (2018)

17. Safrin, R., Sharmila, K.R., Shri Subangi, T.S., Vimal, E.A.: Sentiment analysis on online product review. Int. Res. J. Eng. Technol. (IRJET) **4**(04) (2017)

18. Rahat, A.M., Kahir, A., Masum, A.K.M.: Comparison of naive bayes and SVM algorithm based on sentiment analysis using review dataset. In: 2019 8th International Conference System Modeling and Advancement in Research Trends (SMART), pp. 266–270. IEEE (2019)

19. Hasib, K.M., Towhid, N.A., Alam, M.G.R.: Online review based sentiment classification on bangladesh airline service using supervised learning. In: 2021 5th International Conference on Electrical Engineering and Information Communication Technology (ICEEICT), pp. 1–6. IEEE (2021)

20. Chung, J., Gulcehre, C., Cho, K., Bengio, Y.: Empirical evaluation of gated recurrent neural networks on sequence modeling. arXiv preprint arXiv:1412.3555 (2014)

21. Miller, A., Fisch, A., Dodge, J., Karimi, A.-H., Bordes, A., Weston, J.: Key-value memory networks for directly reading documents. arXiv:1606.03126 (2016)

22. Vaswani, A., et al.: Attention is all you need. In: Advances in Neural Information Processing Systems, vol. 30 (2017)

23. Sabour, S., Frosst, N., Hinton, G.E.: Dynamic routing between capsules. In: Advances in Neural Information Processing Systems, vol. 30 (2017)

24. Wang, Q., Xu, C., Zhou, Y., Ruan, T., Gao, D., He, P.: An attention-based BI-GRU-CapsNet model for hypernymy detection between compound entities. In: 2018 IEEE International Conference on Bioinformatics and Biomedicine (BIBM), pp. 1031–1035. IEEE (2018)

25. Patel, A., Oza, P., Agrawal, S.: Sentiment analysis of customer feedback and reviews for airline services using language representation model. Procedia Comput. Sci. **218**, 2459–2467 (2023)

Implementation of Digital Healthcare for Improving Maternal Care: A Systematic Review

Sarika Kumari Shaw[ID] and Jayati Lahiri Dey[(✉)][ID]

Department of Computer and Information Science, Raiganj University, Raiganj 733134, India
ld.jayatii@gmail.com

Abstract. In this paper we have reviewed the impact of digital healthcare for improving maternal care in West Bengal. To reduce the rate of mortality the Indian Government implemented various types of funding schemes which provide cash incentive to a pregnant woman. In this paper we have studied the effectiveness of two such schemes, namely Janani Suraksha Yojana (JSY), and Pradhan Mantri Matra Vandana Yojana (PMMVY) along with impact of digital healthcare in improving the mother and neonatal healthcare. We conducted this study in all the 23 districts of West Bengal mainly in rural areas. The data is being taken from all the round of the sample registration system to understand the pattern of maternal healthcare in West Bengal. All four National Family Health Survey [1998–99, 2005–06, 2015–16, and 2020–21] were used to analyze the data. It has been seen that this scheme seems to be a very beneficial reducing 77% reduction in the rate of maternal mortality while the world's average is only 43%. The schemes seem to be very beneficial in increasing institutional birth and antenatal and postnatal visit as 20.5%, 3.8%, and 6% growth respectively. The rate of maternal and child mortality decreases gradually that is the result of those schemes along with growing technologies in healthcare.

Keywords: Sample Registration System · National Family Health Survey · Emergency obestric care · Digital Health · Janani Suraksha Yojana · Pradhan Mantri Matra Vandana Yojana

1 Introduction

Digital health is used to improve accessibility, affordability, timeliness, efficiency, inclusiveness and safety. National digital health mission is announced on 15th August, 2020 on the 74th independence day to provide universal access of healthcare. Previously the government of India invests in different policies to upgrade EmOC facilities but more important is to make different strategies so that highest maternal deaths rating poor people could able to utilize those facilities, [2, 6, 32]. As up gradation of EmOC facilities helps to reduce the death rate, but poor women were facing different types of barriers mainly economic to access the facilities of EmOC, so they were still remained at the highest risk [3, 16, 29]. There was the highest need to prioritize poor and low-income people's maternal health. In 2005 National Rural.

M. Majumder et al. (Eds.): ICCTE 2023, CCIS 2376, pp. 251–262, 2025.
https://doi.org/10.1007/978-3-031-81935-3_22

Health Mission (NRHM) was established to help rural poor people. Janani Suraksha Yojana (JSY), a conditional cash incentive scheme was launched under NRHM to help poor and low- income people to access facilities of healthcare. The main goal of this scheme was to increase institutional delivery. JSY helps to reduce the mortality rate by nearly 68–72% while the global average was 42%. This extreme success in reducing the mortality rate in India is also appreciated by World Health Organization [1, 4, 19]. In 2017 Pradhan Mantri Matra Vandana Yojana, which was earlier known as Indira Gandhi Matra Vandana Yojana (2012) was launched that provides cash incentives of INR 5000 to lactating mothers for their first live child [9, 10, 28]. On the other hand new emerging digital technologies have the potential to work in low resource environment to help maternal healthcare but they are in their early stages of development mainly in rural areas [13, 17, 25].

Many research have been done to evaluate the effect of digital technologies and different policies in maternal healthcare in different states of India. But very few works have been done in West Bengal mainly in rural areas. In this study we have reviewed the impact of digital healthcare and two government schemes in improvement of condition of maternal and neonatal health in rural West Bengal by using multiple data sources.

1.1 Objectives

The main objectives of this paper are

i. To analyze the improvement in mothers and child healthcare in rural West Bengal (1998–2022) using multiple data sources
ii. Study the effectiveness of two government schemes (JSY and PMMVY) on the maternal health care status of West Bengal.
iii. Study the advancement of technology and analyze its effect on maternal health care and reducing child mortality.

1.2 Data and Methodology

In this section we have discussed about data and methodologies of the proposed paper.

For doing this study we have taken all 23 districts of West Bengal. Both the cash incentive schemes JSY & PMMVY have implemented in all these districts. Secondary data is used for the study using different quantitative methods. All sample registration systems are used for data collection that is further analyzed by four rounds of the National Family Health Survey which were done in the year 1998–99 (NFHS-2), 2005–06 (NFHS-3), 2015–16 (NFHS-4), and 2020–21, (NFHS-5) [11, 12, 29]. The study was also conducted by a systematic review of literature. In this study, we considered only those studies which were published after the 2005, implementation of JSY. We mainly used Google Scholar, Pub Med & Research Gate for collecting related information.

We established two probit equations. One equation is used for finding the place of delivery of the mother. It might be a public sector institution or private. The second equation used to find the mortality ratio of the child who died below five years of the concerned mother. So as an outcome variable, we are getting two variables, the place of delivery (DH), & the mortality of the child (MC). So these two equations are being estimated.

$$DH = \alpha 0 + \alpha 1 Xc + uc + ec \tag{1}$$

$$MC = \beta 0 + \beta 1 Xh + uh + eh \tag{2}$$

Here delivery in the hospital and mortality of the child is denoted by subscript c, h respectively, and constant terms interpretations for these equations are shown as $\alpha 0$, $\beta 0$. Coefficients of different describing variable are denoted by $\alpha 1$, $\beta 1$. XC, Xh act for possible regressors that may significant effects on the requirement of maternal health input and child mortality. Therefore, the binary outcome variables are represented as

$$
\begin{aligned}
DH = 1 \quad & \text{if delivery taken place in the hospital} \\
0 \quad & \text{if delivery is not in the hospital}
\end{aligned} \tag{3}
$$

$$
\begin{aligned}
MC = 1 \quad & \text{if the child below 5 years is dead} \\
0 \quad & \text{if no child below 5 years is dead}
\end{aligned} \tag{4}
$$

Here un (n = c, h) is denoting an error term which handles overlooked diverseness generated by mother. So un ~ N (0, $\mu 1^2$); $\mu 1^2$ is negligible factors variance and $\in$ is used for all other remaining variation.

1.3   Results and Discussions

Digitization of Healthcare (2000–2020), Between Those Periods Two Government Schemes Were Operational in West Bengal
ICT has been widely used in the healthcare industry for the last two decades and played a major role in the healthcare sector. In healthcare different types of ICT technologies are being used as hospital management and information system, keeping huge amounts of health-related data electronically. PAX is used for the archival and retrieval of all digital imaging systems and bar- coding [14, 27, 34]. There are two very beneficial perspectives of ICT in healthcare telemedicine and telehealth which are developing rapidly. Health

Table 1. Digital Technologies in Rural Healthcare

Health Information digital technologies	Functions
Electronic health Record(EHR)	Keeping all records electronically instead of paper
Networks	Secure networks for communication
Telehealth	Application for enhancing providers Accessibility
Health Portal	Confidential and secure platform
Electronic test result	Provide More accuracy
Mobile devices and tablets	Real time update of patient record

information technology (HIT) mainly working in low resources rural area providing different facilities (Table 1).

Webel (West Bengal Electronics Industry Development Corporation Limited) started working rapidly on the project of telemedicine and telehealth. In West Bengal, WEBEL is going to implement the National Optic Fiber Network (NOFN), which would provide huge bandwidth up to rural areas, and then an enormous amount of telemedicine could be possible. In West Bengal, about 16,000 telemedicine centers are going to be set up within a few years [5, 15]. Health information technology (HIT) mainly working in low resources rural area providing different facilities.

JSY increases use of maternal healthcare facilities by providing cash incentives. During the pregnancy period, the nutritional requirement of woman increases to deliver a healthy child, that increases their out of pocket expenses. PMMVY yojana helps pregnant and lactating mother by providing a cash incentive of 5000/- in three installments to fulfill their daily nutritional need [9, 10, 18, 20].

Antenatal Care. All the woman who gave birth to their child five years prior to the survey, their data is being taken and analyzed by all rounds of NFHS. It has been seen that the urban woman had a slightly higher rate of 4 antenatal care visit percentage than rural woman [12, 22, 30].

Postnatal Care. Postpartum healthcare is very much crucial for reducing mother mortality and protecting their health. It is recommended that the mother should have a postnatal check-up within 2 days of delivery as recommended as shown in the Fig. 1.

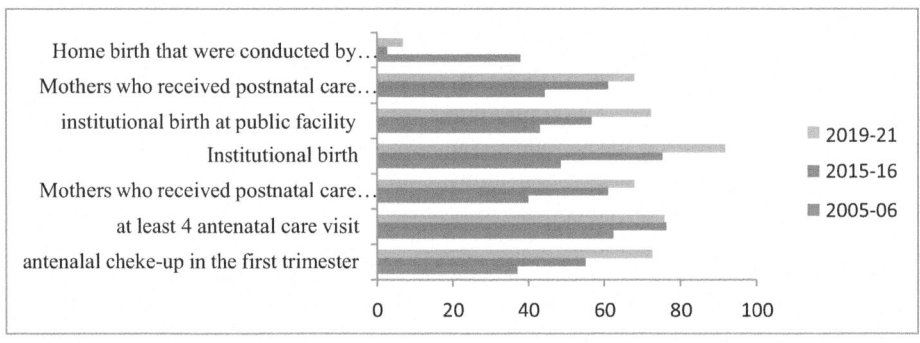

Fig. 1. Facility utilized by mother [Sources – NFHS3, NFHS4, NFHS5]

Abbreviation used in the paper:

EmOC   Emergency obestric care
NFHS   National family health survey
HMIS   Health management information system
PMMVY   Pradhan mantri matru vandana yojana

A Comparative Study of Janani Suraksha Yojana (JSY) and Pradhan Mantri Matru Vandana Yojana (PMMVY)

(See Table 2, Fig. 2).

Table 2. Comparative Study of two cash incentive schemes for maternal care in India.

	Area	Target Group	Type of Program	Incentive for	Payment mechanism	Expected Outcomes	Current Status
JSY	Nationwide (studied here West Bengal)	All mothers	Conditional cash transfer(CCT) by State to mother	Institutionaldelivery	In low performing state mother will get INR 1400 for rural area and 1000for urban area. In high performing state mother will get INR 700for rural area and INR 600 for urban areas	Reducing the mother and infant mortality rate by promoting the institutional delivery among poor women	In the financial year 2022–23 the eligible citizens were only 38 lakh which was 47 percent less than previous year. The partial reason behind this decline is the new guidelines of making aadhar number compulsory for eligibility for benefits
PMMVY	Nationwide (studied here West Bengal)	All mothers (Pregnant/Lactating)	Direct benefit transfer(DBT) By Central Gov	To improve health care behavior in pregnant or lactating mother by providing compensation in wage loss	INR 5000 is givenin 3 installment 1) INR 1000 After registering pregnancy INR 2000 after 6 month of pregnancy & ANC visit INR2000 after child registration and doing some essential vaccination	Reducing wage loss in pregnant and lactating (PM/LM) Mother and support in receiving medical facilities during and after pregnancy	Between thefinancial year 2019–20 and 2020–21 the institutional deliveries also decreases from 202.28 lakh to 193.75 lakh

1.4 Findings

After comparing and analyzing data from different sources viz. All NFHS, and other sources as mentioned in the data collection methodology, different types of result in table forms as given below [7, 8, 11, 12].

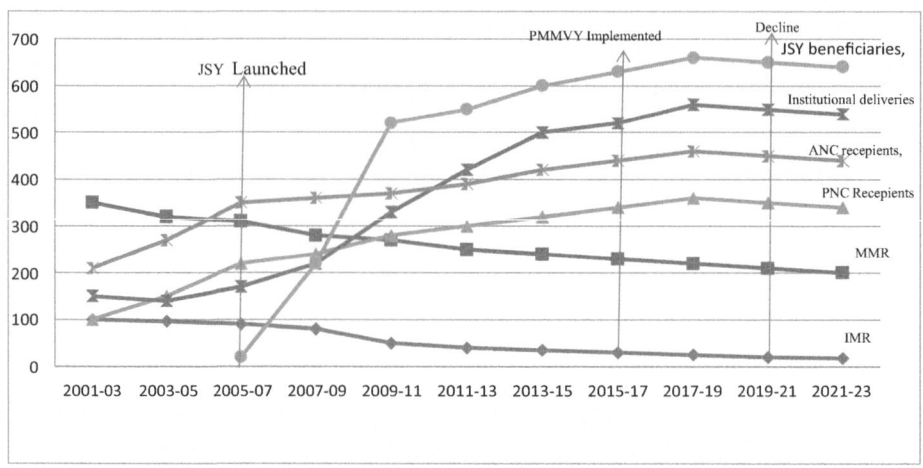

Fig. 2. Comparative analysis of JSY AND PMMVY (Sources HMIS, NFHS(1–4))

District wise distribution of percentage of mother getting all types of antenatal and delivery care for their recent most live births in the 5 years prior to the survey is shown in Table 3 and maternal healthcare services received according to background characteristics is shown in Table 4.

Table 3. Indicators of antenatal & delivery care District Wise

Name of district	4 antenatal visit by mother (%)	In first trimester, antenatal visit by the mother (%)	Iron & Folic acid consumption by mother for 100 days (%)	Two or more tetanus (TT) injection taken by mother (%)	Mother Received all type of antenatal care (%)	Mother Having MCP card (%)	The total number of pregnant woman	Registered pregnancies
Bankura	89.4	65.6	46.2	93.6	40.7	97.2	163	158
Bardhaman	83.4	67.1	35.8	82.4	28.9	96.9	359	350
Birbhum	77.9	51.2	22.5	92.9	17.1	99.6	164	155
Cooch Bihar	74.6	57.9	20.1	87.9	18.0	96.9	132	118
Dakshin Dinajpur	68.9	53.2	39.2	56.2	25.4	99.2	69	66
Darjiling	66.2	80.9	42.6	96.7	32.9	93.4	67	62
Howrah	87.1	73.8	38.5	60.1	32.2	94.3	218	202
Hooghly	77.1	52.6	32.9	93.8	24.8	97.9	240	233
Jalpaiguri	80.8	56.9	24.3	86.9	20.1	96.8	161	148
Kolkata	84.6	76.0	41.9	91.3	33.1	91.9	145	123
Malda	53.0	43.1	18.9	87.9	11.8	95.8	232	199

(continued)

Table 3. (*continued*)

Name of district	4 antenatal visit by mother (%)	In first trimester, antenatal visit by the mother (%)	Iron & Folic acid consumption by mother for 100 days (%)	Two or more tetanus (TT) injection taken by mother (%)	Mother Received all type of antenatal care (%)	Mother Having MCP card (%)	The total number of pregnant woman	Registered pregnancies
Murshidabad	71.9	48.9	22.8	96.9	16.8	99.0	454	426
Nadia	92.0	67.9	35.9	95.6	34.1	98.5	278	273
North 24 Pargana	80.2	51.9	22.1	95.8	17.1	96.5	454	426
Paschim Mednipur	83.9	50.2	25.9	95.1	21.3	97.8	278	273
Purba Mednipur	79.1	42.0	30.2	95.3	22.8	98.1	275	255
Purulia	69.0	51.9	35.2	78.8	23.6	99.9	161	158
South 24 Pargana	76.2	50.2	23.8	92.1	15.9	98.0	499	484
Uttar Dinajpur	42.9	37.2	5.8	84.4	4.5	96.6	183	156
West Bengal	76.4	55.2	27.9	90.8	22.0	97.5	4,462	4,212

Table 4. Based on the background characteristics of West Bengal percentage of mother getting all types of antenatal and delivery care for their recent most live birth in the 5 years prior to the survey is shown

Characteristics referring background	4 antenatal visit by mother (%)	First trimester antenatal visit by mother(%)	Two or more tetanus (TT) injection taken by mother (%)	Protected newborn from neonatal tetanus (%)	Iron & Folic acid brought or provides to mother (%)	Iron & Folic acid consumption for 100 days by mother (%)	Mother Received all type of antenatal care (%)	Intestinal Parasite drug taken by the mother (%)	Total Number of Woman
Mother'age At the time of delivery									
<20	79.2	56.2	93.0	96.3	91.2	25.2	21.1	16.4	1,142
20–35	76.5	54.6	92.8	95.8	92.3	30.2	23.5	20.3	3,327
36–50	61.3	46.7	91.6	94.6	88.8	34.6	25.4	19.2	102
Number of children									
1	83.7	61.2	93.4	96.2	93.5	32.4	26.3	19	2,116
2–3	75.0	52.3	91.5	94.9	90.6	27.6	20.4	20	2,029
4+	52.2	40.6	88.6	92.8	82.4	22.4	14.5	14	319

(*continued*)

Table 4. (*continued*)

Characteristics referring background	4 antenatal visit by mother (%)	First trimester antenatal visit by mother(%)	Two or more tetanus (TT) injection taken by mother (%)	Protected newborn from neonatal tetanus (%)	Iron & Folic acid brought or provides to mother (%)	Iron & Folic acid consumption for 100 days by mother (%)	Mother Received all type of antenatal care (%)	Intestinal Parasite drug taken by the mother (%)	Total Number of Woman n
Place of living									
Urban	77.9	59.2	91.4	94.9	90.9	32.2	26.4	21.2	1,284
Rural	76.2	52.8	82.2	96.1	90.5	27.0	21.6	19.3	3,177
Education									
No education	61.2	46.2	89.0	94.3	87.1	27.1	17.2	18.0	725
<5 years of schooling	73.2	52.5	94.1	96.4	90.2	27.3	20.5	19.2	602
5–10 years of schooling	80.1	54.5	93.5	96.9	92.7	26.4	21.3	20.4	1,960
11–12 years of schooling	81.5	57.4	90.2	97.0	94.6	33.4	28.2	19.8	587
>12 years	87.6	73.2	93.7	96.2	95.2	37.6	31.5	18.6	591
Religion									
Hindu	78.8	54.6	90.8	94.6	92.7	29.9	23.6	17.2	2,906
Muslim	66.5	54.3	89.7	94.1	85.9	22.7	16.7	16.8	1,382
Christian	81.6	79.4	94.6	94.0	94.2	26.8	13.9	2.8	17
Other	97.4	32.9	92.4	93.7	91.5	15.9	15.7	45.6	163
Caste/ Tribe									
General	86.4	64.3	94.1	95.2	93.9	32.2	26.7	24.9	2,320
OBC	82.6	62.2	92.3	94.8	93.1	30.7	25.4	23.7	2,282
SC/ST	78.6	54.9	90.0	93.9	94.5	28.6	22.6	20.4	1,486
Other	75.2	54.2	92.4	94.2	89.7	26.8	21.5	18.3	123
Total	77.2	55.8	92.6	96.7	90.4	27.8	22.4	19.2	4,458

We have calculated two probit equations (Eq. 1, Eq. 2) with two output variables HD, MC are shown in Tables 5 and 6.

Delivery in Hospital (DH): Mother's age had no remarkable effect in that but mother's education had great effect. Primary education has (8%) and secondary education has (15%) in choosing place of delivery as hospital (Table 5) [21, 23, 24].

Mortality of Child (MC): According to regression analysis hospital delivery remarkably reduces child mortality. So mother who delivered their child in hospitals were 12% less risk of child mortality (Table 6) [19, 26, 31, 33].

Table 5. Delivery in Hospital (DH) (Calculated using Eq. 1 & 2)

	Effects on Hospital Delivery	Marginal Effects of the respective covariate
Prenatal Care	0.417 (0.39)	–
Place of residence	1.053*** (9.85)	0.2441
Age of mother	0.000496** (2.60)	0.0001
Religion (ref. Group of Hindu)		
Muslim	−0.581*** (-6.54)	−0.1344
Living Standard (ref. Group of economical)		
Good	0.758*** (4.93)	0.1768
Very good	0.508** (2.83)	0.1178
Access to Information	0.00627**(3.22)	0.0013
Mother's education (ref. Group of Illiterate)		
Primary level	0.304** (2.80)	0.070
Secondary level	0.603*** (5.12)	0.1398
Birth order of the child	−0.191*** (-4.26)	−0.0442
Constant	−1.576 (−1.72)	

Table 6. Mortality of Child (MC) (Calculated using Eq. 1 & 2)

	Result on Child Mortality	Insignificant Effects of the respective covariate
Place of delivery	−0.693* (−2.08)	−0.1293
Age of the mother	−0.000400* (−1.97)	−0.0001
Religion (ref. Group Hindu)		
Muslims	−0.256* (−2.48)	−0.0477
Christian	−1.036* (−2.19)	−0.1933
Mother's occupation (ref. group unemployed)		
Manual	0.320** (2.63)	0.0596
Say in household decision making	0.0129* (2.56)	0.0024
Birth order of the child	0.376*** (7.12)	0.0704
Constant	−1.103 (−1.33)	

t statistics in parentheses. $*p < 0.05$, $**p < 0.01$, $***p < 0.001$.

1.5 Conclusion

In this paper, we studied about how rural areas of West Bengal is gradually adopting digital technologies and its impact in improvement of mother and neonatal healthcare. We also analyzed the effectiveness of two large-scale cash incentive schemes for maternal care. This analysis would helps policymakers in our country and outside of India where these types of financing programs are currently working or planning to launch a relevant program. The study also helps to find the area that needs improvement and helps to suggest some better plans after strengthens the current working schemes. In West Bengal the cash incentive schemes help in better maternal care, lowering the mortality rate, the rise of institutional birth in the public sector, and reducing the out-of-pocket expenses of the mother. Despite these benefits, mothers are getting only partial financial support from JSY because it supports mainly institutional delivery, and antenatal and postnatal care provisions are very less. So there was a need for some more flexible financial support for the mother. The integration of JSY and PMMVY schemes together helps to support overall maternity care and the sustainabilityof maternal care could be ensured. Secondary data is used for the evaluation of these schemes and it found a remarkable growth in institutional delivery, antenatal and postnatal care. It has been also seen that technological development and advancement of ICT were playing a significant role in this positive result.

References

1. Alkema, L., Chou, D., Hogan, D., Zhang, S., Moller, A.B., Gemmill, A., et al.: Global, regional and national levels and trends in maternal mortality between 1990 and 2015, with scenario-based projections to 2030: a systematic analysis by the UN maternal mortality estimation inter-agency group. Lancet **387**(10017), 462–474 (2016)
2. Bandyopadhyay, S., Pal, D., Dasgupta, A., Datta, M., Paul, B.: Status of maternal health care services: an assessment study in slums of Kolkata. J. Family Prime Care **9**(1), 4861–4881 (2020)
3. Brauw, A., Peterman, A.: Can conditional cash transfers improve maternal healthcare? Evidence from El Salvador's "Comunidades Solidaries Rurales" program. Health Econ. **29**(6), 700–715 (2020)
4. Dash, S., Aarthy, R., Mohan, V.: Telemedicine during COVID- 19 in India-a new policy and its challenges. J. Public Health Pol. **42**, 501–509 (2021). https://doi.org/10.1057/s41271-021-00287-w
5. Davis, J., Vyankandondera, J., Luchters, S., Simon, D., Holmes, W.: Male involvement in reproductive, maternal and child health: a qualitative study of policymaker and practitioner perspective in the pacific. Reprod. Health **13**(1), 1–11 (2016)
6. Deo, K.K., et al.: Barriers to utilization of antenatal care services in Eastern Nepal. Front. Public Health **3**, 1–7 (2018)
7. Glassman, A., et al.: Impact of conditional cash transfers on maternal and newborn health. J. Health Popul. Nutr. **31**(4) (2013). https://doi.org/10.3329/jhpn.v31i4.2359
8. Govil, D., Purohit, N., Gupta, S.D., Mohanty, S.K.: Out-of-pocket expenditure on prenatal and natal care post Janani Suraksha Yojana: a case from Rajasthan, India. J. Health Popul. Nutr. **3**(1), 15–35 (2016)
9. Gautam, A.: A critical evaluation of Pradhan Mantri Matru Vandana Yojana. Jindal J. Public Policy **4**(1), 46–62 (2020)

10. Gupta, S.K., et al.: Impact of Janani Suraksha Yojana on institutional delivery rate and maternal morbidity and mortality: an observational study in India. J. Health Popul. Nutr. **30**(4), 64–71 (2012)
11. Jana, A., Basu, R.: Examining the changing health care seeking behavior in the era of health sector reforms in India: evidences from the national sample surveys 2004 & 2014. Glob. Health Res. Policy **2**(1), 6 (2017)
12. Joshi, S.R., George, M.: Healthcare through community participation role of ASHAs. Econ. Pol. Wkly **47**(10), 70–76 (2012)
13. Kesterton, A.J., Cleland, J., Sloggett, A., Ronsmans, C.: Institutional delivery in rural India: the relative importance of accessibility and economic status. BMC Pregnancy Childbirth **10**(1), 30 (2010)
14. Mahmood, S.S., Amos, M., Hoque, S., Mia, M.N., Chowdhury, A.H., HaniFi, S.M.A., et al.: Does healthcare voucher provision improve utilization in the continuum of maternal care for poor pregnant women? Experiences from Bangladesh. Glob. Health Action **12**(1), 1701324 (2019)
15. Mandal, B.: Demand for maternal health inputs in West Bengal- Inference from NFHS 3 in India. Econ. Bull. **35**(4), 2685–2700 (2015)
16. Marco, J. Haenssgen, A.: The struggle for digital inclusion: phones, healthcare, and marginalization in rural India. In: International Conference on Communication System and Networks, Banglore, India, vol. 104, pp. 358–374 (2018). https://doi.org/10.1016/j.worlddev.2017.12.023
17. Ministry of Women and Child Development: Pradhan Mantri Matru Vandana Yojana (PMMVY): Scheme implementation guidelines, pp. 1–3 (2017)
18. Mukhopadhyay, D.K., Mukhopadhyay, S., Mallik, S., Nayak, S., Biswas, A.K., Biswas, A.B.: A study on utilization of Janani Suraksha Yojana and its association with institutional delivery in the state of West Bengal, India. Indian J. Public Health **1**, 60–118 (2017)
19. Murray, S.F., Hunter, B.M., Bisht, R., Ensor, T., Bick, D.: Demand-side financing measures to increase maternal health service utilization and improve health outcomes: a systematic review of evidence from low and middle – income countries. JBI Database Syst. Rev. Implement Rep. **10**, 4165–4567 (2012)
20. Murray, S.F., Hunter, B.M., Bisht, R., Ensor, T., Bick, D.: Effects of demand-side financing on utilization, experiences and outcomes of maternity care in low and middle income countries: a systematic review. BMC Pregnancy Childbirth **17**, 14–30 (2014)
21. Modugu, H.R., Kumar, M., Kumar, A., Millett, C.: State and socio- demographic group variation in out-of-pocket expenditure, borrowings and Jannani Suraksha Yojana (JSY) programme use for birth deliveries in India. BMC Public Health **12**(1), 1048 (2012)
22. National Family Health Survey (NFHS-3), 2005–06, 2007. International Institute for Population Sciences (IIPS) and Macro International, India
23. Paneru, D.P.: Pattern of institutional delivery in Dadeldhura District of Nepal: a cross sectional study. J. Sci. Soc. **41**(2), 94–100 (2014)
24. Panja, T.K., Mukhopadhyay, D.K., Sinha, N., Saren, A.B., Sinhababu, A., Biswas, A.B.: Are institutional deliveries promoted by Janani Suraksha Yojana in a district of West Bengal, India? Indian J. Public Health **56**, 69–72 (2015)
25. Powell, J.T., Mazumdar, S., Mills, A.: Financial incentives in health: new evidence from India's Janani Suraksha Yojana. J. Health Econ. **154** (2015)
26. Randive, B., Sebastian, M., Decosta, A., Lindholm, L.: Inequalities in institutional delivery uptake and maternal mortality reduction in the context of cash incentive program, Janani Suraksha Yojana: results from nine states in India. Soc Sci Med **123**, 1–6 (2014)
27. Rajaranjan, K., Kumar, S.G., Kar, S.S.: Proportion of beneficiaries and factors affecting Janani Suraksha Yojana direct cash transfer scheme in Pondicherry, India. J. Family Med. Prim. Care **5**(4), 817–821 (2016)

28. Sachin, C., et al.: Technology driven rural healthcare practices for the villagers of Sarai Nooruddinpur, Uttar Pradesh. In: Kumar, A., Paprzycki, M., Gunjan, V. (eds.) ICDSMLA 2019. LNEE, vol. 601, pp. 1438–1449.Springer, Singapore (2020). https://doi.org/10.1007/978-981-15-1420-3_153

29. Sibiya, M.N., Ngxongo, T.S.P., Bhengu, T.J.: Access and utilization ofantenatal care services in a rural community of Thekwini district in KwaZulu-Natal. Int. J. African Nurs. Sci. **8**, 1–7 (2018)

30. Silali, M., Owino, D.: Factors influencing accessibility of maternal & child health information on reproductive health practices among rural women in Kenya. Family Med. Med. Sci. Res. **05**(01), 1–7 (2016)

31. Sidney, K., Tolhurst, R., Jehan, K., Diwan, V., Decosta, A.: "The money is important but all women anyway go to hospital for childbirth nowadays" – a qualitative exploration of why women participate in a conditional Cah transfer program to promote institutional deliveries in Madhya Pradesh, India. BMC Prgnancy Childbirth 16–47 (2016)

32. Singh, P.K.: India has achieved groundbreaking success in maternal mortality. WHO Regional Director for South-East Asia, New Delhi (2018)

33. Vellakkal, S., Reddy, H., Gupta, A., Chandran, A., Fledderjohann, J.: A qualitative study of factors impacting accessing of institutional delivery care in the context of India's cash incentive program. Soc Sci Med **17**, 55–65 (2017)

34. Wilunda, C., Scanagatta, C., Putoto, G., Montalbetti, F., Segafredo, G., Takahashi, R.: Barriers to utilization of antenatal care services in south Sudan: a qualitative study in Rumbek North Country. Reprod. Health **14**(1), 1–10 (2017)

Video Content Analysis and Classification Based on Human Activity Recognition

K. C. Hari$^{(\boxtimes)}$ ⓘ, Manish Pokharel, and Sushil Shrestha

Department of Computer Science and Engineering, School of Engineering, Kathmandu
University, Dhulikhel, Nepal
harikc@wrc.edu.np

Abstract. Video content analysis holds significant potential in computer vision
tasks for the automated understanding and interpretation of human motion and
activities. This paper aims to introduce an intelligent framework to recognize
human motion and activities within video data. By leveraging advanced techniques
in feature extraction, temporal modeling and deep learning, a comprehensive approach that aims to accurately categorize and interpret various human activities is
presented in this study. The combination of Convolution layers and Bidirectional
Long Short-Term Memory Layers is used to extract the spatial and temporal features of the video enabling the detection of motion patterns and activity sequences.
Video classification and Activity Recognition (VCHAR) is developed and trained
using the University of Central Florida 101 Classes (UCF101) dataset. This dataset
consists of 101 classes of activities such as archery, surfing, typing, swing and so
on. The model is tested on day-to-day human activities videos. The study provides
the classification of videos based on the activities within the video. The result of
this study is compared with previous benchmark results and it showed better accuracy of 87.67%. The finding of this study contributes to improving surveillance
security, sports activity, intrusion detection and robotics. Sport media companies,
automation Industry, education sector and researchers will be benefitted from this
type of study.

Keywords: Human motion · Activity recognition · Video content analysis ·
CNN · Bi-LSTM · UCF101

1 Introduction

In recent years, the exponential growth of digital video content has led to an increased
interest in developing intelligent systems capable of understanding and interpreting
human activities and motions within videos. This endeavor holds profound implications
for numerous fields, ranging from surveillance and security to healthcare and entertainment. The ability to automatically recognize and analyze human motion and activities
not only enhances our understanding of complex behaviors but also enables the creation
of applications that can aid in various aspects of human life [1]. Activity recognition has
the ability to detect and recognize more specific activities in video frames. The process

M. Majumder et al. (Eds.): ICCTE 2023, CCIS 2376, pp. 263–274, 2025.
https://doi.org/10.1007/978-3-031-81935-3_23

of extracting meaningful insights from video data, known as video content analysis, represents a multidisciplinary challenge that merges computer vision and machine learning. By deciphering the intricate patterns and interactions embedded within video frames, researchers uncover valuable information about the behavior, actions and intentions of individuals or groups depicted in the videos. Human motion and activity recognition, as a subset of video content analysis, holds particular significance due to its potential to revolutionize artificial intelligence industry.

Video Content Analysis and Classification Based on Human Activity Recognition refers to a process in the field of computer vision where videos are analyzed and categorized by recognizing and classifying human activities depicted in the video content. Human activity recognition is a fundamental task within the domain of computer vision. The primary goal is to develop algorithms and techniques that enable computers to automatically identify and interpret various human activities, gestures, and motions from video sequences. This task has garnered substantial attention due to its applications in diverse fields. In the realm of surveillance, accurate human activity recognition facilitates the design of advanced security systems that can promptly detect suspicious behaviors in crowded spaces, airports, and public events. Moreover, healthcare professionals are increasingly interested in utilizing motion recognition technologies to monitor patients' movements and assess their rehabilitation progress remotely [2]. In the entertainment industry, motion recognition forms the basis for interactive gaming experiences, where players' motions are translated into in-game actions. The complexity of human motion recognition arises from the inherent variability in movements across individuals, environmental conditions, and contexts. This necessitates the development of sophisticated algorithms capable of handling these challenges while delivering reliable results.

The UCF101 dataset has been widely used by computer vision research community to develop and evaluate action recognition algorithms and methods, especially those based on deep learning techniques [3]. Researchers often use this dataset to benchmark the performance of the models and compare them against state-of-the-art approaches. It contains a diverse collection of 101 action categories, spanning a diverse range of activities such as sports, dancing, cooking, playing musical instruments, and more. Each category consists of video clips of total 13320 videos extracted from YouTube, making the dataset quite diverse and representative of real-world scenarios. This paper delves into the methodologies and advancements in video content analysis such as CNN and Bi-LSTM for human motion and activity recognition using UCF101 dataset. The fusion of two powerful neural network architectures leverages CNNs' ability to capture spatial features and Bi-LSTMs' proficiency in modeling temporal dependencies. Convolutional Neural Networks are well-suited for extracting spatial features from images and video frames. They use convolutional layers to learn hierarchical representations of visual patterns. In the context of human motion recognition, a CNN can process individual frames and capture information about body poses, object interactions, and other relevant spatial cues. Bidirectional Long Short-Term Memory Networks are designed to model sequential data and capture temporal dependencies over time. They are effective in handling time-series data, making them suitable for analyzing video sequences. Bi-LSTMs can capture the dynamics of motion and actions across multiple frames, which is essential for recognizing complex activities that unfold over time. It involves accurately detecting, classifying, and tracking human actions and motions within video sequences. Despite significant advancements in recent years, several challenges persist when deploying human motion and activity recognition systems in real-world environments. Human activities are characterized by inherent complexities such as viewpoint variations and cluttered backgrounds. These factors pose substantial obstacles to achieving accurate and robust recognition. Furthermore, the diverse range of activities ranging from simple gestures to intricate interactions requires methods capable of handling various levels of spatial and temporal details [4]. This paper addresses these challenges by proposing combined approaches of CNN and Bi-LSTM techniques that enhance the accuracy, robustness, and efficiency of the human motion and activity recognition model. Hence the objectives of this study are as follows:

a) To develop the human action recognition and classification model combining CNN and Bi-LSTM.
b) To recognize the human activities in videos.

The main contribution of this study is to develop an intelligent framework to recognize human activities and classify the videos based on activities within the videos.

2 Related Works

The history of video action recognition dates back to the early days of computer vision where the goal was to classify the videos. However, with the advent of image and video technology, the need for human activity recognition became increasingly important. This led to the development of deep learning techniques for video activity recognition and classification. Previously hand-crafted feature-based approaches [4, 5] were popularly used but nowadays deep learning models are in hype.

The paper [6] explained the combination of traditional classifiers, SVM and CNN for activity recognition. Resnet CNN was used as a feature generator and multiclass SVM as a classifier. The model was trained and evaluated on MSR Daily activity dataset. It is a daily activity dataset. There are 16 activity types such as drink, eat, read book, call cellphone etc. The study by authors [7] applied a deep learning technique known as deep recurrent neural network (DRNN) for high throughput human activity recognition. They developed the model with training dataset of 432 trials. Against the test data of 108 segmented trials, each with a single activity type, the maximum recognition rate was 95.42%. The throughput of the recognition per unit time for the DRNN was only 1.347ms which was far better than the traditional method with 11.031ms. For evaluation, HASC corpus dataset was used and distributed by HASC. The cell phone dataset (HAR dataset) from the UCI Machine Learning Repository was utilized to test the method's generalizability. Authors [8] presented a flexible video framework for action learning in videos. They formulated the model known as a temporal segment network with segment-based sampling and an aggregation module. Deep-learned features from the video were extracted by the model from the video. The model was deployed and tested on both trimmed and untrimmed videos. The datasets used in the study were HMDB51, THUMOS14 and Activity Net.

Additionally, co-occurrence features were given priority by authors [9] for action recognition. They employed end to end convolution learning framework with a hierarchical methodology. Both spatial and temporal features are integrated into the model. They exploited multi-person feature fusion strategies and CNN for learning global co-occurrences. NTU, SBU and MMD datasets were used for evaluating the model. The model greatly improves performance on both action recognition and detection tasks with an accuracy of 98.6%, according to experiments on three benchmark datasets. Further, the challenge to represent action captured from different viewpoints was overcome by the authors [10]. They introduced a method that automatically determines the virtual observation viewpoint. Two adaptive neural networks for view adaptive such as VA-RNN and VA-CNN were deployed based on LSTM and CNN. The model focused on action-specific features and the fusion of two networks. The experiment was evaluated on five benchmark datasets NTU RGB+D Dataset, SYSU 3DSet, UWA3D Multiview Activity II Dataset, Northwestern-UCLA and SBU Dataset. With VA-CNN, highest accuracy of 95.7 is obtained and with VA-RNN, 97.5 on SBU dataset. One of the main factors for the human action recognition is the dynamics of human body skeleton.

Authors [11] explained the Spatial Temporal Graph Convolution Network (ST-GCN) that overcome the limited expressive power and generalizations. The network was trained and evaluated using two dataset, Kinetics and NTU-RGBD. The model focused on pose estimation of human in video. The accuracy of 88.3% was obtained with given datasets. Hence ST-GCN can capture motion information in dynamic skeleton which is complementary to RGB modality. Research on human activity recognition from sensor data in smart healthcare environment using deep learning by authors [12] provide significant decision-making in healthcare. They applied the multisensory framework with deep learning methods gated recurrent units and simple recurrent units. The dataset used was MHealth dataset and evaluation metrics used were confusion metrics and F-score. The overall F-score obtained was 0.996. People's well-being, health, and lifespan are being improved by sensor-based user behavior and health status monitoring. Further, a recurrent convolutional attention model was formulated for imbalance activity recognition [13]. They solved the issue of multimodal sensor data and class imbalance. The co-training framework was used to balance the latent patterns of activity data and recurrent convolutional attention models classifiers exploited unlabeled samples. The experiments were performed on four datasets namely MHealth, PAMAP2, UCI HAR and MARS. One of the challenging tasks in video understanding is temporal action detection. The graph convolutional network (GCN) model was developed by authors [14] to incorporate multi-level semantic context into video data. Video clips were used as nodes in the graph, correlations between individual clips as edges, and graph convolution as the process operation. The algorithms used were interpolation and rescaling in SGAlign. The model was trained and evaluated on ActivityNet dataset and Thumos-14 dataset with metrics mean average precision (34.09) and Intersection over Union (23.42). Hence author explained about the graph convolution network that extracted semantic information from videos for action recognition.

3 Framework, Methods and Tools

3.1 Framework

The framework for this study is depicted in Fig. 1.

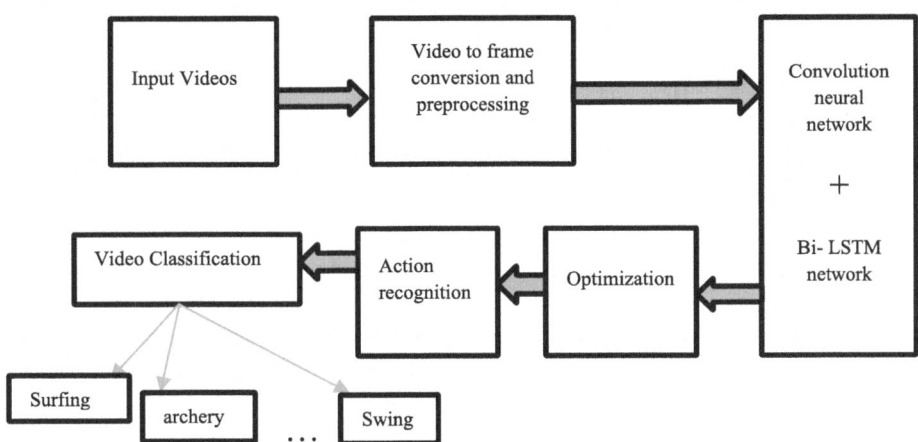

Fig. 1. Framework for action recognition and video classification

In the above figure, Video is inputted which contains various information such as created date, device information and other technical information. Video taken for an experiment is 24 frames per second. These frames are ordered and can be found by the frame number. Then using the number of frames per second, and the time in the video, the frame number can be easily calculated and extracted from the video. Finally, those frames are saved for further pre-processing and activity recognition. Since action tags in dataset are non-numerical, label encoding is used as preprocessing which will reduce training overhead.

3.2 VCHAR Model Function

The model includes two parts: The first part is a convolution neural network. This network will extract spatial information and features from the video (Fig. 2).

It consists of convolution layers, maxpooling, dropout and a fully connected layer. Fully connected layers are used for the final stage of a CNN architecture where they take the flattened or vectorized features extracted from earlier layers and map them to the desired output format for example: class probabilities in classification. Here, Resnet50 CNN is used to extract the feature. It is trained with images from the ImageNet dataset [15] and video frames. Resnet50 architecture includes 50 layers where there are 48

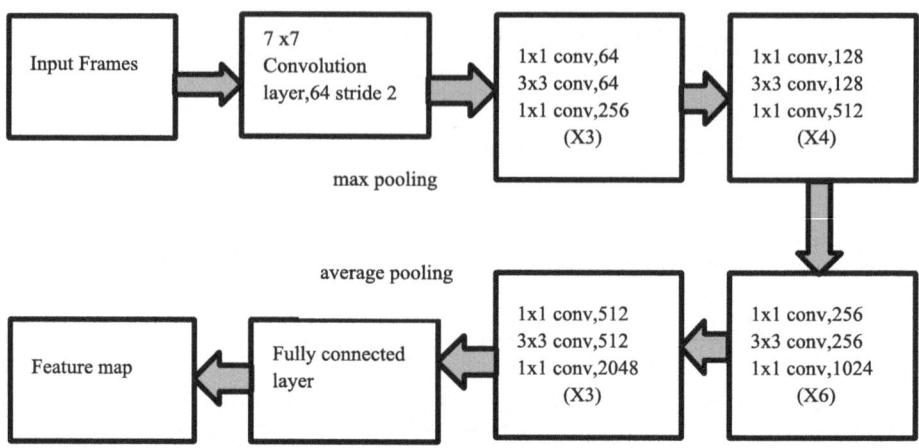

Fig. 2. Resnet50 architecture

convolution layers and 1 pooling layer and 1 fully connected layer. At first the image size of 224 × 224 × 3 is reduced and processed in four stages of convolution as shown in figure. Finally, the feature vector is produced at output of Resnet50 CNN. In each stage, activation function and batch normalization are used. Then the resulting feature map or feature vector(X) is inputted to the Bi-LSTM. In Bi-LSTM input is given from both right to left and from left to right directions (Fig. 3).

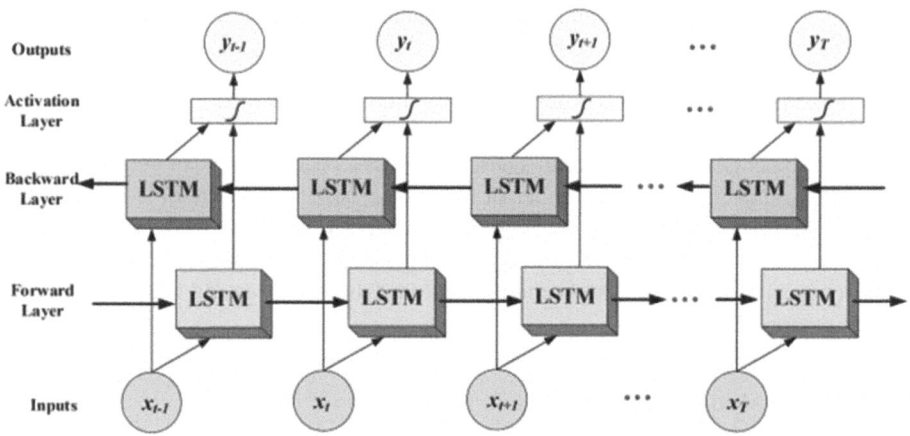

Fig. 3. Bidirectional LSTM

Bi-LSTM can process the video frames sequentially, capturing the temporal evolution of visual information. It handles the long-term dependencies and captures more complex motion patterns and interactions. The forward layer of the LSTM cell is responsible for processing the input sequence in the forward, while the backward layer processes the sequence in the reverse direction. The forward and backward layers work in parallel, capturing both past and future context information for each time step. Bi-LSTM cells have input gates, forget gates, candidate cell states, cell state updates, hidden states, and output gates for each LSTM cell. Information flow into and out of the cell state is controlled by the input gate and forget gate, respectively. They choose the appropriate balance between adding new information and keeping existing knowledge. The candidate cell state represents potential additions to the cell state in the form of new candidate values. By merging the input and the candidate cell state using the input and forget gates, the cell state is updated. The hidden state at the current time step is determined by the output gate. Utilizing the output gate and cell state, the hidden state is calculated. Activation functions are applied at different stages within the LSTM cell to introduce non-linearity and enable the cell to model complex relationships in sequential frames. The activation functions used in a Bi-LSTM cell include the sigmoid activation function. Sigmoid activation functions are used for gating mechanisms in the LSTM cell. Finally, the Adam optimizer is used to optimize and provide a better performance of the model.

3.3 Dataset

UCF101 video dataset is The UCF101 dataset is a popular video dataset for action recognition and classification in the field of computer vision [3]. It contains a wide variety of human activities captured in videos. UCF101 consists of a total of 13,320 video clips, covering 101 action categories. UCF101 covers a diverse range of action categories such as sports actions, domestic activities and dance styles (Table 1).

Table 1. Video Categories

ApplyEyeMakeup	Billiards	CricketBowling	Hammering	HammerThrow
ApplyLipstick	BlowDryHair	CricketShot	HandstandPush	HeadMassage
Archery	BlowingCandles	CuttingInKitchen	HandstandWalk	Haircut
BabyCrawling	BodyWeightSquats	Diving	HighJump	Kayaking
BalanceBeam	Bowling	Drumming	HorseRace	Knitting
BandMarching	BoxingPunchingBag	Fencing	HorseRiding	LongJump
BaseballPitch	BoxingSpeedBag	FieldHockeyPenalty	HulaHoop	Lunges
Basketball	BreastStroke	FloorGymnastics	IceDancing	MilitaryParade
BasketballDunk	BrushingTeeth	FrisbeeCatch	JavelinThrow	Mixing
BenchPress	CleanAndJerk	FrontCrawl	JugglingBalls	MoppingFloor
Biking	CliffDiving	GolfSwing	JumpingJack	Nunchucks

(continued)

Table 1. (*continued*)

PizzaTossing	PlayingCello	PlayingDaf	JumpRope	ParallelBars
PlayingDhol	PlayingFlute	PlayingGuitar	PlayingPiano	PlayingSitar
PlayingTabla	PlayingViolin	PoleVault	PommelHorse	PullUps
Punch	PushUps	Rafting	RockClimbingIndoor	RopeClimbing
Rowing	SalsaSpin	ShavingBeard	Shotput	SkateBoarding
Skiing	Skijet	SkyDiving	SoccerJuggling	SoccerPenalty
StillRings	SumoWrestling	Surfing	Swing	TaiChi
TennisSwing	ThrowDiscus	TrampolineJumping	TableTennisShot	Typing
UnevenBars	VolleyballSpiking	WalkingWithDog	WallPushups	WritingOnBoard
YoYo				

4   Result

For an experiment, the dataset is divided into training and testing datasets. 80% dataset is taken for training set and 20% data is taken for testing dataset. Here is the outcome of the model.

4.1   Activity Recognition and Video Classification

The different types of activity recognition outputs are as follows (Fig. 4):

Fig. 4. Action recognition outputs

The different classifications of video samples output are as follows (Fig. 5):

A) PushUps or Not PushUps

B) Archery or Not Archery

C) Surfing or Not Surfing

Fig. 5. Video Classification outputs

5   Discussion and Limitation

5.1   Accuracy and Loss

From the graph, it is clearly seen that the training accuracy of model is 92.76% and testing accuracy of 87.67%. It means the model has learned to fit the patterns present in the training data to accurately recognize action and classify videos. Similarly, the model loss is 0.23 for training and 0.63 for testing. Lower values of loss indicate better fit to the training data. The graph indicates that the model is learning and generalizing to some extent. The training and testing accuracy values suggest that the model is performing reasonably well in given datasets (Fig. 6).

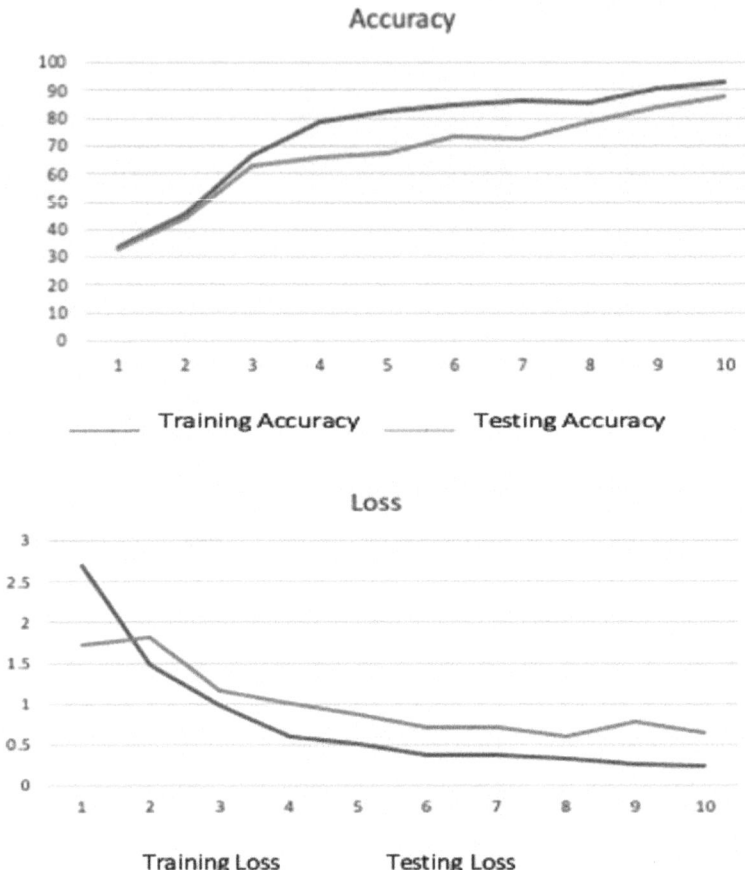

Fig. 6. Accuracy and Loss Curve

5.2 Comparison with Previous Methods Using the UCF101 Dataset

The accuracy is better than other benchmark results as mentioned in the above table but there are certain limitations. The limitation of this study is that the model will not work

Table 2. Accuracy comparison

Methods	UCF-101 accuracy
CNN+Motion Vector [16]	87.5%
RLSTM [17]	86.9%
3D Convolution Network [18]	85.2%
VCHAR (Our method)	87.67%

properly in the case of varied contexts and environments due to the limitation of the video classes. Occlusions may degrade the performance of the model. Only trimmed and short videos are allowed but not preferable for untrimmed and lengthy videos (Table 2).

6 Conclusions

The advancement of video content analysis and classification based on human activity recognition has marked a new era of understanding and interpreting visual data. The exploration of deep learning techniques in this paper such as CNNs and Bi-LSTM plays an important role in deciphering the complex dynamics of human activities. Through the experiment, it has become evident that the fusion of temporal modeling and feature extraction is essential for distinguishing various human actions. Techniques like Bi-LSTM networks have exhibited the power in capturing long-range dependencies, allowing for the recognition of complicated sequences of motions and gestures. This understanding of temporal patterns, combined with the ability to extract meaningful spatial features, has propelled the accuracy and robustness of activity recognition systems. From enhancing surveillance systems to enabling personalized healthcare monitoring, video content analysis and human activity recognition have permeated various facets of lives. The accurate identification of actions, whether routine or exceptional, empowers intelligent systems to make informed decisions, thereby enabling the system to make video classifications task easier.

References

1. Gkioxari, G., Girshick, R., Dollar, P., He, K.: Detecting and recognizing human-object interactions. In: 2018 IEEE/CVF Conference on Computer Vision and Pattern Recognition (2018). https://doi.org/10.1109/cvpr.2018.00872
2. Zhou, X., et al.: Deep-learning-enhanced human activity recognition for internet of healthcare things. IEEE Internet Things J. 7(7), 6429–6438 (2020). https://doi.org/10.1109/jiot.2020.2985082
3. Khurram, S., Zamir, A.R., Shah, M.: UCF101: a dataset of 101 human action classes from videos in the wild. IEEE Access 1, 817–830 (2013). https://doi.org/10.1109/ACCESS.2013.2270168
4. Cutler, R., Turk, M.: View-based interpretation of real-time optical flow for gesture recognition. In: Proceedings Third IEEE International Conference on Automatic Face and Gesture Recognition. https://doi.org/10.1109/afgr.1998.670984
5. Wang, H., Schmid, C.: Action recognition with improved trajectories. In: 2013 IEEE International Conference on Computer Vision (2013). https://doi.org/10.1109/iccv.2013.441
6. Basly, H., Ouarda, W., Sayadi, F.E., Ouni, B., Alimi, A.M.: CNN-SVM learning approach based human activity recognition. In: Proceedings of the Image and Signal Processing: 9th International Conference, ICISP 2020, Marrakesh, Morocco, 4–6 June 2020, vol. 9, p. 271281. Springer, Cham (2020)
7. Inoue, M., Inoue, S., Nishida, T.: Deep recurrent neural network for mobile human activity recognition with high throughput. Artif. Life Robot. 23, 173–185 (2018)
8. Wang, L., et al.: Temporal segment networks for action recognition in videos. IEEE Trans. Pattern Anal. Mach. Intell. 41(11), 2740–2755 (2019). https://doi.org/10.1109/tpami.2018.2868668

9. Li, C., Zhong, Q., Xie, D., Pu, S.: Co-occurrence feature learning from skeleton data for action recognition and detection with hierarchical aggregation. In: Proceedings of the Twenty-Seventh International Joint Conference on Artificial Intelligence (2018). https://doi.org/10.24963/ijcai.2018/109

10. Zhang, P., et al.: View adaptive recurrent neural networks for high-performance human action recognition from Skeleton Data. In: 2017 IEEE International Conference on Computer Vision (ICCV) (2017). https://doi.org/10.1109/iccv.2017.233

11. Yan, S., Xiong, Y., Lin, D.: Spatial-temporal graph convolutional networks for skeletonbased action recognition. In: Proceedings of the AAAI Conference on Artificial Intelligence, vol. 32, no. 1 (2018). https://doi.org/10.1609/aaai.v32i1.12328

12. Gumaei, A., Hassan, M.M., Alelaiwi, A., Alsalman, H.: A hybrid deep learning model for human activity recognition using multimodal body sensing data. IEEE Access 7, 99152–99160 (2019). https://doi.org/10.1109/access.2019.2927134

13. Chen, K., et al.: A semisupervised recurrent convolutional attention model for human activity recognition. IEEE Trans. Neural Netw. Learn. Syst. 31(5), 1747–1756 (2020). https://doi.org/10.1109/tnnls.2019.2927224

14. Xu, M., Zhao, C., Rojas, D.S., Thabet, A., Ghanem, B.: G-tad: sub-graph localization for temporal action detection. In: 2020 IEEE/CVF Conference on Computer Vision and Pattern Recognition (CVPR) (2020). https://doi.org/10.1109/cvpr42600.2020.01017

15. Deng, J., Dong, W., Socher, R., Li, L.-J., Li, K., Fei-Fei, L.: ImageNet: a large-scale hierarchical image database. In: 2009 IEEE Conference on Computer Vision and Pattern Recognition, Miami, FL, USA, pp. 248–255 (2009). https://doi.org/10.1109/CVPR.2009.5206848

16. Zhang, B., Wang, L., Wang, Z., Qiao, Y., Wang, H.: Real-time action recognition with deeply transferred motion vector CNNs. IEEE Trans. Image Process. 27(5), 2326–2339 (2018). https://doi.org/10.1109/tip.2018.2791180

17. Mahasseni, B., Todorovic, S.: Regularizing long short term memory with 3D human-skeleton sequences for action recognition. In: 2016 IEEE Conference on Computer Vision and Pattern Recognition (CVPR) (2016). https://doi.org/10.1109/cvpr.2016.333

18. Tran, D., Bourdev, L., Fergus, R., Torresani, L., Paluri, M.: Learning spatiotemporal features with 3D convolutional networks. In: 2015 IEEE International Conference on Computer Vision (ICCV) (2015). https://doi.org/10.1109/iccv.2015.510

Machine Learning Based Expert System for Breast Cancer Prediction (MLESBCP)

Akhil Kumar Das[1]([✉])(ID), Saroj Kr. Biswas[2](ID), Ardhendu Mandal[3](ID), Arijit Bhattacharya[1](ID), and Debasmita Saha[4](ID)

[1] Department of Computer Science, Gour Mahavidyalaya, Mangalbari, Malda 732142, West Bengal, India
dasakhi@gmail.com
[2] Department of Computer Science and Engineering, National Institute of Technology, Silchar, Silchar 788010, Asaam, India
[3] Department of Computer Science and Technology, University of North Bengal, Darjeeling 734013, West Bengal, India
[4] Department of Computer Science, University of Gour Banga, Malda 732101, West Bengal, India

Abstract. Today, Breast Cancer (BC) fatality risks are increasing dramatically. The detection of BC takes a long time due to the restricted nature of traditional systems. Despite the lack of treatments for breast cancer, the chances of survival are greatly influenced by early identification and diagnosis. A significant development and anticipated trend in the future of medicine is the use of machine learning (ML) for accurate medical diagnosis. Research on BC has greatly advanced our understanding of the disease over the past two decades, leading to safer and more effective treatment options. Despite this, numerous researchers have developed expert systems for BC early diagnosis. However, most expert systems usually fall short in successfully dealing with the class imbalance issue, as well as in performing systematic feature selection and efficient data pre-processing. To address these shortcomings, this study presents a "Machine Learning Based Expert System for Breast Cancer Prediction (MLESBCP)" for better BC prediction utilising ML analytics. The proposed system employs the KMeansSMOTE oversampling technique to balance the WBC dataset. The chi-square feature selection technique is utilised to identify the most essential feature of the WBC dataset to increase the model's accuracy. The proposed model's accuracy, F1-score, recall, and precision are compared to those of the previous single classifier models. The outcomes demonstrate that the MLESBCP achieved a maximum accuracy of 97.32%.

Keywords: Breast Cancer · KMeansSMOTE · Chi-square · Machine Learning

1 Introduction

BC is presently the most prevalent cause of death and the second deadliest cancer in the world, after lung cancer. It is a horrible disorder which brings both psychological and physical suffering to women all over the world. Though it can affect both sexes, women

M. Majumder et al. (Eds.): ICCTE 2023, CCIS 2376, pp. 275–286, 2025.
https://doi.org/10.1007/978-3-031-81935-3_24

are more likely to be affected by BC [1]. Around 0.5 to 1.0% of breast cancers happen in men. Breast cancer is a form of cancer that begins in the breast cells. These cells grow abnormally in the breast. Normally, BC can start in any part of the breast. Usually, breast lobules or ducts are where BC develops. The term "duct" refers to the milk-transporting ducts, whereas "lobules" refers to the milk-producing glands. Nevertheless, BC has spread to many bodily parts, including as the lungs, bones, and brain.

In 2013, over 2,32,340 women in the United States were diagnosed with BC; of these, 39,620 women lost their lives as a result of the disease [2]. This deadly illness claimed the lives of 6,27,000 females in 2018. There are over three million BC survivors in the US, according to the American Cancer Society (ACS). According to ACS's 2019 release, 2,68,500 (approx.) women have been found to have invasive BC, whereas 62,920 (approx.) women have been found to have non-invasive BC. By 2020, 2.30 million women will have received a BC diagnosis, and 6,85,100 (approx.) people will have died from the disease globally, predicts the WHO. The WHO estimates that 2.10 million girls are impacted by it annually. About 15% of female fatalities are attributable to BC [3]. Based on CDC study, 2,50,000 persons are diagnosed with BC annually on average. In India, 1.50 lakh women receive a breast cancer diagnosis each year; 70,000 of them pass away as a result, according to the ICMR. In the United Kingdom, BC now affects 1 in 12 females between the ages of 1 and 85 [4]. BC affects one out of every eight persons in 2021.

There are some familiar methods for diagnosing BC, such as mammography [5], Ultrasound, and dynamic MRI [6]. Despite the demonstration of various modalities, none of them are able to produce an accurate and reliable outcome. Traditional procedures are also time-consuming and, in certain cases, incorrectly detect the condition. Experienced physicians can normally identify diseases with an accuracy of 78%, whereas ML approaches can detect them with an accuracy of up to 97% [7]. This will allow patients to receive vital therapies when they are needed.

Clinical data may be utilised to build a prediction system that uses ML-based expert systems to identify BC early on. Therefore, this type of diagnostic method is less expensive, safer, faster, and easier than other alternative surgical procedures, unneeded harsh therapies, and excessive treatment costs [8]. Even though many researchers have created expert systems to identify BC. However, the majority of expert systems usually fall short in handling the issue of class imbalance correctly, as well as in performing efficient data preparation and methodical feature selection. This paper suggests an expert system, "Machine Learning Based Expert System for Breast Cancer Prediction (MLESBCP)" to predict the BC using a K-NN classifier in order to address these issues. To balance the WBC dataset, the proposed system uses the KMeansSMOTE [27–29] oversampling approach. The chi-square [30] feature selection strategy is utilised to identify the most relevant feature of the WBC dataset to increase the model's accuracy. The suggested prediction system produces very accurate findings for predicting diseases in patients.

The article was organised as follows: Sect. 2 introduces literature survey. Proposed methodology was given in Sect. 3. Section 4 discussed about the performance metrics employed in the study. Result and discussion was done in Sect. 5. Finally, in Sect. 6, the research was concluded.

2 Literature Survey

Multiple studies are carried out on the BC dataset, employing a variety of classifiers and feature selection strategies. The literature contains a lot of information about BC datasets. Many of them have high categorization accuracy. Asri et al. [7] used four ML methods like k-NN, SVM, NB, and C4.5 to predict BC using the WBC dataset with 699 instances. They used the 10-CV approach as well as the WEKA software tool. They sought to examine the usefulness and utility of different algorithms based on accuracy, sensitivity, specificity, and precision values in order to establish the ideal classification accuracy. Of the aforementioned techniques, the SVM classifier had the highest accuracy, at 97.13%. Ahmed et al. [9] presented five ML methods, such as NB, SVM, MLP, J48, and RF, to diagnose and detect BC using the WBC dataset with 699 instances and 10 features. They employed the WEKA tool. They compared the results on the basis of accuracy, recall, precision, F-score, MCC, PRC area, kappa statistic, and ROC area. The NB classifier produced the best accuracy of 97.42% using 10 CV among the mentioned methods. Sivakami K. [10] described the treatment information about the BC and proposed a hybrid model named DT-SVM to diagnose, and detect the BC using the WBC dataset with 699 instances collected from the UCI repository. The suggested DT-SVM model is built using the Weka tool on the Java platform. The suggested approach compared the accuracy to other classifier algorithms like SMO, IBL, and NB. They got an accuracy of 91% for DT-SVM, 85.23% for IBL, 72.56% for SMO, and 89.48 for NB, respectively. Telsang et al. [11] analysed five ML methods like NB, RF, SVM, and k-NN to detect the BC using the WBC dataset with 699 instances, which is collected from the UCI repository. They contrasted the outcomes amongst the aforementioned classifiers based on the performance matrix, which includes accuracy, recall, precision, and AUC. When compared to other approaches, the SVM model was the most accurate, with a 96.25% accuracy. Jabbar M.A. [12] proposed a new ensemble model named BN+RPF to diagnose and detect BC utilising the WBC dataset with 699 instances and WEKA tool. This model used two classifiers, like BN and RPF. 10 CV is used to validate the suggested model. They compared the performance in terms of sensitivity, specificity, PPV, NPV, Error rate, FPR, MCC, and accuracy with other models. The experimental findings reveal that the suggested technique beats existing approaches in categorising BC data, with a remarkable accuracy (97.42%). Li et al. [13] presented the five ML methods like LR, RF, DT, SVM, and ANN classifiers to diagnose and predict BC using two BC datasets, such as WBC with 699 instances and BCCD with 116 instances. They compared the results based on accuracy, F-score, precision, recall, and AUC. RF achieved the highest accuracy of 96.1% for the WBC dataset and 74.3% for the BCCD dataset compared to the other four algorithms. Shamrat et al. [14] provided a comparative study that used supervised ML algorithms to diagnose BC patients. They used multiple ML algorithms like RF, DT, NB, KNN, SVM, and LR. The authors compared the result in terms of total accuracy, specificity, sensitivity, and F1-score to the mentioned classifiers. SVM performed best, with the highest accuracy of 97.07%. But the second-best forecast accuracy has been achieved by NB and RF. Shubham et al. [15] presented three ML methods like KNN, SVM, and DT to detect the BC using WBC with 699 instances. They compared the performance based on accuracy, precision, and recall to the mentioned classifiers. SVM achieved the best accuracy of 96% when compared to the other classifiers. Banerjee et al. [16] presented a comparative analysis of various ensemble ML methods to detect the BC using the WBC dataset. They used various ensemble methods like bagging, boosting, and voting-based ensembles. Bagging-based ensembles achieved the highest accuracy of 95.6%.

The following are the major elements of our proposed strategy:

- This paper proposes an expert system named MLESBCP, built using a K-NN classifier for BC detection using the WBC dataset.
- To handle the class imbalance problem using KMeansSMOTE

The suggested approach was contrasted with other cutting-edge techniques.

3   Proposed Methodology

This MLESBCP model has been divided into the following three parts: A. Data Description, B. Data Pre-processing, C. Classification. The data description part discussed about the utilized dataset, whereas the data pre-processing steps were discussed in data pre-processing part. Finally model based classification was done in classification part. Detailing of these phases were given below. Figure 1 displayed the process diagram for the MLESBCP model.

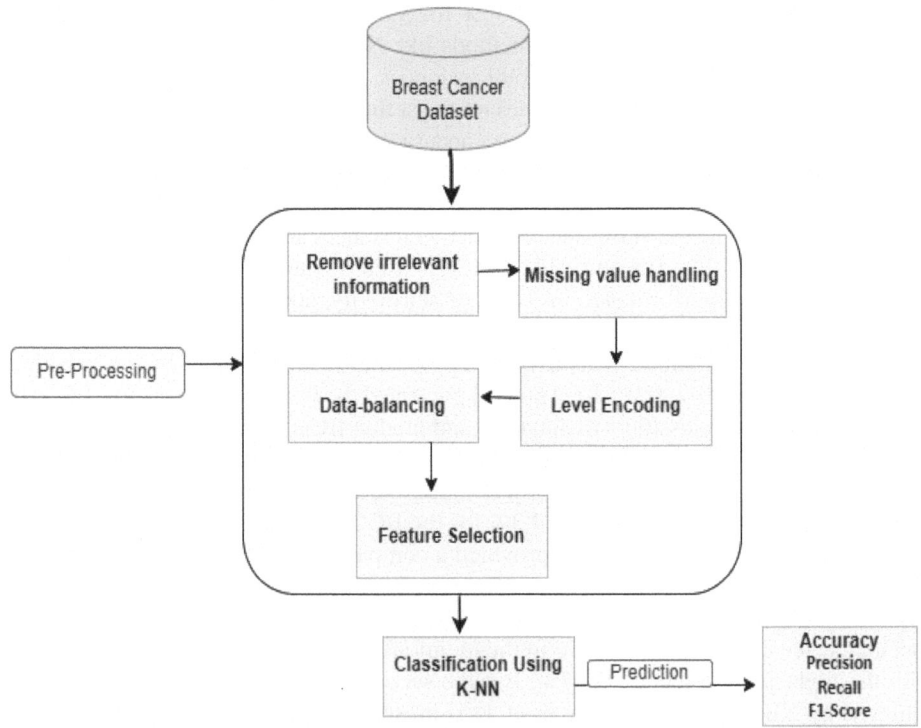

Fig. 1. Machine Learning Based Expert System for Breast Cancer Prediction (MLESBCP)

3.1 Data Description

The WBC dataset represents a common BC dataset for using ML techniques and is widely used in bioinformatics. This dataset has been obtained from the repository at UCI [17]. The suggested model has been validated using this dataset. There are 699 instances with eleven features, including patient ID. There were two types of cases: benign and malignant [17]. There were 458 benign instances and 241 instances of malignancy. Ten features total; nine of them were considered input features such as "lump Thickness", "Uniformity of Cell Size", "Uniformity of Cell Shape", "Marginal Adhesion", "Single Epithelial Cell Size", "Bare Nuclei", "Bland Chromatin", "Normal Nucleoli" and "Mitoses", while final one is considered an output feature. But, the "Bare Nuclei" feature has 16 missing values.

```
In [54]: df.info()

         <class 'pandas.core.frame.DataFrame'>
         RangeIndex: 699 entries, 0 to 698
         Data columns (total 11 columns):
          #   Column                       Non-Null Count  Dtype
         ---  ------                       --------------  -----
          0   id                           699 non-null    int64
          1   Clump Thickness              699 non-null    int64
          2   Uniformity of Cell Size      699 non-null    int64
          3   Uniformity of Cell Shape     699 non-null    int64
          4   Marginal Adhesion            699 non-null    int64
          5   Single Epithelial Cell Size  699 non-null    int64
          6   Bare Nuclei                  699 non-null    object
          7   Bland Chromatin              699 non-null    int64
          8   Normal Nucleoli              699 non-null    int64
          9   Mitoses                      699 non-null    int64
          10  Class                        699 non-null    int64
         dtypes: int64(10), object(1)
         memory usage: 60.2+ KB
```

Fig. 2. Display the WBC data description

3.2 Data Pre-processing

The raw datasets of BC are prepossessed in the data pre-processing step to remove any irrelevant or extraneous features. It consists of the following sub-phases:

- **Remove irrelevant information:** In the BC dataset, "ID" is an irrelevant attribute. "ID" has been removed from the WBC dataset.
- **Missing value handling:** In the WBC dataset, the feature of "Bare Nuclei" is present in 16 rows. In the BC dataset, there are very few missing variables. Thus, all of the missing values are just eliminated. The dataset had 683 records after they were removed, of which 239 instances and 444 instances were malignant and benign, respectively.
- **Level encoding:** In the Fig. 2 above, "Bare Nuclei" is the only characteristic that has a descriptive feature. To make processing easier, Label encoding methods have been

used to partition the feature into six labels (0 to 5). Both malignant ("4") and benign ("2") categories applied to the target attribute class. Furthermore, it was modified to one for aggressive malignancies and zero for benign tumours.

- **Data balancing:** The BC dataset contains 444 benign and 239 malignant incidences, i.e., it is an unbalanced dataset. To balance benign and malignant cases, the KMeansS-MOTE approach is utilised. After balancing the BC dataset, it consists of 447 benign and 447 malignant tumours.
- **Feature selection:** The suggested system uses the Chi-Square (Chi2) to choose the most essential features of the WBC dataset. These are "Clump Thickness", "Uniformity of Cell Size", "Uniformity of Cell Shape", "Marginal Adhesion", "Single Epithelial Cell Size", "Bare Nuclei", "Bland Chromatin", and "Normal Nucleoli".

3.3 Classification

The truncated feature set obtained from the Chi Square technique is utilised by the proposed MLESBCP model to identify BC. The classifiers are trained using the reduced feature set. After that, the suggested model uses the KNN classifier to identify if the tumour is malignant or benign.

K-Nearest Neighbors. It is the most basic and uncomplicated categorization technique. Like the majority of ML algorithms, K-NN simply uses the training data points. It uses the full training dataset as a reference throughout the training phase. When producing predictions, it uses a distance metric like as Euclidean distance to compute the distance between the input data. The method then determines the K closest neighbours to the input data point based on their distances. For various K values, the K-NN result varies. A high value of K results in class overlap, whereas a low value of K results in more computations.

4 Performance Metrics Equation

A Confusion Matrix (CM) represented actual and predicted classifications of the system shown in Table 1.

Table 1. Representation of CM

		Predicted: NO	Predicted: YES
Actual Value	NO	TN	FP
	YES	FN	TP

TP: True Positive; FP: False Positive; TN: True Negative; FN: False Negative

Accuracy, precision, recall, and F-measure are the metrics employed in this study to evaluate the quality of the suggested approach. Table 2 describes the performance metrics [25, 26].

Table 2. Representation different performance metrics

	Formula	Equation No
Accuracy	(TP + TN)/(TP + FP + TN + FN)	(1)
Precision(Pr)	TP/(TP + FP)	(2)
Recall(Rec)	TP/(TP + FN)	(3)
F-Measure	(2 * Pr * Rec)/(Pr + Rec)	(4)

5   Results and Discussion

The suggested MLESBCP model was implemented using Python programming. To evaluate how well the proposed model performed for BC diagnosis, experiments were conducted using the WBC dataset [17] mentioned above. First, using K-Means SMOTE, class imbalance problem was resolved. Then, Chi-Square was employed to get the most important features, which contain eight features out of the ten features. These eight features were used as input to the KNN classifier in the prediction phase. A 10-fold CV [31] was employed to validate MLESBCP results. Confusion Matrix (CM) [32] was utilized to evaluate the performance of the proposed model, MLESBCP. For this model, four statistical performance criteria were used: recall, accuracy, precision, and F1-score.

Table 3 makes comparison of the performance metrics including the accuracy, F1-score, precision, and recall of simple KNN to the suggested model, MLESBCP. Improvement in terms of performance metrics including accuracy of 1.12%, F1-score of 1.65%, precision of 0.81%, and recall of 2.36% were recorded for MLESBCP, which shows the capacity of the model.

Table 3. Comparison of MLESBCP and simple KNN

	Simple KNN	MLESBCP	Increase in Performance
Accuracy (%)	96.2	97.32	1.12
F1-score (%)	95.02	96.67	1.65
Precision (%)	96.99	97.8	0.81
Recall (%)	93.24	95.6	2.36

Figure 3 depicts a graphical depiction of the accuracy, precision, F1-score, and recall comparisons of the proposed model with simple KNN classification methods for better visualisation and understanding.

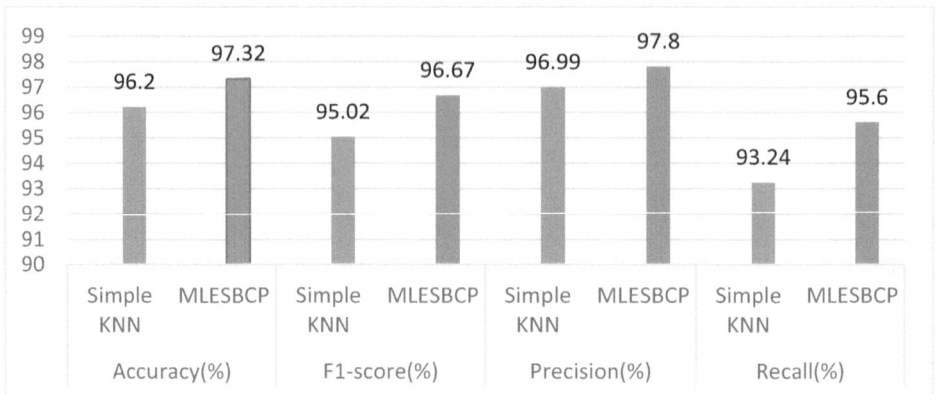

Fig. 3. A graphical representation compares the performance of simple KNN and the proposed system.

From Table 4, the MLESBCP technique outperforms the NB, DT, SVM, and LR methods in terms of accuracy and F1 score, but the precision of NB achieved a better result than DT, SVM, LR, and MLESBCP. The recall of LR outperformed DT, SVM, NB, and MLESBCP.

Table 4. Compares the performance metrics for modified model, base models, and the proposed MLESBCP model.

Model	Accuracy (%)		F1-score (%)		Precision (%)		Recall (%)	
	Simple-Model	Modified-Model	Simple-Model	Modified-Model	Simple-Model	Modified-Model	Simple-Model	Modified-Model
NB	96.34	96.75	94.92	95.72	98.32	98.53	91.79	93.25
DT	93.71	94.97	91.71	93.29	92.58	94.32	92.85	92.77
SVM	96.64	97.09	95.63	96.39	96.79	97.26	94.59	95.58
LR	96.35	97.09	95.33	96.13	96.17	96.31	94.68	95.98
Proposed Model	**97.32**		**96.67**		97.80		95.6	

Figure 4 depicts a graphical depiction of the accuracy, F1-score, precision, and recall comparisons of the proposed model with NB, DT, SVM, and LR methods for better visualization and understanding.

Comparing the performance of MLESBCP with several models found in the literature, ensemble models, and single-classifier models.

Compares the performance to current cutting-edge models. It has been observed that previous models did not appropriately apply feature selection and data preprocessing procedures in order to generate greater insights. As demonstrated in Table 5, the proposed framework MLESBCP outperforms several models identified in the literature, ensemble models, and single-classifier-based models in terms of accuracy.

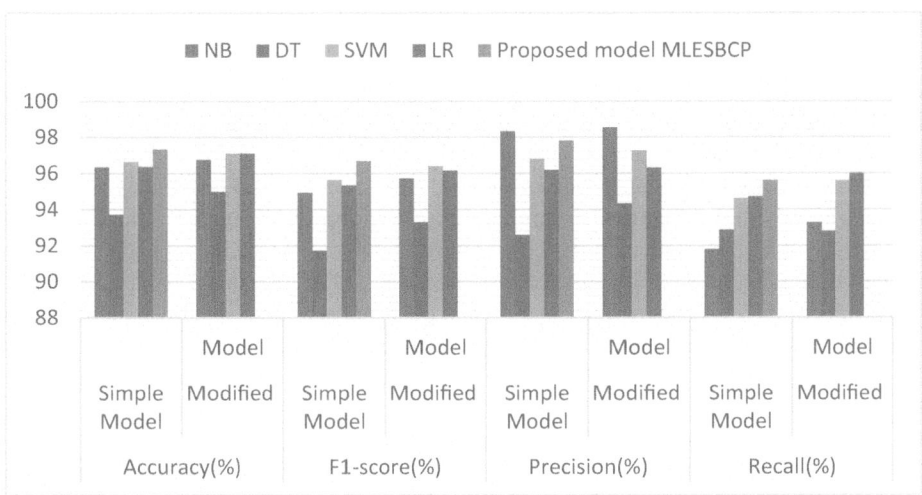

Fig. 4. Comparison of accuracy, F1-score, precision, and recall graphically represented.

Table 5. Comparison with Previous Works.

Reference	Methodology	Best accuracy (%)
Single-classifier models		
[7], 2016	SVM	97.13
[11], 2020	SVM	96.25
[14], 2020	SVM	97.07
[15], 2022	SVM	96.00
[19], 2020	K-NN	97.00
[21], 1995	GA-based FS + ANN	94.72
[22], 2016	SVM	97.00
[23], 2019	SMO + SVM	97.00
[24], 2017	NB	96.85
Ensemble models		
[13], 2018	RF	96.10
[16], 2017	Bagging based ensemble method	95.60
Various models present in literature		
[10], 2015	DT –SVM model	91.00
[18], 2012	RBFNN-EKF	96.43
[20], 2020	Ensemble model(FS + SVM)	97.14
[25], 2022	ESBCP system	94.01

Accuracy of proposed Model MLESBCP is 97.32% that is better from the above works.

6 Conclusion

MLESBCP employed the SMOTE based KMeansSMOTE data oversampling method, the Chi-square feature selection algorithm, and the KNN classification technique. In order to balance the samples of benign and malignant cases in the WBC dataset, the KMeansSMOTE approach was employed. Chi2 was utilised to identify the WBC dataset's most crucial features to improve classification performance. The suggested MLESBCP was compared with the conventional KNN, NB, DT, SVM, and LR classifier based models, as well as with the modified NB, DT, SVM, and LR based models. The results of the experiment demonstrate that the proposed model is substantially more effective in terms of accuracy, F1 measure, and other metrics. The accuracy of the suggested MLESBCP is also contrasted with standalone-classifier models, ensemble models, and other models found in the study. According to the experimental findings, the suggested MLESBCP model has much higher accuracy.

Therefore, it may be concluded that the proposed method for categorising breast cancer yields incredibly encouraging findings. The suggested MLESBCP model can be highly useful for physicians seeking a second opinion and making a final choice. More emphasis should be placed in future studies on how to enhance the present research by employing some of the optimisation strategies to increase the accuracy and other performance metrics of BC detection.

Acknowledgement. This research received no specific grant from any funding agency in the public, commercial, or not-for-profit Sectors.

Conflicts of Interest/Competing Interests. The authors declare that they have no known competing financial interests or personal relationships that could have appeared to influence the work reported in this paper.

References

1. India against cancer 2019. "Breast Cancer", National Institute of Cancer Prevention and Research, 12 November 2019. http://cancerindia.org.in/breast-cancer/
2. Akram, M., Iqbal, M., Daniyal, M., Khan, A.U.: Awareness and current knowledge of breast cancer. Biol. Res. **50**(1), 1–23 (2007). https://doi.org/10.1186/s40659-017-0140-9
3. Jaikrishnan, S.V.J., Chantarakasemchit, O., Meesad, P.: A breakup machine learning approach for breast cancer prediction. In: 2019 11th International Conference on Information Technology and Electrical Engineering (ICITEE), pp. 1–6. IEEE (2019). https://doi.org/10.1109/ICITEED.2019.8929977
4. Han, S.J., et al.: Prognostic significance of interactions between ER alpha and ER beta and lymph node status in breast cancer cases. Asian Pac. J. Cancer Prev. **14**(10), 6081–6084 (2013). https://doi.org/10.7314/APJCP.2013.14.10.6081
5. Mori, M., et al.: Diagnostic accuracy of contrast-enhanced spectral mammography in comparison to conventional full-field digital mammography in a population of women with dense breasts. Breast Cancer **24**, 104–110 (2017). https://doi.org/10.1007/s12282-016-0681-8
6. Nagashima, T., et al.: Dynamic-enhanced MRI predicts metastatic potential of invasive ductal breast cancer. Breast Cancer **9**(3), 226–230 (2002)

7. Asri, H., Mousannif, H., Al Moatassime, H., Noel, T.: Using machine learning algorithms for breast cancer risk prediction and diagnosis. Procedia Comput. Sci. **83**, 1064–1069 (2016). https://doi.org/10.1016/j.procs.2016.04.224

8. Sharma, D., Kumar, R., Jain, A.: Breast cancer prediction based on neural networks and extra tree classifier using feature ensemble learning. Measur.: Sens. **24**, 100560 (2022). https://doi.org/10.1016/j.measen.2022.100560

9. Ahmed, M.T., Imtiaz, M.N., Karmakar, A.: Analysis of wisconsin breast cancer original dataset using data mining and machine learning algorithms for breast cancer prediction. J. Sci. Technol. Environ. Inform. **9**(2), 665–672 (2020). https://doi.org/10.18801/jstei.090220.67

10. Sivakami, K., Saraswathi, N.: Mining big data: breast cancer prediction using DT-SVM hybrid model. Int. J. Sci. Eng. Appl. Sci. (IJSEAS) **1**(5), 418–429 (2015)

11. Telsang, V.A., Hegde, K.: Breast cancer prediction analysis using machine learning algorithms. In: 2020 International Conference on Communication, Computing and Industry 4.0 (C2I4), pp. 1–5. IEEE (2020). https://doi.org/10.1109/C2I451079.2020.9368911

12. Jabbar, M.A.: Breast cancer data classification using ensemble machine learning. Eng. Appl. Sci. Res. **48**(1), 65–72 (2021). https://doi.org/10.14456/easr.2021.8

13. Li, Y., Chen, Z.: Performance evaluation of machine learning methods for breast cancer prediction. Appl. Comput. Math. **7**(4), 212–216 (2018). https://doi.org/10.11648/j.acm.20180704.15

14. Shamrat, F.J.M., Raihan, M.A., Rahman, A.S., Mahmud, I., Akter, R.: An analysis on breast disease prediction using machine learning approaches. Int. J. Sci. Technol. Res. **9**(02), 2450–2455 (2020)

15. Kumar, S., Kamalraj, D.R.: Breast cancer detection using machine learning algorithms. Int. J. Adv. Eng. Manag. (IJAEM) **4**(3), 987–994 (2022). https://doi.org/10.35629/5252-0403987994

16. Baneriee, C., Paul, S., Ghoshal, M.: A comparative study of different ensemble learning techniques using Wisconsin breast cancer dataset. In: 2017 International Conference on Computer, Electrical & Communication Engineering (ICCECE), pp. 1–6. IEEE (2017). https://doi.org/10.1109/ICCECE.2017.8526215

17. Breast Cancer Wisconsin (Original) Data Set. https://archive.ics.uci.edu/dataset/15/breast+cancer+wisconsin+original. Accessed 18 June 2023

18. Senapati, M.R., Panda, G., Dash, P.K.: Hybrid approach using KPSO and RLS for RBFNN design for breast cancer detection. Neural Comput. Appl. **24**, 745–753 (2014). https://doi.org/10.1007/s00521-012-1286-6

19. Kumar, V.: Evaluation of computationally intelligent techniques for breast cancer diagnosis. Neural Comput. Appl. **33**(8), 3195–3208 (2021). https://doi.org/10.1007/s00521-020-05204-y

20. Dhanya, R., Paul, I.R., Akula, S.S., Sivakumar, M., Nair, J.J.: F-test feature selection in stacking ensemble model for breast cancer prediction. Procedia Comput. Sci. **171**, 1561–1570 (2020). https://doi.org/10.1016/j.procs.2020.04.167

21. Kermani, B.G., White, M.W., Nagle, H.T.: Feature extraction by genetic algorithms for neural networks in breast cancer classification. In: Proceedings of 17th International Conference of the Engineering in Medicine and Biology Society, vol. 1, pp. 831–832. IEEE (1995). https://doi.org/10.1109/IEMBS.1995.575385

22. Bazazeh, D., Shubair, R.: Comparative study of machine learning algorithms for breast cancer detection and diagnosis. In: 2016 5th International Conference on Electronic Devices, Systems and Applications (ICEDSA), pp. 1–4. IEEE (2016). https://doi.org/10.1109/ICEDSA.2016.7818560

23. Bayrak, E.A., Kırcı, P., Ensari, T.: Comparison of machine learning methods for breast cancer diagnosis. In: 2019 Scientific Meeting on Electrical-Electronics & Biomedical Engineering

and Computer Science (EBBT), pp. 1–3. IEEE (2019). https://doi.org/10.1109/EBBT.2019.
8741990

24. Kumar, U.K., Nikhil, M.S., Sumangali, K.: Prediction of breast cancer using voting classifier
technique. In: 2017 IEEE International Conference on Smart Technologies and Management
for Computing, Communication, Controls, Energy and Materials (ICSTM), pp. 108–114.
IEEE (2017). https://doi.org/10.1109/ICSTM.2017.8089135

25. Das, A.K., Biswas, S.K., Mandal, A.: An expert system for breast cancer prediction (ESBCP)
using decision tree. Indian J. Sci. Technol. **15**(45), 2441–2450 (2022). https://doi.org/10.
17485/IJST/v15i45.756

26. Bhattacharya, A., Biswas, S.K., Mandal, A.: Credit risk evaluation: a comprehensive study.
Multimed. Tools Appl. **82**(12), 18217–18267 (2023). https://doi.org/10.1007/s11042-022-
13952-3

27. Last, F., Douzas, G., Bacao, F.: Oversampling for Imbalanced Learning Based on K-Means
and SMOTE (2017). https://arxiv.org/abs/1711.00837

28. Fonseca, J., Douzas, G., Bacao, F.: Improving imbalanced land cover classification with
K-Means SMOTE: detecting and oversampling distinctive minority spectral signatures.
Information **12**(7), 266 (2021). https://doi.org/10.3390/info12070266

29. Xu, Z., Shen, D., Nie, T., Kou, Y., Yin, N., Han, X.: A cluster-based oversampling algorithm
combining SMOTE and k-means for imbalanced medical data. Inf. Sci. **1**(572), 574–589
(2021). https://doi.org/10.1016/j.ins.2021.02.056

30. Williamson, S., Vijayakumar, K., Kadam, V.J.: Predicting breast cancer biopsy outcomes
from BI-RADS findings using random forests with chi-square and MI features. Multimed.
Tools Appl. **81**(26), 36869–36889 (2022). https://doi.org/10.1007/s11042-021-11114-5

31. Das, A.K., Biswas, S.K., Mandal, A., Bhattacharya, A., Sanyal, S.: Machine learning based
intelligent system for breast cancer prediction (MLISBCP). Expert Syst. Appl. **25**, 122673
(2023). https://doi.org/10.1016/j.eswa.2023.122673

32. Paul, R., Biswas, S.K., Boruah, A.N., Das, A.K., Reshmi, S., Purkayastha, B.: Expert system for breast cancer prediction using ensemble learning. In: 2022 International Conference
on Machine Learning, Big Data, Cloud and Parallel Computing (COM-IT-CON) 26 May
2022, vol. 1, pp. 113–118. IEEE (2022). https://doi.org/10.1109/COM-IT-CON54601.2022.
9850678

ARU NET: Follicle Segmentation from Ultrasound Images of Ovaries Using Attention Residual U-NET Model

Debasmita Saha[1]([✉]) [ID], Ardhendu Mandal[2] [ID], Saroj Kr. Biswas[3] [ID],
Arijit Bhattacharya[4] [ID], and Akhil Kumar Das[4] [ID]

[1] Department of Computer Science, University of Gour Banga, Malda, West Bengal, India
debasmita.saha.cs@gmail.com
[2] Department of Computer Science and Technology, University of North Bengal, Siliguri,
West Bengal, India
[3] Department of Computer Science and Engineering, National Institute of Technology, Silchar,
Silchar, Assam, India
saroj@cse.nits.ac.in
[4] Department of Computer Science, Gour Mahavidyalaya, Malda, West Bengal, India

Abstract. The ovary is a vital component of the female reproductive system, housing numerous spherical structures known as follicles. These follicles play a crucial role in diagnosing various female health conditions such as infertility, Polycystic Ovarian Syndrome (PCOS), and Ovarian Cancer. Presently, the evaluation of follicle size, shape, and count relies on the manual examination of ultrasound images of the ovaries, a process prone to errors, labour-intensiveness, and time consumption. To enhance this monitoring process performed by radiologists and doctors, researchers have proposed several segmentation methods over the past few decades. Nevertheless, there is still room for improvement in the segmentation process, specifically in accurately isolating the Region of Interest (ROI) in terms of shape, position, and count. Moreover, as accuracy rates are improved, it becomes critical to minimize the occurrence of false identifications. Addressing these challenges necessitates the extraction of a comprehensive set of features from the target region while maintaining a sharp focus on the ROI. In this context, this research introduces an automated approach that leverages deep learning techniques for the segmentation of ovarian follicles. The foundation of this method is a U-Net model, strengthened by the incorporation of attention mechanisms and residual connections to amplify its performance. To validate the effectiveness of this approach, experiments were conducted using the USOVA 3D dataset. The experimental results validate the superiority of the proposed method, referred to as the Attention Residual U-Net (ARU Net), compared to conventional U-Net models, Attention U-Net models, and state-of-the-art models. Notably, the Attention Residual U-Net achieved an impressive accuracy rate of 97.65% using a rigorous 5-fold cross-validation. Additionally, the model exhibited outstanding performance with Recall, Precision, and Dice Coefficient scores of 89.45%, 92.13%, and 69.26%, respectively.

Keywords: Follicle Segmentation · Deep Neural Network · Ovarian Ultrasound Image · Computer Aided Diagnosis

M. Majumder et al. (Eds.): ICCTE 2023, CCIS 2376, pp. 287–298, 2025.
https://doi.org/10.1007/978-3-031-81935-3_25

1 Introduction

An Ovarian Follicle is a cellular aggregation set located in the ovaries, and it holds substantial significance in the female reproductive system. Monitoring the size, shape, and number of follicles is essential in diagnosing various conditions such as Polycystic Ovarian Disease (PCOD), Infertility, Cancer of Ovary etc. [1]. Follicle segmentation from ultrasound image of ovaries is a critical job in the field of gynecological imaging [2]. It entails identifying and isolating follicles from surrounding tissue in the ovaries. The size and number of follicles can provide significant information about a woman's reproductive health and contributes in the diagnosis process of various fertility related difficulties [3]. As a result, accurately segmenting the ovary and ovarian follicles in ultrasound imaging is critical for therapeutic purposes. Now-a-days, follicle segmentation in ultrasound images is carried out manually by experts, a process that is both time-consuming and susceptible to human errors [4]. Therefore, there is a need to develop an automated and precise technique for accurately segmenting the ovary and ovarian follicles, which would offer significant benefits in the field of computer aided diagnosis of ovarian diseases.

Image segmentation is the process of automatically classifying pixels within an image. In the field of medical image segmentation, the primary objective is to improve the visibility of specific structures in medical images. Utilizing computer vision-based automated detection techniques provides several benefits, including enhanced efficiency, reduced expenses, and improved reliability. These advancements hold great promise for computer-assisted diagnosis in the medical domain. With the advent of deep learning, automated methods have been developed to perform this task, reducing the dependency on human expertise and increasing the speed and accuracy of the segmentation process. Several approaches based on deep learning have made new development in medical image segmentation field in recent years. U-Net [5] is one of the most widely used deep learning models for medical image segmentation tasks because of its simple architecture and ability to capture fine details in images. However, U-Net exhibits a few limitations, such as the loss of spatial information caused by extensive down sampling and vulnerability to over fitting, especially when dealing with small datasets. Consequently, various modified versions of the U-Net model have emerged, aiming to address these limitations like Attention U-Net [6], R2U-NET [7], MU-NET [8], Res-UNet [9], etc.

In this research, an endeavor was made to improve the efficiency of follicle segmentation by refining the architecture of the fundamental U-Net model. The paper introduces a novel deep neural network model called "Attention Residual U-Net (ARU Net)," which integrates both the attention mechanism and residual block into the U-Net framework. This integration aims to enhance its performance and mitigate the constraints associated with the traditional U-Net model, specifically for the task of segmenting follicles in ultrasound images of ovaries.

2 Related Work

In quest of isolating the Region of Interest (ROI) covering follicular structures, a diverse array of methods have been employed, encompassing segmentation, thresholding, and region growing techniques [10–13]. Additionally, researchers have harnessed various machine learning approaches, including clustering [4, 14, 15] and classification [16, 17], to achieve this goal. Artificial Neural Networks (ANN) [18] and Convolutional Neural Networks (CNN) [1, 8, 19, 20] became more and more adept at tasks involving image segmentation and classification as the field developed. Particularly, the field of deep learning attracted a lot of attention [21] because of its aptitude for handling enormous data volumes while simultaneously identifying pertinent information that can later be used to guide the construction of creative models. Wanderly et al. [1] adopted a supervised approach to evaluate the performance of the Faster R-CNN, YOLOv3 and RetinaNet models for detecting ovarian structures, specifically the follicles and ovary, within B-mode ultrasound images. Among these models, Faster R-CNN achieved the most favorable outcome, boasting 95.5% precision rate and 94.7% recall rate. Additionally, RetinaNet also achieved impressive results, with both recall and precision rates exceeding 90%. In 2022, a team of researchers led by Zhong Chen from Ain Shams University introduced a novel approach for automatic follicle segmentation using a combination of deep neural network and edge information [20]. This method harnesses the strengths of deep neural networks alongside edge data. The researchers integrated a Canny operator to detect edges simultaneously in both the labeled and predicted images, enabling the calculation of edge loss. Their proposed model outperformed classical models like FCN, Segnet, and U-Net. The experiment involved 1,050 images, and model performance was evaluated using metrics including Recall, Precision, Jaccard, and Dice coefficient. Saha et al. (2022) proposed a novel model namely MU-net [8], a 2D segmentation network that combined U-Net and MobileNetV2 to accurately segment follicles from ultrasound images of ovaries. The model was evaluated on the USOVA3D Training Set 1. The model successfully achieved 98.4% accuracy, despite facing challenges such as low image contrast commonly found in ultrasound scans.

3   Proposed Method

Given the promising track record of the traditional U-Net model in the field of image segmentation, we initiated our experiment by training the U-Net model using our dataset. Following this, we added two components, namely the Attention Gate and Residual Block, into the U-Net architecture for the purpose of overall performance improvement of the proposed framework. The attention gate (see Fig. 1) will focus towards crucial regions while fading feature activation in irrelevant areas. This greatly boosts the model's capacity to represent information effectively without imposing a substantial rise in computational cost. The Residual Blocks (see Fig. 2) will enhance the effectiveness of deep neural networks by accommodating additional neural layers while minimizing the error rate. By utilizing skip connections, the integration of outputs from earlier layers into subsequent layers becomes possible, facilitating the training of notably deeper networks and leading to improved segmentation outcomes.

Before feeding to the model, the images underwent resizing and feature normalization. Afterwards, the dataset was distributed into training, validation, and test sets, and then the models were trained. (see Fig. 3). The subsequent section discusses the data preparation, implementation specifics, evaluation metrics, experimental findings, and a comparison with state-of-the-art methods.

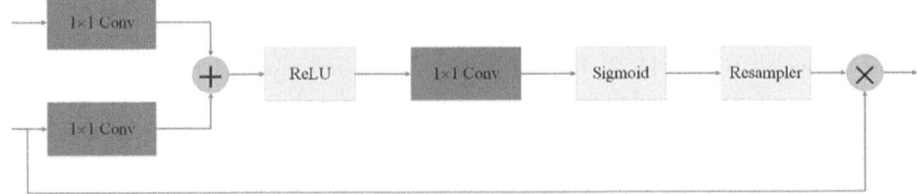

Fig. 1. Attention mechanism

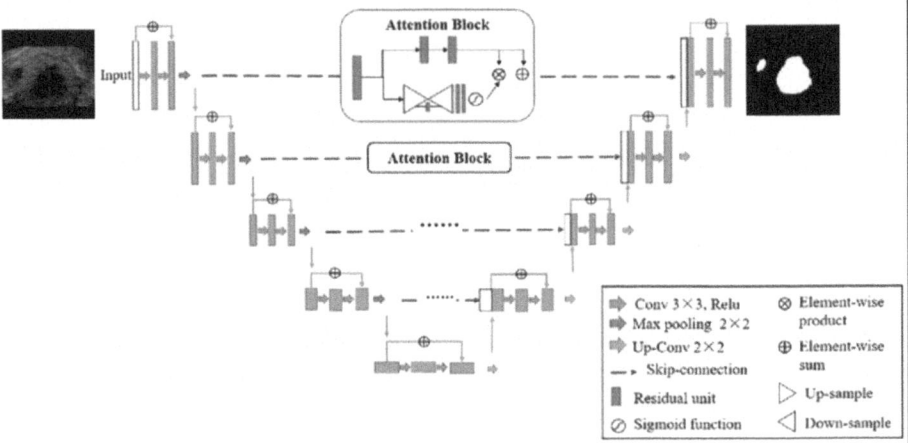

Fig. 2. Architecture of Attention Residual U-Net

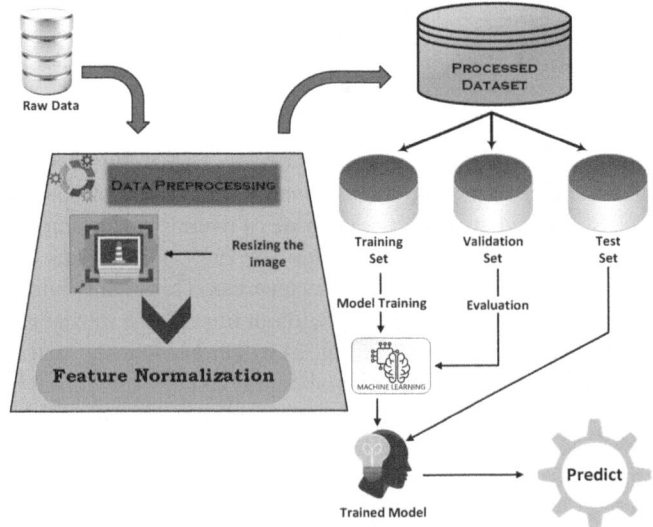

Fig. 3. System workflow

3.1 Attention Gate

The Attention gate functions by focusing on particular areas of interest while reducing the activation of features in irrelevant regions. Here the gating signal is denoted by "G $\in \mathbb{R}^{C' \times H \times W}$" and input feature map by X, which captures contextual information at a granular scale. Attention gate employs improved attention to derive the gating coefficient. Firstly, the gating signal and input X go through linear transformations, projecting them into a dimensional space "$\mathbb{R}^{F \times H \times W}$". The spatial attention weight map "$S \in \mathbb{R}^{1 \times H \times W}$" is generated by compressing the output within the channel domain in a subsequent step. This entire procedure is expressed as follows (see Fig. 4):

$$S = \sigma\big(\varphi\big(\delta\big(\varphi_x(X) + \varphi_g(G)\big)\big)\big)Y = SX$$

Linear transformations φ, ϕ_x, and ϕ_g are represented as 1×1 convolutions. The Attention gate serves as a powerful mechanism that directs the model's focus towards crucial regions, simultaneously inhibiting feature activation in less significant areas. It significantly augments the model's depictive capabilities without incurring a substantial increase in computational expense or an expansion in the model parameters numbers, thanks to its efficient design. Moreover, its modular and general nature renders it easily adaptable for use in a variety of Convolutional Neural Network (CNN) models.

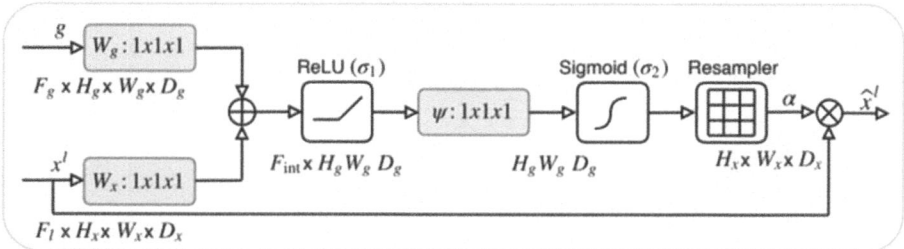

Fig. 4. Mechanism of Attention Gate

3.2 Residual Block

In conventional Convolutional Neural Networks (CNNs), as network layers are stacked, the accumulation of multi-layer features can become excessive. However, the straight-forward stacking of network layers leads to the issue of gradient vanishing, impeding the model's convergence. To enhance the accuracy of environmental recognition and miti-gate gradient vanishing, the introduction of a residual network into the network structure was considered. This addition aids in training and addresses the degradation problem. The residual network is primarily constructed as a sequence of residual blocks featuring a shortcut connection. This shortcut connection is akin to incorporating a direct pathway within the network. The residual units are represented as follows:

$$y_i = F(x_i, w_i) + h(x_i),$$

$$x_i + 1 = f(y_i)$$

The representation for the input of the residual unit in layer i is x_i, whereas the network parameters for the identical unit are denoted as w_i. The residual function is symbolized as $F(\cdot)$, the identity mapping function is denoted as $h(\cdot)$, and activation function as $f(\cdot)$. The following figure shows the architecture of Residual Block (see Fig. 5). Residual neural network contains two units:

Identity Mapping Unit. The main purpose of this section is to merge the input with the output that has undergone processing by the residual component. This procedure improves the incorporation of subsequent feature information.

Residual Unit. This section typically consists of several inverted neural networks, activation functions and normalization layers.

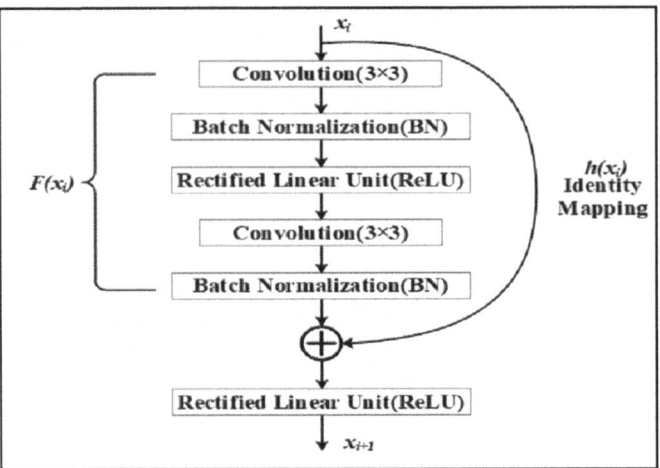

Fig. 5. Residual Block

4 Experiment

4.1 Dataset Preparation

The assessment of various models was conducted using the USOVA3D Training Set 1 [22]. This dataset includes 16 3D ultrasound images of ovaries in VTK format. For model training, the 3D images were sliced to generate a substantial number of 2D images. This slicing yielded 3,419 axial images for use in our experiments. The following figure (see Fig. 6) shows axial slices from a 3D ultrasound image of the ovary and corresponding annotated images labelled by domain experts. The dataset has been divided into three distinct sets: an 80% training set, a 10% validation set, and an additional 10% testing set. Out of the total number of images, 2,735 have been assigned for training purposes, while the remaining images have been set aside for validation and testing.

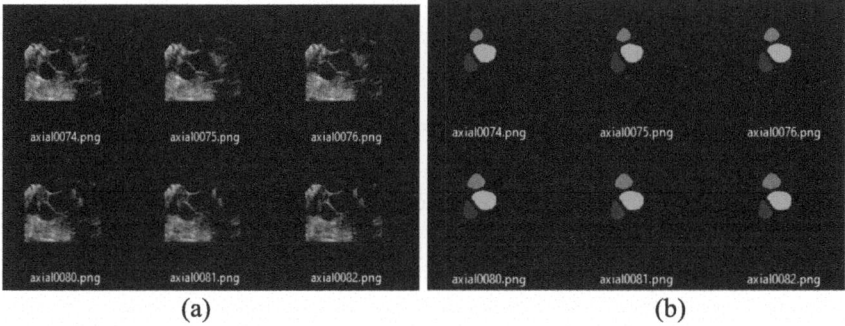

(a)   (b)

Fig. 6. Axial slices of 3D ultrasound images of the ovary, (a) Original Images, (b) Annotated Images

4.2 Implementation Details

The models undergo 100 epochs of training keeping the batch size 16. Here the Adam optimizer is used and the initial learning rate is 1e−4 (see Table 1).

Table 1. Values of parameters for model training.

Parameters	Value
Learning rate	1e−4
Epochs	100
Batch size	16
Optimizer	Adaptive Moment Estimation (Adam)
Total Parameters	39,090,377
Trainable Parameters	39,068,871

4.3 Evaluation Metrics

To conduct a quantitative analysis of experimental outcomes, various evaluation metrics are employed. These metrics include Accuracy, Precision, Recall, Dice Co-efficient, and Jaccard similarity. These measures are founded on four essential concepts: "True Positive" (TP), "True Negative" (TN), "False Positive" (FP), and "False Negative" (FN). Here, "GT" indicates ground truth, which represents actual correct segmentation, while "SR" represents the predicted segmentation mask. The formulas for computing Accuracy, Recall, and Precision are as follows [18, 19]:

$$\text{Accuracy} = \frac{TP + TN}{TP + TN + FP + FN}$$

$$Recall = \frac{TP}{TP + FP}$$

$$Precision = \frac{TP}{TP + FP}$$

$$Dice = 2|GT \cap SR|/(|GT| + |SR|)$$

$$Jaccard = |GT \cap SR|/|GT \cup SR|$$

4.4 Experimental Result

Table 2 shows the results for segmentation using three different models: U-Net, Attention U-Net, and our proposed Attention Residual U-Net (ARU-Net). These results are measured in terms of evaluation metrics including Accuracy, Recall, Precision, Jaccard Similarity, and Dice Coefficient. The findings clearly indicate that the ARU-Net model surpasses the other two models according to the Accuracy, Precision, Recall, and Dice coefficient values. This favorable outcome indicates that our innovative model is capable in achieving improved segmentation results from ultrasound images of ovaries. The diagram below presents a visual comparison of various components, including Loss, Accuracy, Recall, Precision, Jaccard Similarity, and Dice Coefficient, for the three models (see Fig. 7).

Quantitative Result

Table 2. Outcomes of segmentation by employing various components.

Models	Loss	Accuracy	Recall	Precision	Jaccard Similarity	Dice co-efficient
U-Net	0.3540	0.8865	0.9032	0.9098	0.1909	0.6349
Attention U-Net	0.3186	0.8857	0.8853	0.9012	0.1831	0.6814
Attention Residual U-Net (ARU-Net)	0.3073	0.9765	0.8945	0.9213	0.2189	0.6926

Qualitative Result

In segmentation-related challenges, qualitative assessment holds the same significance

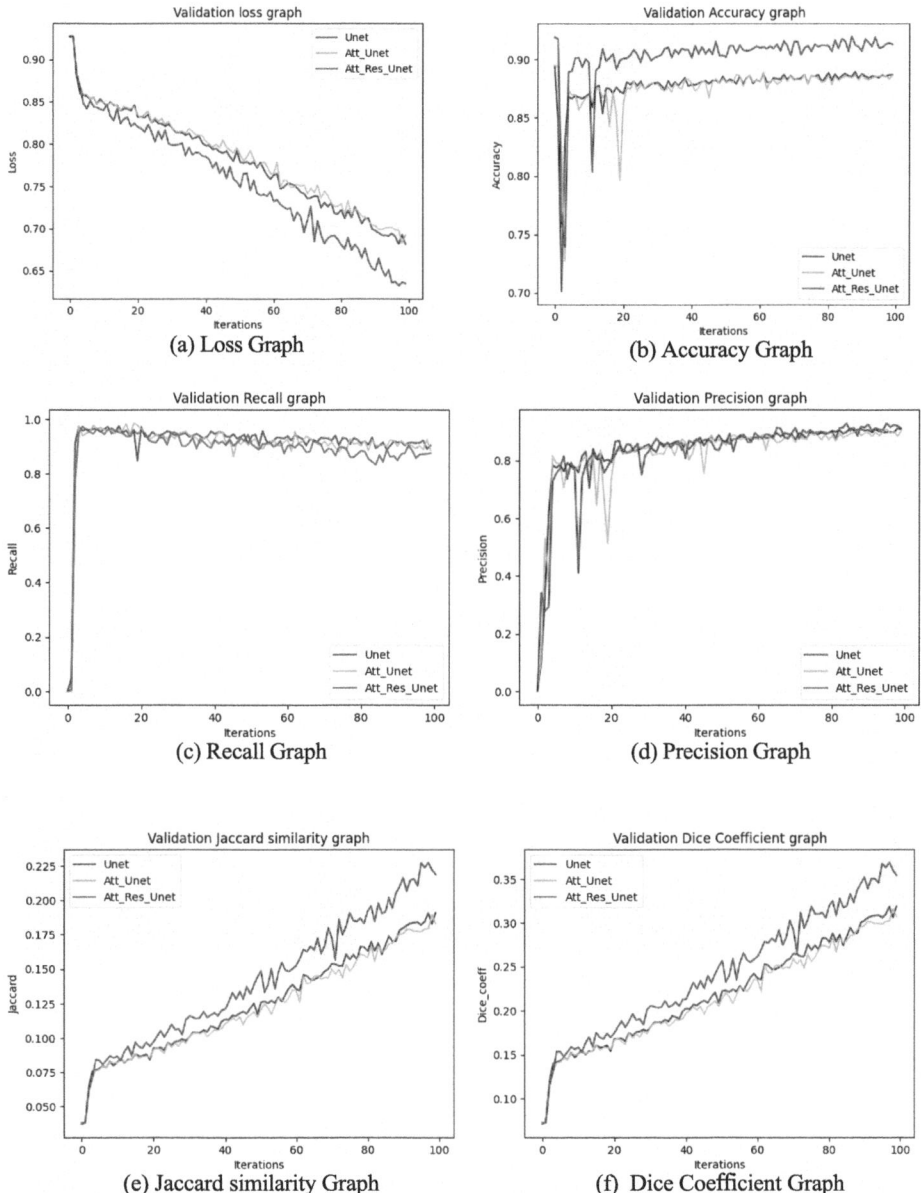

Fig. 7. Comparison of segmentation results for the three models using different components.

as quantitative evaluation. The segmentation results obtained by the three models i.e., U Net, U Net with Attention Mechanism (Unet + AM) and U Net with attention mechanism and Residual Connection (Unet + AM + RC) are depicted in the following figure, showcasing a visual contrast among predicted segmentation and actual segmentation (see Fig. 8).

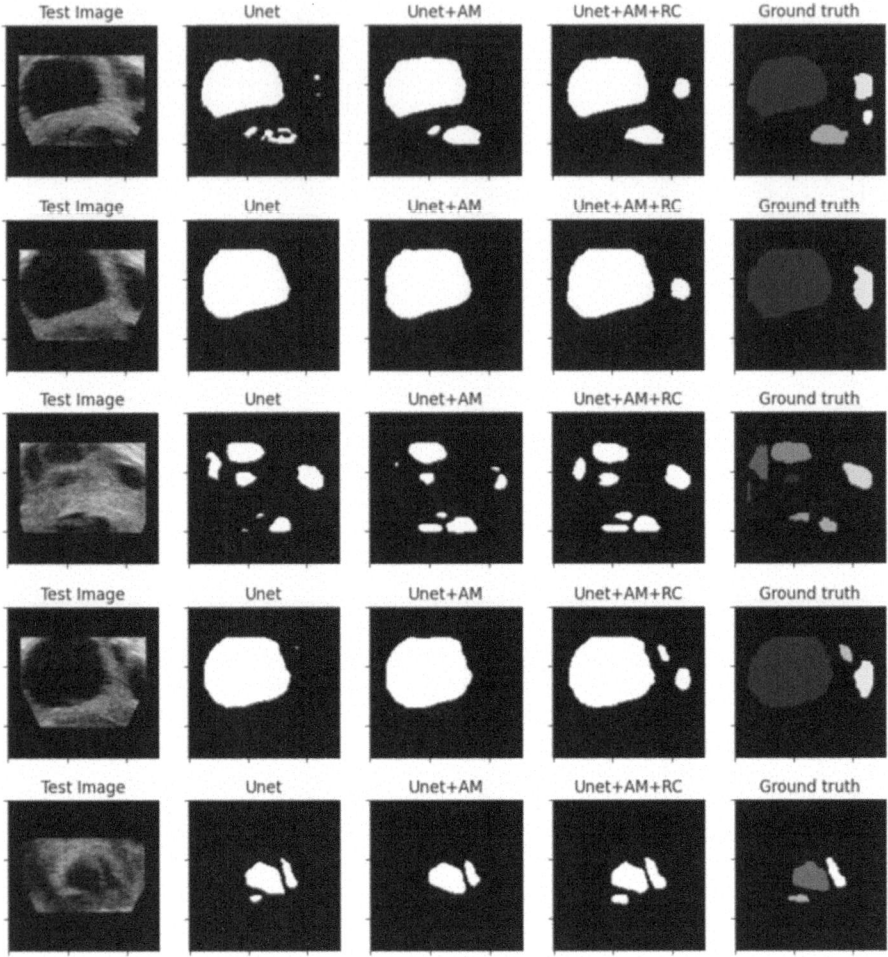

Fig. 8. Comparison of segmentation results by U Net, U Net with Attention Mechanism and U Net with attention mechanism and Residual Connection.

4.5 Comparison with Existing Work

Comparison between the performances of the proposed model (ARU-Net) and two contemporary state-of-art methods is presented in Table 3. It is evident that the proposed model exhibits higher Recall and Precision values compared to the first model [20]. Furthermore, in comparison to the second model [8], our proposed approach achieves notably improved Recall and Dice coefficient values, while maintaining commendable levels of Accuracy and Precision.

Table 3. Comparison of segmentation results with existing models.

Reference	Year	Method	Size of dataset	Performance				
				Accuracy	Recall	Precision	DSC	Jaccard
[20]	2022	Unet+EI+AM	1050	-	0.7773	0.8444	0.8095	0.6799
[8]	2022	MU Net	3419	0.984	0.638	0.983	0.674	-
Proposed ARU-Net			3419	0.9765	0.8945	0.9213	0.6926	0.2189

5 Conclusion

The present study proposes an automated framework for the precise segmentation of follicles from ovarian ultrasound images which integrates a deep neural network with both attention mechanisms and residual connections. The incorporation of an attention mechanism within the network's skip-connections allows for heightened focus on the target area, ultimately enhancing the accuracy of the segmentation process. The outcomes of the experiments provide clear evidence of the superiority of the approach when contrasted with the latest techniques in terms of segmentation performance. The proposed architecture excels in precisely delineating ovarian follicles within ultrasound images, yielding well-defined boundaries and achieving intricate segmentation.

References

1. Wanderley, D., Ferreira, C., Campilho, A., Silva, J.: Ovarian structures detection using convolutional neural networks. Procedia Comput. Sci. **196**, 542–549 (2022). https://doi.org/10.1016/j.procs.2021.12.047
2. Denny, A., Raj, A., Ashok, A., Ram, C.M., George, R.: i-HOPE: detection and prediction system for polycystic ovary syndrome (PCOS) using machine learning techniques. In: 2019 IEEE Region 10 Conference (TENCON), TENCON 2019, pp. 673–678 (2019). https://doi.org/10.1109/TENCON.2019.8929674
3. Follicle Detection on the USG Images to Support Determination of Polycystic Ovary Syndrome - IOPscience. https://iopscience.iop.org/article/10.1088/1742-6596/622/1/012027/meta. Accessed 31 July 2023
4. Mandal, A., Saha, D., Sarkar, M.: Follicle segmentation using K-means clustering from ultrasound image of ovary. In: Bhattacharjee, D., Kole, D.K., Dey, N., Basu, S., Plewczynski, D. (eds.) Proceedings of International Conference on Frontiers in Computing and Systems. Advances in Intelligent Systems and Computing, vol. 1255, , pp. 545–553. Springer, Singapore (2021). https://doi.org/10.1007/978-981-15-7834-2_51
5. Marques, S., et al.: Segmentation of gynaecological ultrasound images using different U-Net based approaches. In: 2019 IEEE International Ultrasonics Symposium (IUS), pp. 1485–1488 (2019). https://doi.org/10.1109/ULTSYM.2019.8925948
6. Oktay, O., et al.: Attention U-Net: learning where to look for the pancreas. arXiv (2018). https://doi.org/10.48550/arXiv.1804.03999
7. Alom, Z., Taha, T.M., Asari, V.K.: Recurrent Residual Convolutional Neural Network based on U-Net (R2U-Net) for Medical Image Segmentation

8. Saha, D., Mandal, A., Ghosh, R.: MU Net: ovarian follicle segmentation using modified U-Net architecture. Int. J. Eng. Adv. Technol. **11**(4), 30–35 (2022). https://doi.org/10.35940/ijeat.D3419.0411422

9. Diakogiannis, F.I., Waldner, F., Caccetta, P., Wu, C.: ResUNet-a: a deep learning framework for semantic segmentation of remotely sensed data. ISPRS J. Photogramm. Remote Sens. **162**, 94–114 (2020). https://doi.org/10.1016/j.isprsjprs.2020.01.013

10. Rihana, S., Moussallem, H., Skaf, C., Yaacoub, C.: Automated algorithm for ovarian cysts detection in ultrasonogram. In: 2013 2nd International Conference on Advances in Biomedical Engineering, pp. 219–222 (2013). https://doi.org/10.1109/ICABME.2013.6648887

11. Mehrotra, P., Chakraborty, C., Ghoshdastidar, B., Ghoshdastidar, S., Ghoshdastidar, K.: Automated ovarian follicle recognition for Polycystic Ovary Syndrome. In: 2011 International Conference on Image Information Processing, Shimla, Himachal Pradesh, India, pp. 1–4. IEEE (2011). https://doi.org/10.1109/ICIIP.2011.6108968

12. Potocnik, B., Zazula, D.: Automated ovarian follicle segmentation using region growing. In: Proceedings of the First International Workshop on Image and Signal Processing and Analysis. In Conjunction with 22nd International Conference on Information Technology Interfaces. (IEEE Cat. No.00EX437), IWISPA 2000, Pula, Croatia: Univ. Zagreb, pp. 157–162 (2000). https://doi.org/10.1109/ISPA.2000.914907

13. Deng, Y., Wang, Y., Shen, Y.: An automated diagnostic system of polycystic ovary syndrome based on object growing. Artif. Intell. Med. **51**(3), 199–209 (2011). https://doi.org/10.1016/j.artmed.2010.10.002

14. Rao, M.J., Kumar, D.R.K.: Follicle detection in digital ultrasound images using adaptive K-means clustering algorithm. **14**(2), 7 (2019)

15. Raj, A.: Ovarian follicle detection for polycystic ovary syndrome using fuzzy C-means clustering. Int. J. Comput. Trends Technol. **4**(7), 4 (2013)

16. Gopalakrishnan, C., Iyapparaja, M.: Detection of polycystic ovary syndrome from ultrasound images using SIFT descriptors. Bonfring Int. J. Softw. Eng. Soft Comput. **9**(2), 26–30 (2019). https://doi.org/10.9756/BIJSESC.9017

17. Tegnoor, J.R.: Automated ovarian classification in digital ultrasound images using SVM. Int. J. Eng. Res. **1**(6), 17 (2012)

18. Dewi, R.M., Adiwijaya, Wisesty, U.N., Jondri: Classification of polycystic ovary based on ultrasound images using competitive neural network. In: Journal of Physics: Conference Series, vol. 971, p. 012005 (2018). https://doi.org/10.1088/1742-6596/971/1/012005

19. Srivastava, S., Kumar, P., Chaudhry, V., Singh, A.: Detection of ovarian cyst in ultrasound images using fine-tuned VGG-16 deep learning network. SN Comput. Sci. **1**(2), 81 (2020). https://doi.org/10.1007/s42979-020-0109-6

20. Chen, Z., Zhang, C., Li, Z., Yang, J., Deng, H.: Automatic segmentation of ovarian follicles using deep neural network combined with edge information. Front. Reprod. Health 4 (2022). https://www.frontiersin.org/articles/10.3389/frph.2022.877216. Accessed 08 June 2023

21. Luo, D., Zeng, W., Chen, J., Tang, W.: Deep learning for automatic image segmentation in stomatology and its clinical application. Front. Med. Technol. 3 (2021). https://www.frontiersin.org/articles/10.3389/fmedt.2021.767836. Accessed 10 Aug 2023

22. Potočnik, B., et al.: Public database for validation of follicle detection algorithms on 3D ultrasound images of ovaries. Comput. Methods Programs Biomed. **196**, 105621 (2020). https://doi.org/10.1016/j.cmpb.2020.105621

d-RIMNet: RIMNet with Depthwise Separable Convolutional Layer for Retinal OCTA Image Segmentation

Farhana Sultana[1] , Abu Sufian[1(✉)] , and Paramartha Dutta[2]

[1] Department of Computer Science, University of Gour Banga, Malda, India
sufian@ieee.org
[2] Department of Computer and System Sciences, Visva-Bharati, Santiniketan, India

Abstract. Optical Coherence Tomography Angiography (OCTA) is a recently developed noninvasive imaging technique capable of capturing detailed images from distinct layers of the retinal vascular complexes. This imaging modality extracts images from the capillary level, consisting of thin, tiny, and high-complexity vascular information. Therefore, segmentation of the retinal microvasculature in OCTA images is challenging. To accurately segment the vessels in OCTA images, we have applied an improved version of RIMNet. RIMNet is a convolutional neural network (CNN) based image magnification architecture. In this study, we have used a depthwise separable convolutional layer instead of the standard convolutional layer in the RIMNet model dubbed as d-RIMNet and experimented on the Optical Coherence Tomography Angiography Retinal Scans and Segmentations (OCTA-SS) dataset. We have noticed that using depthwise separable convolutional layers in d-RIMNet reduces the parameter and complexity of the model by ≈20% than RIMNet without significantly decreasing the segmentation performance. A thorough performance analysis of the state-of-the-art models and our model on the OCTA-SS dataset demonstrates that with a small dataset, smaller network sometimes outperforms large complex networks.

Keywords: Convolutional Neural Network · Depthwise Separable Convolution Layer · Image Magnification Network · Medical Image Segmentation · OCTA · Retinal Blood Vessel Segmentation

1 Introduction

The circulatory system connects retinal blood vessel with a person's entire body. Hence, the retinal vessels get affected by diseases affecting the other parts of the body. Recent studies show that detection of constructional and functional differences of the retinal vessel is crucial in diagnosing various ocular diseases (e.g. retinal vein and artery occlusions), conditions related to diabetes (e.g. diabetic retinopathy, diabetic macular edema, diabetes mellitus, glaucoma) [6,10], cardiovascular diseases (e.g. hypertensive retinopathy) [11], chronic kidney diseases

M. Majumder et al. (Eds.): ICCTE 2023, CCIS 2376, pp. 299–310, 2025.
https://doi.org/10.1007/978-3-031-81935-3_26

[15] and neurodegenerative diseases (e.g. Alzheimer and mild cognitive impairment) [1,7]. Therefore, proper examination of the retina can portray about a person's condition on the whole.

Fundus photography and fluorescein angiography are popular imaging modalities used to capture retinal microvasculature for a long time, but they have some limitations. The former can not capture thin complex vessels and capillaries from deep layers of the retina, and the latter requires dye injection, which may affect a person's health severely. In contrast, optical coherence tomography angiography(OCTA) is a recently emerging non-invasive imaging modality which can capture high-resolution images from different layers of the retinal vascular complexes. This imaging modality extracts images from the capillary level, consisting of thin, tiny, and high-complexity vascular information, which is very useful for diagnosing various diseases mentioned above, particularly neurovascular [7,16] ones.

The initial step in diagnosing a disease from a retinal OCTA image is to segment the vessels accurately. This task is time-consuming, tiring and challenging for an ophthalmologist or optometrist. Automatic image segmentation comes here to handle the issue. This study proposed d-RIMNet (RIMNet with depthwise separable convolution layer) to segment retinal vessels in OCTA images.

RIMNet [13] used convolutional neural network as a magnifying tool for image segmentation. It has a three-step up-sampling magnification path followed by a three-step down-sampling demagnification path. The former path magnifies the feature maps, and the later path contracts the same. Despite a small network with only o.42 million parameters, RIMNet retains the minute details of an image and shows comparative performance on the retinal fundus image segmentation task. However, it suffers from a downfall in having a large number of floating point operations (Flops) due to feature magnification, leading the model to have high-model complexity.

To reduce the model complexity, we have used the depthwise separable convolution layer instead of the standard convolution layer in d-RIMNet and evaluated the performance of the model on the Optical Coherence Tomography Angiography Retinal Scans and Segmentations (OCTA-SS) dataset. We have noticed that d-RIMNet reduced the model complexity by ≈20% than RIMNet without significantly degrading the segmentation performance. To extend the model's capacity further, we doubled the number of filters applied in each up-sampling and down-sampling step of the model. Then, we thoroughly analysed the d-RIMNet and state-of-the-art models on the OCTA-SS dataset. Our result demonstrates that with a small dataset, smaller network may outperform large complex networks.

Contributions of the Work: The following list shows the key contributions of our study:

– We designed an improved version of the RIMNet model labeled d-RIMNet to segment retinal OCTA image.
– Depthwise separable convolution layer is used instead of standard convolution layer in d-RIMNet to reduce the model complexity.

- To increase the capacity of the model, we doubled the number of filters in each step of the up-sampling and down-sampling path of the model.
- Our model reduced the model complexity by ≈20% without significantly degrading the models segmentation performance.
- It also outperformed some deep, complex models on the image segmentation task of the OCTA-SS dataset.

Structure of the Paper: We structured our paper as follows:

- Section 2: related work.
- Section 3: model description.
- Section 4: details of the dataset.
- Section 5: experimental details.
- Section 6: result presentation and discussion.
- Section 7: conclusion of the study.

2 Related Work

In recent years, CNN-based automatic segmentation networks performed well on various image segmentation tasks. Among them, the performance of U-net [12] mostly dominates image segmentation research. This U-shaped network used a down-sampling contracting way and an up-sampling expanding way for biomedical image segmentation. ResUnet [17] used a residual block instead of a basic convolution block in U-net to maintain a trainable gradient in deep model. An attention gate is used in the short connection of attention U-net [9] to make the model focus on relevant information. MultiResUNet [5] concatenates feature maps generated from multiple consecutive residual layers to capture multiscale lesions in a medical image. IMN [8] is an image magnification network that follows the inverse architectural pattern of U-net. OCT2Former [14] uses the merits of both CNN and transformer to address image segmentation tasks.

3 Model

RIMNet [13] is an image magnification network originally used for segmenting retinal fundus images. It is a CNN based segmentation network. Conventional CNN-based encoder-decoder segmentation network uses a down-sampling path which encodes the feature maps and an up-sampling path which decodes the same to its original resolution. Due to down-sampling at the encoding path, it fails to capture the minute details of the images. RIMNet unconventionally used an up-sampling magnification path and a down-sampling demagnification path, as shown in Fig. 1, to prevent this loss.

The encoder part magnifies the image, and the decoder part contracts it back to its original resolution. The magnification process in the encoder part

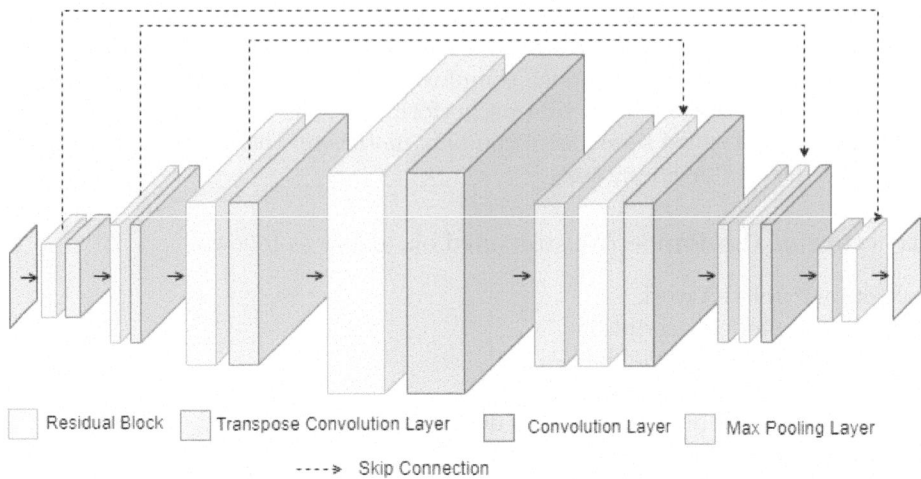

Fig. 1. The RIMNet model [13]

helps the model retain thin and tiny details of the image. The model has three up-sampling blocks, one bottleneck layer and three down-sampling blocks. A single-up sampling block contains a feature extracting residual block [3] followed by a magnifying transpose convolution layer. The bottleneck layer consists of two convolution layers. Each down sampling block contains a single convolution layer, a max pool layer, a concatenation layer, and one residual block consecutively. The concatenation layer concatenates the output feature maps of max pool layer of decoder block and the output feature map of the corresponding residual block of encoder block. Figure 2 depicts different blocks of RIMNet.

Despite being a small network, RIMNet showed state-of-the-art performance in retinal vessel segmentation task. However, as it expands the dimension of feature maps in its encoder path, it suffers from a high number of floating point operations (Flops), which eventually increases the model's complexity and inference time. RIMNet model contains 16 numbers of 2 × 2 standard convolution layers. Among them, 5, 5, 5, and 1 convolution layers have 16, 32, 64, and 128 channels respectively. In this study, we have addressed this problem of model complexity by reducing the amount of parameters. We used a depthwise separable convolution layer following [4] instead of the standard convolution layer in the RIMNet model and named our model d-RIMNet. Figure 3 shows both layers. This approach significantly decreased the model's parameters (≈20%) without any noticeable reduction in the segmentation performance.

Standard convolution operation filters the features and combines them in a single step. Depthwise separable convolution does the same task in two separate layers: 1. depthwise convolution layer and 2. pointwise convolution layer. On each individual channel of input feature map, depthwise convolution layer applies a dedicated filter to extract features. This keeps the number of channels of input

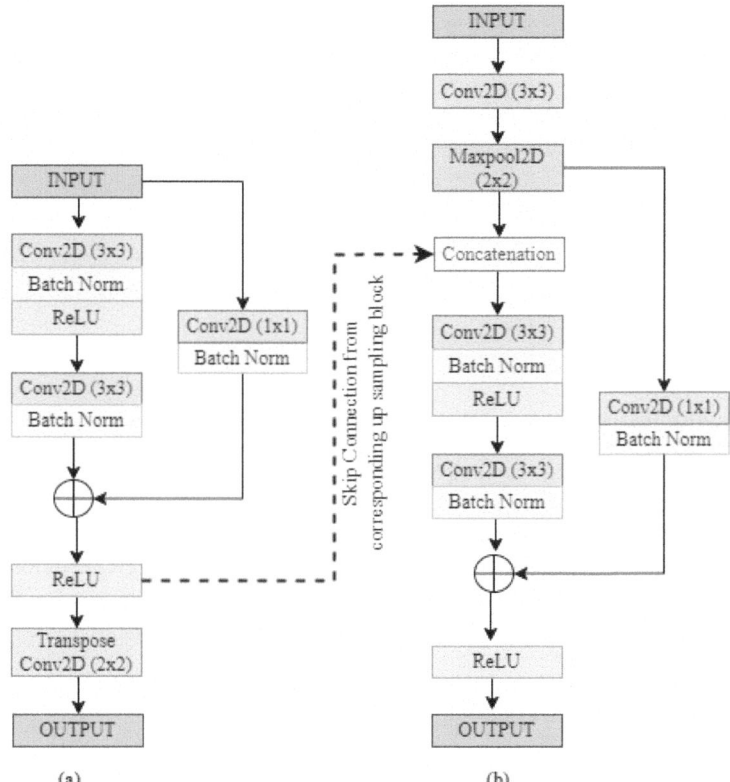

Fig. 2. Up-sampling block (a) and down-sampling block (b) of RIMNet

and output feature maps same. Pointwise convolution layer applies several 1×1 filters on output feature map generated from depthwise convolution layer to combine features extracted from previous layer. Figure 3 shows two different layers of depthwise separable convolution layer.

We assume that size of an input feature map F_i is $H_i \times W_i \times M$ where H_i, W_i, M represents the feature map's height, width and depth (number of channels). On F_i, the standard convolution layer will apply N number of filters of size $H_k \times W_k \times M$ to produce feature map F_o of size $H_o \times W_o \times N$. The spatial dimensions of F_i and F_o are assumed to be square, and F_o has the exact spatial dimensions as F_i. In depthwise separable convolution layer, depthwise convolution layer produces feature map of size $H_d \times W_d \times M$ by applying M number of $H_k \times W_k \times 1$ filters on F_i to get per channel feature map F_d. Then, the pointwise convolution layer applies N number of filters of size $1 \times 1 \times M$ to produce feature map F_o of size $H_o \times W_o \times N$.

If $k(= H_k = W_k)$ is the size of the kernel and $F(= H_i = W_i = H_o = W_o = H_d = W_d)$ is the size of the feature map. Then, assuming single stride and same

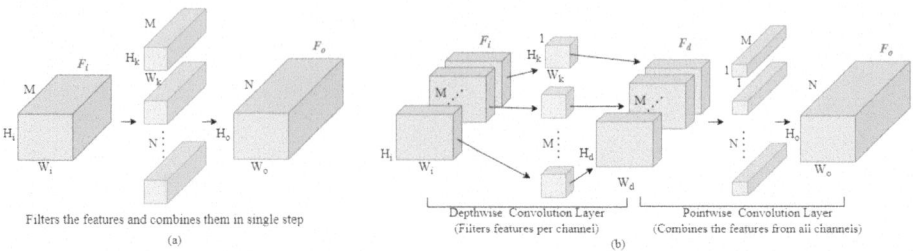

Fig. 3. Standard Convolutional Layer (a) and Depthwise Seperable Convolutional Layer (b)

padding, computational cost for standard convolution will be :

$$k^2 \cdot M \cdot N \cdot F^2 \tag{1}$$

Computational cost for depthwise separable convolution is the summation of the cost of depthwise convolution and pointwise convolution as given below:

$$k^2 \cdot M \cdot F^2 + M \cdot N \cdot F^2 \tag{2}$$

The total reduction of computational cost in case of depthwise separable convolution layer than standard convolution layer will be:

$$\frac{k^2 \cdot M \cdot F^2 + M \cdot N \cdot F^2}{k^2 \cdot M \cdot N \cdot F^2} = \frac{1}{N} + \frac{1}{k^2}$$

We are neither changing number of filters nor size of filters in depthwise separable convolution layer. However, the reduction of computational cost in depthwise convolutional layer will be as effective as we would have change the number of filters (N) or reduced the size of the filter (k) in standard convolution layer.

4   Dataset Used

In this section, we briefly described dataset and data pre-processing techniques.

4.1   Dataset Description

We have used an open-source Optical Coherence Tomography Angiography Retinal Scans and Segmentations (OCTA-SS) [2] dataset provided by Usher Institute, University of Edinburgh, UK. The dataset contains 55 OCTA images collected from five clinical regions (Foveal, Superior, Nasal, Inferior and Temporal) of the retinal parafoveal area of 11 participants aged 44–59, with or without a family history of dementia. It also contains corresponding manually segmented ground truth. All images are of size 91×91. The original images are of **.tif** format, and the ground truths are of **.png** format. The dataset does not contain any training-test division. In this work, we have used 30 original images and corresponding ground truths to train and validate our model and rest are used for testing purpose. Also, there is no information about which images are of dementia patients and which are not. Figure 4 shows a original image and corresponding ground truth of dataset.

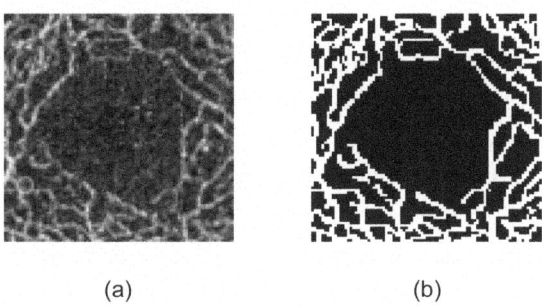

(a)   (b)

Fig. 4. Original image (a) and Ground truth (b) from OCTA-SS dataset

4.2 Data Preprocessing

We have center-cropped the original images and corresponding ground truths of size 80×80 for experimental convenience and then normalized both images and ground truths by dividing each pixel by 255.

5 Experimental Details

In RIMNet, magnification path contains three up-sampling blocks with 16, 32, and 64 filters, consecutively, and the demagnification path also has three down-sampling blocks with 64, 32, and 16 filters, consecutively. The bottleneck block contains 128 number of filters. Dimension and depth of feature maps get doubled in each magnifying stage. In contrast, in each demagnifying stage, both get halved. The model inputs an RGB image and outputs single channeled segmentation mask. There are 16 number of 2×2 convolution layers in total and these layers contribute to the model's parameter. In d-RIMNet, we have used depthwise separable convolution layer, as discussed in Sect. 3, in those 16 layers to reduce the amount of model parameters. To extend the capacity of d-RIMNet further, we have doubled the number of channels in each up-sampling (i.e. 32, 64, 128) and down-sampling block (i.e. 128, 64, 32). We have evaluated our model with state-of-the-art models and discussed the result in Sect. 6.

Optimization: We trained d-RIMNet model using Adam optimizer with a weight decay 0.0001. We have applied poly learning rate with poly power of

0.9 and initialized it to 0.0001. We have trained our model for 100 epochs with early stopping criteria. According to the early stopping criteria, the training will be stopped when validation performance is not improving for ten epochs. A mini-batch of 4 images is used to fit the model. We have used dice loss L_{Dice} described in Eq. (1) as loss function:

$$L_{Dice} = 1 - \frac{2 \sum_{i=1}^{N} p_i q_i + \epsilon}{\sum_{i=1}^{N} p_i{}^2 + \sum_{i=1}^{N} q_i{}^2 + \epsilon} \tag{3}$$

where p_i and q_i denote original image and ground truth respectively. To avoid numerical problem, we used a small positive constant ϵ as a parameter.

Cross-Validation: We used *5-fold* random cross-validation for training and validating d-RIMNet. We have reported mean and standard deviation of evaluation metrics on test set across all five folds in Sect. 6.

Implementation: We experimented our model with PyTorch on single NVIDIA Tesla A100 GPU.

6   Result and Discussion

In this section, we have presented the result obtained from the experiment outlined in Sect. 5. The following evaluation metrics are applied to show the segmentation result of the model:

$$Accuracy = \frac{(TP + TN)}{TP + TN + FP + FN}$$

$$Sensitivity = \frac{TP}{TP + FN}$$

$$Precision = \frac{TP}{TP + FP}$$

$$Specificity = \frac{TN}{TN + FP}$$

$$False\ Detection\ Rate\ (FDR) = \frac{FP}{FP + TP}$$

$$Area\ under\ ROC\ curve\ (AUC) = (Sensitivity + Specificity)/2$$

where TP, FP, TN and FN represent true positive, false positive, true negative, and false negative respectively and

$$Dice\ Similarity\ Coefficient\ (DSC)(A, B) = \frac{2\,A \cap B|}{|A| + |B|}$$

$$Intersection\ over\ Union\ (IoU)(A, B) = \frac{|A \cap B|}{A \cup B|}$$

where A and B represent original image and ground truth respectively. Along with the above-mentioned evaluation metrics, we have also shown amount of parameters (in a million) and floating point operations (Flops (in gigaflops)) of different networks with an input image size of 80×80.

Table 1 shows the comparative performance of RIMNet and d-RIMNet models. We have noticed that using depthwise separable convolution layer decreased our model's parameters by 21.42% and the complexity (Flops) by 21.06% than the RIMNet model with a slight degradation of the segmentation performance. To increase the segmentation performance further, we have doubled the number of filters of d-RIMNet.

Table 1. Segmentation performance of RIMNet and d-RIMNet on OCTA-SS dataset along with number of parameters and Flops. 'M' represents Million parameters and 'G' represents GigaFlops. The + indicates RIMNet with double number of channel, and * represents doubled channel RIMNet with depthwise separable convolution layer.

Methods	ACC	DSC	IoU	AUC	Sensitivity
RIMNet [13]	91.29 ± 1.05	88.78 ± 0.89	88.78 ± 0.89	90.60 ± 0.910	88.14 ± 1.66
RIMNet+	91.26 ± 0.88	89.39 ± 0.76	80.83 ± 1.24	90.95 ± 1.02	89.14 ± 2.92
d-RIMNet	90.19 ± 1.43	86.56 ± 0.94	86.56 ± 0.94	88.93 ± 1.24	84.26 ± 2.78
d-RIMNet*	91.15 ± 1.14	88.89 ± 0.27	88.89 ± 0.27	90.46 ± 0.44	87.85 ± 1.65
Methods	Precision	Specificity	FDR	Param (M)	Flops (G)
RIMNet [13]	89.50 ± 0.70	93.04 ± 0.79	10.20 ± 1.79	0.42	89.29
RIMNet+	89.79 ± 1.79	92.76 ± 1.01	10.20 ± 1.79	1.69	712.32
d-RIMNet	89.15 ± 1.62	93.61 ± 1.03	10.84 ± 1.62	**0.09**	**18.81**
d-RIMNet*	90.10 ± 1.36	93.08 ± 1.08	9.89 ± 1.36	0.33	72.05

In Table 2, we showed comparative performance of d-RIMNet (double channeled) model with seven state-of-the-art models, including six CNN based networks (U-net, ResUnet, Attention u-net, MultiResUNet, IMN, RIMNet) and one transformer-based network (OCT2Former). Among all the models, d-RIMNet has minor parameters (0.33 million) yet outperformed some deep, complex networks with many parameters. In Fig. 5, we have shown different segmentation predictions generated from the model along with the original image and ground truth.

Table 2. Comparative result of d-RIMNet and state-of-the-art model on OCTA-SS dataset. Due to resource limitation, we could not re-implement the OCT2Former in our hyper parameter setting which lead to incomplete information for some metric. Best performance are shown in bold. 'M' represents Million parameters and 'G' represents GigaFlops.

Methods	ACC	DSC	IoU	AUC	Sensitivity
U-net [12]	86.23 ± 1.76	81.69 ± 1.93	69.10 ± 2.78	84.84 ± 1.45	78.48 ± 2.54
ResUnet [17]	89.59 ± 1.32	85.86 ± 0.90	75.24 ± 1.38	88.27 ± 0.85	82.61 ± 1.78
Attention u-net [9]	90.40 ± 0.49	87.00 ± 0.62	77.02 ± 0.98	89.17 ± 0.41	84.62 ± 1.28
MultiResUNet [5]	87.95 ± 1.90	83.90 ± 0.65	72.34 ± 0.95	86.76 ± 1.12	81.75 ± 0.46
IMN [8]	86.39 ± 2.53	80.74 ± 5.26	68.09 ± 7.08	85.42 ± 4.08	80.95 ± 13.7
OCT2Former [14]	-	$\mathbf{90.96 \pm 1.10}$	$\mathbf{83.44 \pm 1.85}$	-	$\mathbf{92.34 \pm 2.63}$
RIMNet [13]	$\mathbf{91.29 \pm 1.05}$	88.78 ± 0.89	79.85 ± 1.44	$\mathbf{90.60 \pm 0.91}$	88.14 ± 1.66
d-RIMNet	91.18 ± 0.75	88.89 ± 0.27	80.00 ± 0.43	90.46 ± 0.44	87.85 ± 1.65
Methods	Precision	Specificity	FDR	Param (M)	Flops (G)
U-net	85.29 ± 1.96	91.19 ± 1.74	14.70 ± 1.96	31.03	5.33
ResUnet	89.50 ± 2.23	$\mathbf{93.93 \pm 0.78}$	10.49 ± 2.23	22.47	5.35
Attention u-net	89.59 ± 0.83	93.73 ± 0.89	10.40 ± 0.83	8.13	4.48
MultiResUNet	86.34 ± 2.22	91.78 ± 1.50	13.65 ± 2.22	13.41	**2.8**
IMN	84.04 ± 9.59	89.89 ± 6.76	15.95 ± 9.59	0.8	513.06
OCT2Former	$\mathbf{92.92 \pm 1.97}$	-	-	49.99	7.34
RIMNet	89.50 ± 0.70	93.04 ± 0.79		0.42	89.29
d-RIMNet	90.10 ± 1.36	93.08 ± 1.08	$\mathbf{9.89 \pm 1.36}$	**0.33**	72.05

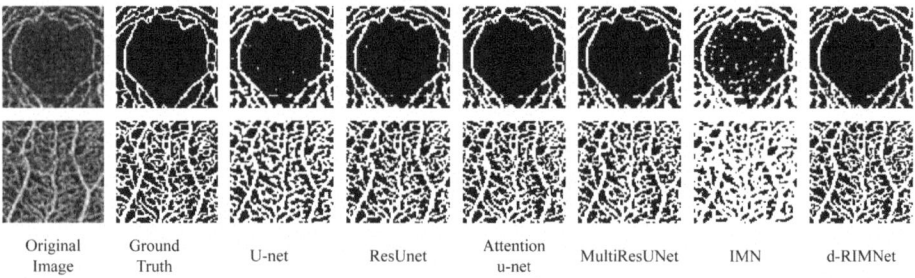

Original Image   Ground Truth   U-net   ResUnet   Attention u-net   MultiResUNet   IMN   d-RIMNet

Fig. 5. Predictions of different models on OCTA-SS dataset

7   Conclusion

We have designed d-RIMNet, which replaces the standard convolution layer of RIMNet with depthwise separable convolution layer. This approach significantly reduces the amount of model parameters and complexity with a slight segmentation performance degradation. To enhance model's performance further, we doubled the number of filters in all six up-sampling and down-sampling steps of the model. These two approaches helped the model achieve state-of-the-art results with fewer parameters which reduces model complexity. This study will direct the deep learning research to acquire better performance with fewer parameters, which will be very helpful, especially in the case of embedded systems.

References

1. Cabrera DeBuc, D., Somfai, G.M., Koller, A.: Retinal microvascular network alterations: potential biomarkers of cerebrovascular and neural diseases. Am. J. Physiol.-Heart Circulatory Physiol. **312**(2), H201–H212 (2017). https://doi.org/10.3390/brainsci13030460

2. Giarratano, Y.: Optical coherence tomography angiography retinal scans and segmentations, [image], University Of Edinburgh, Medical School (2019). https://doi.org/10.7488/ds/2729.

3. He, K., Zhang, X., Ren, S., Sun, J.: Deep residual learning for image recognition. In: The IEEE Conference on Computer Vision and Pattern Recognition (CVPR) (2016). https://doi.org/10.1109/CVPR.2016.90

4. Howard, A.G., et al.: Mobilenets: efficient convolutional neural networks for mobile vision applications. arXiv preprint arXiv:1704.04861 (2017)

5. Ibtehaz, N., Rahman, M.S.: Multiresunet: rethinking the u-net architecture for multimodal biomedical image segmentation. Neural Netw. **121**, 74–87 (2020). https://doi.org/10.1016/j.neunet.2019.08.025

6. Jenkins, A.J., Joglekar, M.V., Hardikar, A.A., Keech, A.C., O'Neal, D.N., Januszewski, A.S.: Biomarkers in diabetic retinopathy. Rev. Diabet. Stud.: RDS **12**(1–2), 159 (2015). https://doi.org/10.1900/RDS.2015.12.159

7. Jiang, H., et al.: Altered macular microvasculature in mild cognitive impairment and Alzheimer disease. J. Neuro-ophthalmol. Official J. North Am. Neuro-Ophthalmol. Soc. **38**(3), 292 (2018). https://doi.org/10.1097/WNO.0000000000000580

8. Li, M., et al.: Image magnification network for vessel segmentation in octa images. In: Pattern Recognition and Computer Vision, pp. 426–435. Springer, Cham (2022). https://doi.org/10.1007/978-3-031-18916-6_35

9. Oktay, O., et al.: Attention u-net: learning where to look for the pancreas. CoRR (2018). http://arxiv.org/abs/1804.03999

10. Onishi, A.C., et al.: Importance of considering the middle capillary plexus on oct angiography in diabetic retinopathy. Invest. Ophthalmol. Vis. Sci. **59**(5), 2167–2176 (2018). https://doi.org/10.1167/iovs.17-23304

11. Poplin, R., et al.: Prediction of cardiovascular risk factors from retinal fundus photographs via deep learning. Nat. Biomed. Eng. **2**(3), 158–164 (2018). https://doi.org/10.1038/s41551-018-0195-0

12. Ronneberger, O., Fischer, P., Brox, T.: U-net: convolutional networks for biomedical image segmentation. In: Navab, N., Hornegger, J., Wells, W.M., Frangi, A.F. (eds.) MICCAI 2015. LNCS, vol. 9351, pp. 234–241. Springer, Cham (2015). https://doi.org/10.1007/978-3-319-24574-4_28

13. Sultana, F., Sufian, A., Dutta, P.: Rimnet: image magnification network with residual block for retinal blood vessel segmentation. In: 2022 IEEE Region 10 Symposium (TENSYMP), pp. 1–6 (2022). https://doi.org/10.1109/TENSYMP54529.2022.9864467

14. Tan, X., et al.: Oct2former: a retinal oct-angiography vessel segmentation transformer. Comput. Methods Programs Biomed. **233**, 107454 (2023). https://doi.org/10.1016/j.cmpb.2023.107454

15. Vadalà, M., Castellucci, M., Guarrasi, G., Terrasi, M., La Blasca, T., Mulè, G.: Retinal and choroidal vasculature changes associated with chronic kidney disease. Graefes Arch. Clin. Exp. Ophthalmol. **257**, 1687–1698 (2019). https://doi.org/10.1007/s00417-019-04358-3

16. Van De Kreeke, J.A., et al.: Optical coherence tomography angiography in preclinical Alzheimer's disease. Br. J. Ophthalmol. **104**(2), 157–161 (2020). https://doi.org/10.1136/bjophthalmol-2019-314127

17. Zhang, Z., Liu, Q., Wang, Y.: Road extraction by deep residual u-net. IEEE Geosci. Remote Sens. Lett. **15**(5), 749–753 (2018). https://doi.org/10.1109/LGRS.2018.2802944

Multi-modal Biometric Authentication: Harnessing Human Gait and Keystroke Dynamics for Enhanced Security

Sandip Dutta[1]([⊠]) [iD], Utpal Roy[1], and Soumen Roy[2] [iD]

[1] Department of Computer and System Sciences, Visva-Bharati,
Santiniketan 731235, West Bengal, India
`duttasandip5824@gmail.com`
[2] Department of Computer Science and Engineering, University of Calcutta, Acharya
Prafulla Chandra Roy Siksha Prangan, JD - 2, Sector - III, Saltlake City,
Kolkata 7000106, West Bengal, India

Abstract. In an era of escalating cyber threats, user authentication has become paramount in safeguarding sensitive information. Traditional methods like passwords and PINs are proving inadequate in the face of sophisticated attacks. This article proposes a cutting-edge approach that leverages the unique biometric signatures of human gait and keystroke dynamics for robust user authentication. Through a combination of sensor technologies and advanced machine learning algorithms, this method offers a multi-layered security solution that promises to revolutionize the authentication landscape.

A group of fifty-two volunteers with varying ages, professions, and levels of education was chosen for the preparation of the study database. The keystroke pattern and gait of each individual were captured utilizing built-in smartphone sensors in four different sessions. To capture human keystroke patterns four different fixed-texts are considered. Afterward, both keystroke and gait patterns underwent comprehensive analysis employing multiple machine learning algorithms, both in isolation and in combination. The findings indicate that the concurrent utilization of keystroke and gait patterns resulted in a markedly elevated degree of precision in user identification and authentication when compared to the individual application of either of these patterns.

The research pathways for the development and integration of this dual-factor authentication system have been outlined properly. It is expected that this comprehensive and multi-tiered authentication approach will substantially enhance security in both physical and digital spaces while ensuring a layer of convenience for the users.

Keywords: Behavioral Biometric · Computer Security · Human Gait · Keystroke Dynamics · Smartphone Sensors

1 Introduction

User authentication represents a foundational facet of security within the contemporary interconnected landscape. Ensuring that individuals accessing systems or physical spaces are who they claim to be is essential for safeguarding sensitive information and assets. Conventional user authentication mechanisms, exemplified by passwords and personal identification numbers (PINs), exhibit inherent vulnerabilities susceptible to exploitation by malicious entities. In contrast, biometric authentication methods present a compelling alternative, relying on distinctive physiological or behavioral attributes that pose significant challenges for unauthorized replication or theft.

Over the past few years, S. Roy along with his research group have published a number of interesting articles [1–5] centered around user identification, authentication, and prediction using either keystroke dynamics or the analysis of human gait. This builds upon our prior research endeavors, now integrating keystroke dynamics and human gait patterns to bolster the security of contemporary computing devices.

The principal objective of this investigation is to determine whether combining keystroke and gait patterns leads to enhanced user authentication accuracy in comparison to using keystroke or gait alone. Moreover, the efficacy of diverse machine learning algorithms in the authentication of users utilizing their gait and keystroke patterns will be systematically evaluated. The algorithm exhibiting superior performance in this assessment will be selected for subsequent implementation.

1.1 Significance of Multi-modal Biometrics

In the past few years, the confluence of biometrics and machine learning has facilitated the emergence of innovative methodologies in the domain of user authentication. One such promising avenue is the integration of human gait patterns with keystroke dynamics for authentication purposes. This multidisciplinary field offers a plethora of research opportunities with implications for security, healthcare, and human-computer interaction. In this article, we delve into the potential avenues for exploration and innovation in this dynamic research area.

1.2 Human Gait Analysis

Gait analysis involves the measurement and analysis of human walking patterns. Research [2] has shown that each individual possesses a unique gait signature, influenced by factors such as body structure, weight distribution, and muscle movement. Advances in computer vision and machine learning have enabled the extraction of meaningful features from gait data for authentication purposes.

1.3 Keystroke Dynamics

Keystroke dynamics focus on the manner and timing of key presses during text input. Each person has a distinctive typing rhythm, influenced by factors like finger length, typing habits, and motor coordination. This biometric trait has been employed in various applications, from access control to continuous authentication.

1.4   Advantages of Combined Analysis

By merging gait and keystroke patterns, we create a multi-modal biometric system that benefits from the complementary strengths of both traits. This fusion increases the complexity of the identification process, making it more resistant to spoofing attacks or impersonation. The major advantages of merging keystroke dynamics with human gait are as follows:

1. **Enhancing Security and Access Control:** The primary motivation behind combining gait and keystroke dynamics is to bolster security measures. Researchers can delve into developing more robust algorithms and models for authenticating users. This includes exploring advanced feature extraction techniques, refining machine learning models, and investigating methods to counter potential attacks or spoofing attempts.
2. **Behavioral Biometrics for Continuous Authentication:** The integration of gait and keystroke dynamics facilitates continuous authentication, representing a departure from the conventional paradigm of one-time access verification. This integration provides avenues for research into the development of algorithms capable of adapting to subtle variations in an individual's gait or typing patterns over time. Continuous authentication holds the promise of substantially augmenting security measures in dynamic environments.
3. **Human-Computer Interaction and Accessibility:** Beyond security applications, the combined analysis of gait and keystrokes holds promise in improving human-computer interaction. Researchers can explore how this technology can be harnessed to create more intuitive and accessible interfaces for individuals with mobility or dexterity challenges. This research avenue has the potential to empower a wider range of users in the digital landscape.

2   Literature Survey

In the realm of literature, both keystroke dynamics and human gait have been employed for user authentication purposes [6–19]. While these techniques are not novel, they have gained increasing popularity in recent research due to their efficacy in data collection, cost-effectiveness, easy accessibility, and adaptability to existing computer security systems with minimal adjustments.

The exploration of keystroke dynamics has a historical trajectory spanning several decades. Indeed, the formal investigation into keystroke dynamics traces its origins back to the late nineteenth century, specifically when Bryan and Harter [6] immersed themselves in this domain during a period marked by significant attention to the modernization of telegrams. In a 1975 IBM technical report, Spillane [7] postulated that human keystroke patterns could serve as a distinctive means of identifying individuals. Subsequent to this assertion, a series of inquiries conducted in the late 1990s and early 2000s [8–12] aimed to substantiate the viability of keystroke dynamics as a method for authenticating an individual's unique identity. Consequently, research endeavors in the realm of keystroke dynamics have endured and persisted into the present day.

Conversely, human gait analysis has conventionally found application primarily in neural stress analysis, typically involving the monitoring of an individual's body movements through video imagery. However, this approach incurs substantial costs and is predominantly conducted within controlled laboratory environments. In response to these limitations, Morris et al. [13] introduced an innovative methodology in 1973, entailing the incorporation of an accelerometer sensor within the human body to discern body movements. Nonetheless, it became evident that relying solely on data from an accelerometer sensor resulted in heightened errors, especially with variations in movement speeds. To address this challenge, Dejnabadi et al. [14] in 2006 and Favre et al. [15] in 2008 advocated for the integration of both accelerometer and gyroscope sensors. This dual-sensor approach has since been embraced by numerous subsequent studies, as it has demonstrated efficacy in mitigating errors and enhancing the accuracy of user identification and authentication.

There has been a scarcity of research dedicated to combining keystroke dynamics and gait analysis for the purpose of fortifying computer system security. [16–19] are some recent studies that have employed keystroke dynamics and gait analysis to enhance the security of contemporary computer systems.

3   Methodology

This research aligns with the inductive approach to research, where specific observations lead to the formulation of theories at the conclusion of the research process. In the initial phase of our study, we initiated the development of a web-based application employing the Hyper Text Markup Language (HTML), Cascading Style Sheets (CSS), and JavaScript programming languages. This application was designed to accurately capture both keystroke patterns and human gait patterns, aiming to facilitate user authentication grounded in the principles of keystroke dynamics and human gait analysis. The subsequent stages of the process encompass *Data Collection* and *Data Analysis*. The steps of the model implementation have been discussed below.

3.1   Dataset Preparation

To collect data for keystroke dynamics and human gait analysis, we introduced two distinct data acquisition protocols. In this regard, we utilized the Android 10.0 platform, primarily making use of sensors such as the accelerometer, rotation, gyroscope, and gravity sensors. A total of fifty-two volunteers were engaged in the data collection process. From each of these volunteers, keystrokes and gait patterns were recorded across four separate sessions in a controlled environment. Detailed explanations about our research objectives were provided to each participant. The collected data was meticulously structured and stored in .CSV format, organized within two Excel files to facilitate subsequent analysis. The ensuing graphical representations delineate the distribution of subjects within their respective classes (Figs. 1 and 2).

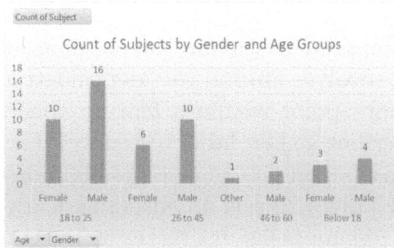

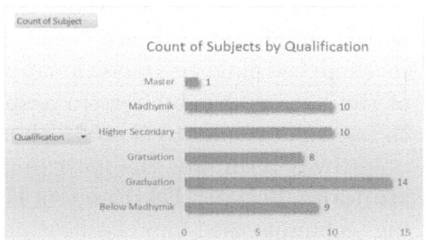

Fig. 1. Age and Gender Distribution

Fig. 2. Academic Qualification of Subjects

During the collection of keystroke patterns, participants were instructed to input a predefined set of commonly used keywords into the web-based application. Specifically, they were asked to type "Kolkata", "Facebook", "Computer" and "123456". Through this process, the web-based application extracted various distinct feature vectors, including keypress time, key release time, key hold duration, and latencies between two consecutive key presses, among others. These extracted feature vectors were subsequently combined and examined to generate a distinctive signature specific to each individual.

To conduct human gait analysis, the smartphone holding the web-based application has been positioned in the volunteer's trouser pocket. Participants were then instructed to walk naturally for approximately 1 min and 30 s on a plane surface. As the individual commenced movement, the smartphone in their pocket moved in tandem, causing the sensors within the device to generate data in response to shifts in its orientation. This method effectively captured the natural patterns of the person's gait, making it suitable for distinctive user identification and authentication.

3.2 Data Analysis

For analysis and implementation purposes, we have used the R statistical language and WEKA version 3.9 data mining tool. Both are open-source software available over the Internet. The partitioning of datasets into judiciously allocated training and testing subsets has been performed employing the 10-fold cross-validation methodology. Our analysis encompasses the scrutiny of the amassed datasets, both individually and collectively, utilizing six well-established supervised machine learning algorithms:

1. **Attribute Selected Classifier:** An Attribute-Selected Classifier (ASC) is a machine-learning method that integrates feature selection and classification into a unified process. It starts by identifying a subset of pertinent features from the dataset and subsequently employs a classifier on this reduced set of features. This approach enhances the efficiency and accuracy of models by eliminating irrelevant or redundant features prior to classification.

2. **Instance-Based k-Nearest Neighbors:** This algorithm constitutes a widely employed instance-based learning approach within the domain of machine learning. Its operational paradigm revolves around the identification of the k-nearest neighbors to a specified data point within a feature space. In tasks pertaining to classification, IBk ascribes a class label to a novel data point by determining the most prevalent class among its k-nearest neighbors.

3. **Random Forest:** The Random Forest (RF) algorithm is an ensemble learning technique used for classification and regression tasks. It creates an ensemble of decision trees during training, with each tree formulated using a randomized subset of training data and features, enhancing resilience and mitigating overfitting.

4. **Filtered Classifier:** A filtered classifier (FC) is a machine learning technique used in data preprocessing for classification tasks. It combines a feature selection or extraction method (filter) with a classification algorithm.

5. **Random Committee:** In machine learning, a Random Committee (RC) is an ensemble learning approach that combines predictions from multiple base models, or "committee members", to formulate a definitive prediction. Each base model is trained on distinct subsets of data or uses diverse algorithmic approaches, enhancing the ensemble's predictive diversity and robustness.

6. **Support Vector Machine:** A Support Vector Machine (SVM) is a robust supervised machine learning algorithm used for classification and regression tasks. It operates by identifying the optimal hyperplane in the feature space to effectively separate different classes and maximize classification accuracy. The algorithm achieves this by identifying support vectors, crucial data points, and determining the hyperplane that maximally separates these vectors. This approach enhances the SVM's ability to discern and categorize distinct classes within a given dataset.

4   Results and Discussions

4.1   Performance Evaluation Metrics

Classification accuracy serves as a ubiquitous metric within the domain of machine learning, acting as an evaluative criterion to assess the effectiveness of a classification model. This metric precisely quantifies the ratio of accurately classified instances to the total instances encompassed within the dataset, thereby providing a comprehensive measure of the model's performance in categorizing data points. Mathematically, it is defined as:

$$Accuracy = \frac{Number\ of\ Accurately\ Classified\ Instances}{Total\ Number\ of\ Instances}$$

For instance, in the scenario where a model accurately classifies 90 instances out of a total of 100 instances, the corresponding accuracy would be quantified as 90%.

While accuracy serves as a direct evaluation of a model's comprehensive performance, its suitability may be limited, particularly in scenarios characterized by imbalanced class distribution. In such instances alternative metrics, namely precision and recall, can furnish a more nuanced evaluation of the efficacy of the model.

Precision: Precision, within the context of binary classification, manifests as a quantitative metric delineated by the ratio of true positive predictions to the aggregate of true positive and false positive predictions. In its essence, precision serves to quantify the proportion of predicted positive instances that concord with the actual positive instances present within the dataset. Mathematically, it is defined as:

$$\text{Precision} = \frac{\text{True Positives}}{\text{True Positives} + \text{False Positives}}$$

High precision signifies that a model excels in minimizing false positives. This metric is of significance in scenarios where the cost associated with false positives is substantial.

Recall: Recall, also referred to as sensitivity or True Positive Rate (TPR), is a quantitative measure that delineates the proportion of correctly predicted positive observations relative to the entire set of actual positive observations. It serves as a metric assessing the efficacy of correctly predicting instances designated as positive within the actual dataset. Mathematically it is defined as:

$$\text{Recall} = \frac{\text{True Positives}}{\text{True Positives} + \text{False Negatives}}$$

Elevated recall signifies the model's proficiency in effectively detecting and capturing the entirety of actual positive instances within the dataset.

We have also considered the root mean square error (RMSE) value of the predicted and actual numerical instances of our databases. A low RMSE value indicates that the predicted values are very close to the actual values. The RMSE value and the model prediction accuracy are inversely proportional to each other. The mathematical formulation of RMSE is as follows:

$$\text{RMSE} = \sqrt{\frac{1}{n} \sum_{i=1}^{n} (y_{\text{actual}_i} - y_{\text{predicted}_i})^2}$$

where:

- n n is the total number of data points.
- y_{actual_i} is the actual value of the target variable for the ith data point.
- $y_{\text{predicted}_i}$ is the predicted value of the target variable for the ith data point.

4.2   Result and Discussion

The gait database and the keystroke database were examined using both WEKA version 3.9 and RStudio. This analysis was conducted independently as well as in combination, considering the six appropriate supervised machine learning algorithms mentioned earlier. The outcomes of these assessments are detailed in Table 1.

Table 1. The comparative evaluation of six distinct machine learning algorithms.

Algorithms	Attributes	Accuracy	Precision	Recall	RMSE
Attribute Selected Classifier	**Keystroke Dynamics**	**94.71%**	**95.80%**	**94.70%**	**0.0451**
	Human Gait	80.28%	81.90%	80.30%	0.0843
	Combined	**93.26%**	**94.00%**	**93.30%**	**0.0509**
Random Committee	Keystroke Dynamics	71.63%	72.90%	71.60%	0.0908
	Human Gait	36.53%	40.80%	36.50%	0.127
	Combined	**75.48%**	**78.10%**	**75.50%**	**0.0833**
Filtered Classifier	Keystroke Dynamics	78.84%	78.00%	78.80%	0.0795
	Human Gait	77.88%	78.30%	77.90%	0.0847
	Combined	**87.01%**	**89.50%**	**87.00%**	**0.0599**
IBK	Keystroke Dynamics	66.82%	68.20%	66.80%	0.1043
	Human Gait	60.09%	62.70%	60.10%	0.1136
	Combined	**77.40%**	**79.20%**	**77.40%**	**0.0877**
Random Forest	Keystroke Dynamics	77.88%	78.60%	77.90%	0.0924
	Human Gait	47.59%	45.60%	47.60%	0.1179
	Combined	**86.53%**	**87.50%**	**86.50%**	**0.0908**
SVM with RBF Kernel	Keystroke Dynamics	49.04%	47.20%	45.40%	0.156
	Human Gait	43.27%	42.70%	42.10%	0.198
	Combined	**61.54%**	**60.20%**	**61.40%**	**0.0985**

Note: Combined in Table 1 refers to the fusion of keystroke and human gait patterns.

Table 1 provides clear evidence that integrating keystroke patterns with gait patterns leads to improved accuracy, precision, recall, and reduced RMSE values when compared to using either keystroke or human gait data alone. This trend is observed across most algorithms, with the exception of the ASC algorithm, where the classification accuracy for the KD dataset and the combined dataset is nearly identical.

Among these six algorithms, it is evident that ASC stands out as the most effective in authenticating users based on their gait and keystroke patterns. This effectiveness can be attributed to its ability to handle our imbalanced dataset, which includes more male participants than females and others. During the data preprocessing step, this classifier employs techniques such as dimensionality reduction, attribute selection, bias mitigation, addressing data sparsity, and enhancing interpretability, leading to its superior performance. Similarly, the support vector machine's performance suffers due to its limited capability to effectively manage imbalanced datasets. The inverse relation between model user authentication accuracy and RMSE value is shown in Fig. 3. It visually illustrates that as the accuracy bar becomes taller, the corresponding RMSE bar becomes shorter, indicating an inverse relationship.

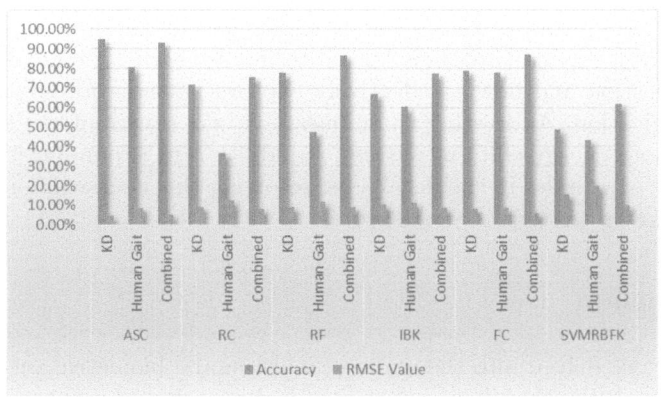

Fig. 3. Inverse relation between prediction accuracy and the RMSE value

Considering that ASC consistently achieves the best results in authenticating users based on their gait and keystroke patterns, we have chosen ASC as the preferred model for our system.

5   Comparison with Existing Systems

In order to contextualize our findings, a comparative analysis was conducted by benchmarking our results against a relevant study denoted as [19]. This investigation involved the acquisition of gait and keystroke samples from 20 subjects across varied scenarios, employing diverse machine learning classifiers. The experimental results revealed a significant accuracy rate of 99.11% when utilizing a Multilayer Perceptron classifier. The associated false acceptance rate, false rejection rate, and equal error rate values were observed to be 0.684%, 7%, and 1%, respectively.

Contrastingly, studies referenced [16–18] involved fewer than 30 subjects, and their outcomes indicated an inability to achieve an equal error rate (EER) below 1%.

Through this comparative analysis, we infer that, to date, no extant study has integrated gait and keystroke patterns from a cohort exceeding 50 individuals, particularly utilizing cost-effective smartphone sensors, while concurrently achieving an identification accuracy comparable to our attained accuracy level.

6   Limitation and Future Work

One significant drawback associated with smartphone authentication involves the substantial power drain incurred when utilizing hardware resources for extended periods of multi-sensor feature-level fusion. A study [19] has mentioned that monitoring gyroscopes and accelerometers at a rate of 16 Hz results in an energy consumption overhead. Another study [20] has reported that re-authentication consumes an additional 2.4% of the device's battery. Consequently, collecting sensory data over extended periods of time imposes a considerable energy expenditure. It is logical to inquire about the ideal time frame for data collection and how to effectively capture patterns for continuous user identity verification. Addressing these questions will contribute to the development of a more energy-efficient system. In addition to expanding the quantity and diversity of samples in the databases, resolving this issue will form the foundation for our future research endeavors.

7   Conclusion

In this study, we delved into the realm of multi-modal biometric authentication, leveraging the distinctive characteristics of human gait and keystroke dynamics. Our comprehensive analysis and experimentation demonstrated the potential of this combined approach for bolstering security measures.

ASC emerged as a standout performer, showcasing remarkable accuracy, precision, and recall rates. Its proficiency in handling imbalanced datasets, particularly those skewed towards male participants, proved invaluable. Through techniques like Reduced Dimensionality, Bias Mitigation, and Improved Generalization, we achieved enhanced performance and interpretability. The inverse relationship observed between accuracy and RMSE underscores the significance of a balanced model. As the accuracy bar rose, the RMSE bar saw a corresponding reduction, highlighting the importance of precision in authentication systems.

In conclusion, our findings advocate for the adoption of a multi-modal approach, incorporating gait and keystroke dynamics, to fortify biometric authentication systems. ASC stands as a testament to the potential of this methodology, offering a promising avenue for heightened security in authentication processes. This research paves the way for future advancements in biometric authentication, ensuring robust and reliable systems for safeguarding sensitive information.

Acknowledgement. One of the authors S.D. acknowledges the financial support of DST as an INSPIRE fellow.

References

1. Roy, S., Pradhan, J., Adhikary, D.R.D., Roy, U., Sinha, D., Pal, R.K.: A systematic literature review on latest keystroke dynamics based models. IEEE Access **10**, 92192–92236 (2022). https://doi.org/10.1109/ACCESS.2022.3197756
2. Ghosh, D., Roy, S., Roy, U.: Gait identity verification using equipped smartphone sensors. In: National Conference on Emerging Trends on Sustainable Technology and Engineering Applications (NCETSTEA) (2020). https://doi.org/10.1109/NCETSTEA48365.2020.9119955. ISBN 978-1-7281-4362-0
3. Roy, S., Roy, U., Sinha, D.D.: Efficacy of typing pattern analysis in identifying soft biometric information and its impact in user recognition. In: Battiato, S., Farinella, G.M., Leo, M., Gallo, G. (eds.) ICIAP 2017. LNCS, vol. 10590, pp. 320–330. Springer, Cham (2017). https://doi.org/10.1007/978-3-319-70742-6_30
4. Roy, S., Roy, U.: Enhanced knowledge-based user authentication technique via keystroke dynamics. Int. J. Eng. Sci. Invention (IJESI), 41–48 (2014)
5. Roy, S., Roy, U., Sinha, D.D.: Password recovery mechanism based on keystroke dynamics. In: Mandal, J.K., Satapathy, S.C., Sanyal, M.K., Sarkar, P.P., Mukhopadhyay, A. (eds.) Information Systems Design and Intelligent Applications. AISC, vol. 339, pp. 245–257. Springer, New Delhi (2015). https://doi.org/10.1007/978-81-322-2250-7_24
6. Thornton, M.A.: Retrieved from Encyclopedia of Cryptography, Security and Privacy, Keyboard Dynamics, pp. 1–6 (2021)
7. Spillane, R.: Keyboard apparatus for personal identification. In: IBM Tech, 62 Disclosure Bulletin, vol. 17, no. 3346 (1975)
8. Forsen, G.E., Nelson, M.: Personal attributes authentication techniques, Technical report RADC-TR77-333, Rome Air Development Center (1977)
9. Gaines, R., Lisowski, W.: Authentication by keystroke timing: some preliminary results. In: Rand Report, R-2560- NS, Rand Corporation (1980)
10. Bleha, S.S.: Computer access security systems using keystroke dynamics. IEEE Trans. Pattern Anal. Mach. Intell. **12**(12), 1217–1222 (1990)
11. Joyce, R., Gupta, G.: Identity authorization based on keystroke latencies. Commun. ACM **33**(2), 166–176 (1990)
12. Monrose, F., Rubin, A.D.: Keystroke dynamics as a biometric for authentication. Future Gener. Comput. Syst. **16**(4), 351–359 (2000)
13. Morris, J.R.W.: Accelerometry—a technique for the measurement of human body movements. J. Biomech. 729–736 (1973)
14. Dejnabadi, H., Jolles, B.M., Casanova, E., Aminian, P.F.K.: Estimation and visualization of sagittal kinematics of lower limbs orientation using body-fixed sensors. IEEE Trans. Biomed. Eng. 1385–1393 (2006)
15. Favre, J., Jolles, B.M., Aissaoui, R., Aminian, K.: Ambulatory measurement of 3D knee joint angle. J. Biomech. 1029–1035 (2008)
16. Ross, A., Jaju, A., Nandakumar, K.: Combining keystroke and gait biometrics for user authentication. IEEE Trans. Inf. Forensics Secur. (2008)
17. Ross, A., Kasturi, A.: Gait and keystroke dynamics for person authentication. ACM Trans. Inf. Syst. Secur. (2009)
18. Vidya, T., Arthi, R.: Human authentication using keystroke dynamics and gait patterns: a review. Procedia Comput. Sci. (2015)
19. Lamiche, I., Bin, G., Jing, Y., et al.: A continuous smartphone authentication method based on gait patterns and keystroke dynamics. J. Ambient Intell. Hum. Comput. **10**, 4417–4430 (2019). https://doi.org/10.1007/s12652-018-1123-6

20. Sitova, Z., et al.: HMOG: new behavioural biometric features for continuous authentication of smartphone users. IEEE Trans. Inf. Forensics 3759 Secur. **11**(5,) 877–892 (2016)
21. Lee, W.H., Lee, R.: Implicit smartphone user authentication with 3761 sensors and contextual machine learning. In: Proceedings of the 47th Annual IEEE/IFIP3762 International Conference on Dependable System. Networks (DSN), pp. 297–308 (2017)

Author Index

A

Adhikari, Tinku   II-38
Adhikary, Bhanita   II-185
Adhikary, Dibya Ranjan Das   I-93
Aditya, Vardhan   II-121
Alam, Ekram   I-102

B

Banerjee, Ansuman   II-82
Banerjee, Gouravmoy   I-149
Banerjee, Soham   II-82
Banerji, Nandan   I-127
Barman, Debaditya   I-160
Basak, Rohini   I-184
Bhakta, Puja   I-160
Bharti, J. S.   II-121
Bhattacharjee, Debjyoti   II-206
Bhattacharya, Arijit   I-275, I-287
Bhattacharya, Indrajit   II-26
Biswapati, Jana   II-121
Biswas, Rohan   I-219
Biswas, Rounak   II-228
Biswas, Saroj Kr.   I-275, I-287
Biswas, Swagata   II-155
Borate, Vishal   I-45

C

Chakraborty, Mamata Dalui   II-48
Chakraborty, Samarjit   II-71
Chanda, Debanil   I-64
Chatterjee, Avimita   II-206
Chatterjee, Taniya   I-160
Chattopadhyay, Anupam   II-206

D

Das, Akhil Kumar   I-275, I-287
Das, Arijit   I-219
Das, Bidyut   I-54
Das, Nibaran   I-18
Das, Nilanjana   I-240
Das, Ratul   I-196

Das, Sushovan   II-94, II-109
Das, Swarup   I-136
Datta, Kankana   II-48
Dey, Jayati Lahiri   I-251
Dey, Sumon   I-136
Dhar, Priyodarshini   I-31
Dutta, Kalpita   I-18
Dutta, Paramartha   I-102, I-299
Dutta, Rakesh   I-240
Dutta, Sandip   I-311
Dutta, Subhasis   II-26

F

Faruqui, Nuruzzaman   I-78

G

Ganguly, Kasturi   I-219
Ganguly, Souvik   I-207
Ghosal, Sudipta Kr.   I-78
Ghosh, Arindam   II-130
Ghosh, Indrajit   I-149
Ghosh, Mili   I-160
Ghosh, Sumana   II-82
Ghosh, Swaroop   II-206
Ghosh, Utpal   II-171
Gurung, Suraj   I-229

H

Hamad, Mohammad   II-142
Hari, K. C.   I-263
Hasanujjaman, Md.   II-61

J

Jana, Biswapati   II-48

K

Karmakar, Rahul   I-173
Kaur, Harmandar   II-3, II-14
Khan, Ajoy Kumar   II-38
Khatun, Rubaya   I-3

M. Majumder et al. (Eds.): ICCTE 2023, CCIS 2376, pp. 323–324, 2025.
https://doi.org/10.1007/978-3-031-81935-3

Kour, Satveer   II-3, II-14
Kule, Malay   II-38
Kumar, Rishu   I-54

L
Leo, Marco   I-102
Li, Junde   II-206

M
Mahapatra, Priya Ranjan Sinha   II-26
Majumder, Mukta   II-195
Mali, Yogesh Kisan   I-45
Mandal, Ardhendu   I-275, I-287
Mandal, Jyotsna Kumar   I-78, I-240
Mandal, R. K.   I-64
Mandal, Rakesh Kumar   I-31
Mandal, Sourav   I-184
Mandi, Utpal   II-195
Metta, Ravindra   II-71
Modi, Shabina   I-45
Mohalik, Swarup K.   II-82
Moitra, Moumita   I-207
Mondal, Ritwik   II-26
Mondal, Uttam Kr.   II-94, II-109, II-171
Mondal, Uttam Kumar   I-240
Mukherjee, Pratyush   I-184

N
Nandi, Avishek   II-130
Nandi, Utpal   I-78
Neogi, Sanchita   I-173

P
Pal, Alok Ranjan   I-196
Panda, Sushmita   II-26
Pattanaik, Barsha   I-184
Paul, Bachchu   I-78
Paul, Himadri Sekhar   II-155
Paul, Jaya   I-18
Phalak, Koustubh   II-206
Pokharel, Manish   I-263
Pope, Jeremie   II-206
Prasad, Roshan Kumar   I-229

R
Rai, Pratika   I-229
Rath, Premansu Sekhara   I-93

Roy, Arundhati   I-114
Roy, Bibek Bikash   II-109
Roy, Kaushik   I-18
Roy, Soumen   I-311
Roy, Utpal   I-311, II-228

S
Sadhu, Arnab   II-130
Saha, Amit   II-206
Saha, Basudev   II-195
Saha, Debasmita   I-275, I-287
Saha, Diganta   I-196, I-219
Saha, Sriparna   I-114
Saha, Sujan Kumar   I-207
Sale, Deepali   I-45
Sarangal, Himali   II-3, II-14
Sarkar, Anasua   I-18
Sarkar, Anish   I-78
Sarkar, Arindam   I-136
Sarkar, Arup   I-3
Sarkar, Uditendu   I-149
Sen, Deepanjan   I-136
Shaw, Sarika Kumari   I-251
Sherpa, Lhamu   I-127
Shrestha, Sushil   I-263
Singh, Butta   II-3, II-14
Singh, Manjit   II-3, II-14
Singh, Moirangthem Marjit   I-78
Singh, Vikramjeet   II-3
Sinha, Sharad   I-229
Sourav, Kaity   II-121
Steinhorst, Sebastian   II-142
Subba, Ranjit   I-229
Sufian, Abu   I-102, I-299
Sultana, Farhana   I-299
Swarnakar, Jaydeep   II-185

T
Tripathy, Rudra M.   I-184
Tripathy, Swagatika   I-93

Y
Yeolekar, Anand   II-71

Z
Zaman, J. K. M. Sadique Uz   II-61

The manufacturer's authorised representative in the EU is Springer
Nature Customer Service Centre GmbH, Europaplatz 3, 69115 Heidelberg,
Germany. If you have any concerns regarding our products, please
contact ProductSafety@springernature.com

Printed and bound by CPI Group (UK) Ltd, Croydon, CR0 4YY
29/04/2026
02099544-0007